VICTOR JOHN YANN

ENERGY CRISIS

Danger and Opportunity

WEST PUBLISHING CO.
St. Paul • New York • Boston
Los Angeles • San Francisco

Library of Congress Catalog Card Number: 74–948

ISBN: 0–8299–0013–6

Yannacone—The Energy Crisis

To my son,
Victor.
Thank You.

*

PREFACE

The Chinese ideogram for "crisis" is composed of the symbols for "danger" and "opportunity," and the energy crisis facing most of the industrialized world today also contains the same polar elements—danger and opportunity. It will probably be a number of years before we will be able to decide whether the crisis was in fact a danger, an opportunity, or perhaps both for modern society.

Since our effort to present information on the energy crisis takes the form of a book it is only fitting to consider the limitations of the printed page. As Western man embraced printing—typography—he adopted a linear, cause and effect, mode of thought. The system for communication that we adopted, adapted us, and we began to analyze or breakdown into discrete units what we had formerly perceived as a unified totality: the world around us, our environment and natural resources. We became skilled at manipulating experience recast as sentences, clauses, phrases, words and phonemes. It was a marvelous technique because it allowed us to accomplish similar changes in our world of work.

No longer did the master craftsman, *il maestro,* produce a finely carved cabinet. In his place arose the assembly line that like a great linotype machine orchestrated fragmented, imperfect skills out of which poured finished cabinets, one every 35 minutes ... without fail. Sadly, il maestro was a casualty. Sadder yet, none of the assembly line workers were individually capable of producing a complete cabinet. We had disassembled il maestro and sown his parts along rows of conveyor belts, so that now some men cultivated cabinet doors, or legs, while others grew expert at producing shelves or panels.

The weakness of this technique when applied to acquiring, storing and circulating information is that at the same time it confers upon the user the ability to transmit information beyond the tribal circle and the brief life span of its members, it becomes the tool of the specialists who the deeper they dig, the more they lose contact with the rest of the their fellow men. Many scientists are no longer able to communicate with non-scientists and some cannot even communicate with scientists in other disciplines. The occidental appears totally out of phase with the oriental.

If we are to understand the current energy crisis in sufficient depth to rationally make the public policy and private action decisions that must be made in order to capitalize on the opportunities it offers to society, there can be no substitute for a basic understanding of the physical concepts involved — energy, power and work. Physicists and mathematicians have considered these subjects for centuries and they can be precisely defined and succinctly stated as follows:

$$
\begin{aligned}
W &= \int_0^{S_1} F_s \, dS = \int_0^{S_1} m \frac{d^2 S}{dt^2} \, dS = \int_0^{t_1} m \frac{d^2 S}{dt^2} \frac{dS}{dt} \, dt \\
&= \int^{t_1} m \frac{dv}{dt} v \, dt = \int_0^{t_1} m \frac{d}{dt} \left(\tfrac{1}{2} v^2 \right) dt = \int_0^{v_1} m \, d\left(\tfrac{1}{2} v^2 \right) \\
&= \tfrac{1}{2} m v_1^2 .
\end{aligned}
$$

And while many of the authors might feel comfortable with such a statement, there is an obvious need to furnish a bit more expository discussion. In Chapter 1, Dr. John W. Andrews, a physicist and Professor of Physics considers these fundamental physical concepts in everyday language and without resort to higher mathematics. Even those readers who might already be familiar with the basic physics involved would do well to read this chapter in order to find just the right words and examples to explain what they may already know in precise technical terms to their friends and neighbors who might need a more popular explanation.

Chapter 2 is based on a now legendary study by Dr. M. King Hubbert, the basic conclusions of which were first published as early as 1956. Then in an extended form in 1962, it was released by the National Academy of Sciences/National Research Council, with whose permission portions of the original study have been reprinted. Subsequent revisions in 1969 and 1972, as well as continued input to various federal agencies from his position within the United States Geological Survey, make Dr. Hubbert's studies the fundamental source document on Energy Resources, particularly fossil fuels. Following the format of the original report, current information presented by proponents of various exotic energy sources to various Congressional Committees in hearings on the energy crisis during 1973, as well as updated material from Dr. Hubbert has been included where appropriate. By way of introduction to Chapter 2, a portion of a paper by Dr. Hubbert, entitled, "Energy Resources," which appeared as Chapter 5 in *ENVIRONMENT, Resources, Pollution and Society,* edited by William W. Murdoch and published by Sinauer Associates, Stamford, Connecticut, has been included and represents a more technical discussion of energy, work and power than appeared in Chapter 1, but is still less formidable than a complete mathematical exposition of the concepts.

If fossil fuels are in short supply, some consideration should be given to the other roles they play in our society. In Chapter 3, a substantial amount of material has been gathered from the American Petroleum Institute and other petrochemical manufacturing organizations as well as Congressional hearings on the energy crisis in an attempt to indicate some of the alternative uses for our limited quantities of coal, oil and natural gas. Although this book does not consider the environmental impact of the energy crisis directly, material is included in Chapter 3 on some of the current desulfurization options available to the producers of coal and oil. While we may be running out of cheap fossil fuels, there is still no justification to return to the levels of environmental degradation formerly attributable to the use of fossil fuels, in particular coal and high-sulfur oil.

The principal alternative to fossil fuels as a cheap and available source of energy to power industrialized society is atomic energy. Unfortunately, atomic power has become a subject of deep and abiding controversy over the last ten years. Many citizens are unfamiliar with the complex physical, chemical, biological and medical aspects of the controversy and many of the scientists who have attempted to maintain a relatively moderate position have been damned by extremists on both sides of the issues. The result has been a lack of reliable information which can be used by the general public to evaluate the merits of siting this nuclear power plant here or that nuclear power plant there, or any nuclear power plant anywhere. In an attempt to confront this problem without, at the same time, advocating on behalf of one group or the other, Chapter 4 contains some basic information about atomic power including some of the criticisms raised by opponents to atomic power as well as some of the answers presented by the proponents of atomic power. Much of the controversy over nuclear power and indeed much of the controversy over environmental degradation from commercial and industrial operations is intimately related to the problem of industrial siting, so the balance of the Chapter deals with cost/benefit considerations in site selection. Much of the material for Chapter 4 was obtained from discussions with scientists active in the field.

Chapter 5 has been written for this book by Dr. Victor P. Bond, now the Associate Director of Brookhaven National Laboratory. Much of the material contained in Chapter 5 can be found in other publications by Dr. Bond and other workers cited in the Chapter, however, it will be difficult to find a more objective and accurate treatment of the complex problem: determining and evaluating the public risk attributable to the operation of nuclear facilities such as power plants. That there is some risk associated with the operation of such facilities is conceded by all responsible scientists, however, that risk must be placed in proper perspective when compared with the other risks faced by the human race and individual human beings each day, in particular, the known demonstrable public health hazard from the continued generation of electricity in conventional fossil-fueled plants, particular those burning high-sulfur coal.

Part IV of this book deals with Energy and Society and contains a series of essays which can furnish a basis for informed public consideration of the energy crisis. Chapter 6 is a whimsical look at Philadelphia after its people finally awakened to the dangers of the automobile written by Angelo J. Cerchione, a career Air Force officer and also the only man in the Pentagon at this time with an advanced degree in environmental regional planning. Dr. Large, in Chapter 7, discusses energy conservation from the point of view of the individual and the impact of energy policies upon the individual. Energy conservation measures suitable for application by individuals and families have been culled from testimony at Congressional hearings and added as a guide for the reader who really is concerned about saving energy.

James Lowden, an executive officer of Trizec, one of the largest commercial and industrial builders in Canada discusses the impact of commercial building practices on energy consumption and suggests a number of conservation measures that many local municipalities could easily incorporate into existing building and construction codes, and which the public should be aware of when considering what it needs in the way of convenience at work and in the store.

Is there really an energy crisis? In Chapter 9, Roland W. Comstock of the Northern States Power Company establishes that there is, indeed, a crisis of sorts, and shows how part of that crisis has been brought upon us by ourselves. In the Parable of the Lily Pond, he tolls a warning that all must heed, and then, sets forth a method for technology assessment which equips anyone with a convenient method of analyzing the claims that a new technological breakthrough is only a few months (years?) and a few million (billion?) dollars away.

If you have ever been curious about the Oil Industry and in particular the gasoline business, Mike Morrison in Chapter 10, has the answers to some very perplexing questions: Why are there gas stations on every available corner? Why do (did) we pay less for gasoline than drivers in any other country?

A great deal has been said about protecting the quality of our environment through application of the National Environmental Policy Act of 1969 (NEPA). There is little doubt that NEPA has played some part in the energy crisis and will continue to play a part unless Congress substantially revises the Act or the Courts substantially modify it through interpretation. In Chapter 11, attorney Irving Like, incidentally an active opponent of nuclear power generation in its present form, (counsel for the Lloyd Harbor Study Group) identifies what may be the root cause of the energy crisis—money, in particular the way the Federal Reserve System manipulates money policy outside the constraints of NEPA—so far.

The origins of the National Environmental Policy Act are important if the Act is to be of any use to citizens during the energy cirsis. Of particular concern should be the intention of the Congress that the Act truly be a statement of a national environmental policy which should be observed in spirit as well as to the letter. There is little doubt that the energy crisis will spawn a number of hard choices which must eventually be resolved by litigation or legislation. Among the obvious choices are those between the continued wanton, profligate waste of fossil fuels for burning to make heat to power large numbers of automobiles on already clogged highways, the continued need of a number of the petrochemical industries for coal, oil and natural gas in order to continue to supply society with many of the products which contribute to maintaining our civilization in the form that we now know it.

Chapter 13 is based on another long ignored paper. Written in 1952 as part of a Master's Thesis under Professor Stanley Cain at the University of Michigan, Dr. Nicholas Muhlenberg demonstrates a relationship between Energy and Conservation that certainly exists, but has been little considered and less understood. At this time in history, the relationship between energy utilization and Conservation should be the single most important concern of Conservation (the movement). However, it appears that many of the demands made upon our industrial system in the name of Conservation would, perhaps, force us to return to the "cave and the candle." Dr. Muhlenberg points out that there is a significant relationship between energy consumption and culture. A relationship which can not be ignored in the physical sense, and should not be ignored in any discussion of the energy crisis.

ACKNOWLEDGEMENTS

This book was requested on December 1, 1973, and could never have been released this quickly without the dedicated cooperation of many individuals and organizations. This simple acknowledgement is little enough expression of the gratitude of the editor and the author. At the risk of slighting some of the people who worked so diligently over the Christmas Holiday, names will be mentioned, but the order in which these acknowledgements appear is no indication of the extent of that individual's contribution to the production of this book. It just happens to be the order of our notes and files. Thanks to all, particularly:

. . . my secretary, Muriel Aldrich, her husband Bill and all of her children who shared the Christmas Holidays with this book;

. . . my wife Carol, for keeping house, home and family together during the Holidays and keeping Christmas amid manuscript;

. . . my son, Victor, for helping make a real Christmas for Claire;

. . . Congressman Otis G. Pike, and his extremely efficient staff for furnishing us with Congressional research material assuring that this book would be current and topical;

. . . Herb Kolber, President of Dorset Computype and all his girls who keyboarded from manuscript and put aside other work in order to get this book typeset over the Holidays;

. . . Dr. James Roesser, President of Science Typographers, who developed the computer program which made it possible to typeset Chapter 2 and Chapter 5 in less than a week, and to his wife Rae and all their staff who keypunched the manuscript over Christmas and New Year's weekend;

. . . Angelo J. Cerchione, who gave up a week of well-earned leave from the Pentagon in order to help edit and tie the loose ends together;

. . . Dr. Anne Renouf Headley, whose prescient paper, "Political Granting" originally prepared for delivery at the 1971 Annual Meeting of the American Political Science Association unfortunately could not be included in this volume, but whose tireless efforts on behalf of the editor and the authors at the Environmental Protection Agency did result in substantial contributions to the work;

. . . the National Science Foundation which permitted the reprinting of portions of Dr. Hubbert's prophetic paper;

. . . Monsanto *Enviro-Chem* Systems Inc. and Larry Gillingham, Communications Manager for information on solid waste disposal and resource recovery;

. . . Vance J. Van Laanen, for his advice and comments which contributed greatly to maintaining a balanced consideration of the nuclear power controversy;

. . . and the staff and management at West Publishing Company, which saw a need and met it.

Although many people participated in the preparation of this manuscript and the production of this book and the credit belongs to them, any mistakes, omissions or other causes for complaints are necessarily my responsibility.

Patchoque, New York   Victor J. Yannacone, Jr.
New Year's 1974

ABOUT THE AUTHORS

VICTOR JOHN YANNACONE, jr is an attorney generally considered the founder of Environmental Law as a recognized legal discipline. A working trial lawyer since 1959 he has been instrumental in establishing the Trust Doctrine, the Ninth Amendment of the Constitution and the general principles of equity jurisprudence as the cornerstones of successful environmental litigation. He is the senior author of a comprehensive treatise, *Environmental Rights & Remedies*, and the author of numerous articles on environmental law, trial practice and environmental science, as well as a regular column on Ecology and Real Estate which appears monthly in *National Real Estate Investor*.

Mr. Yannacone was a co-founder of the Environmental Defense Fund and acted as their counsel until September, 1969. Mr. Yannacone was first co-chairman of the Environmental Law Committee of the American Trial Lawyers Association, and served as Chairman of the Environmental Law Committee of the Section of Insurance, Negligence and Compensation Law of the American Bar Association.

At the present time, in addition to writing and practicing law, Mr. Yannacone is working with several groups in applying the techniques of modern systems science to the resolution of environmental controversy.

JOHN W. ANDREWS received his undergraduate education at the Massachusetts Institute of Technology, and was awarded a Ph.D. in high energy particle physics by the University of Notre Dame in 1967. While at Notre Dame and later at Yale University, he was engaged in research in the field of particle physics and published on strong interactions of π and κ mesons.

In 1969 Dr. Andrews came to Southampton College of Long Island University where he was instrumental in establishing a program in environmental studies. Working with local environmental groups he has taken the lead in involving non-technical people in the problems of energy and the environment. Dr. Andrews has even run for local public office on an environmental platform. Bringing to this book extensive experience in teaching complex physical, mathematical and chemical concepts to laymen in a relevant but non-intimidating manner, he provides the concerned citizen with a non-mathematical, but nevertheless, solid basis for understanding the "energy crisis."

M. KING HUBBERT is a research geophysicist with the U.S. Geological Survey in Washington. In 1964, after 20 years of research in petroleum exploration and production for Shell Oil and Shell Development Companies, he accepted dual appointments, one with the Geological Survey, and a second as Professor of Geology and Geophysics at Stanford University, devoting part time to each. Prior to that, he taught geology and geophysics for 10 years at Columbia University, and did geophysical work for the Amerada Petroleum Corporation and for the Illinois State and U.S. Geological Surveys. His scientific education, with a major jointly in geology and physics and a minor in mathematics, was received from 1924 to 1930 at the University of Chicago, where he received both his undergraduate degree and his doctorate. His subsequent researches have been principally in mineral exploration, mechanics of geologic structures, physics of underground fluids, and on the consequences of human exploitation of the earth's mineral energy resources.

VICTOR P. BOND at the present time is Associate Director, Life Sciences and Chemistry, Brookhaven National Laboratory, as well as Adjunct Professor of Medicine, Health Sciences Center, State University of New York at Stony Brook, and Adjunct Professor of Radiology, Columbia University. His professional career has been concerned mainly with the medical sciences, in particular those aspects of hematology which consider the quantitative aspects of cell production and the regulation of cell proliferation. He has done a great deal of research on the effects of radiation as well as research in nuclear medicine. In recent years, as administrative burdens increased, his investigative work has tended to become more evaluative than laboratory oriented, however he has published more than 270 scientific papers in a distinguished scientific career that began with an M.D. degree from the University of California, San Francisco in 1945, and a Ph.D. in Medical Physics from the University of California, Berkeley, in 1952.

ANGELO J. CERCHIONE is a career officer in the United States Air Force presently assigned to the Secretary of the Air Force Office of Information at the Pentagon. He is the first military officer to earn an advanced degree in Environmental Regional Planning, and he is responsible for a number of symposia on land use planning and aviation. He is the editor of *Master Planning the Aviation Environment*, and author of a number of articles which have appeared in such diverse publications as the *Air University Review* (The scholarly journal of the United States Air Force) and *Planning, Environmental Science and Aviation,* the opening volume in a series published by the American Bar Association Section on Insurance Negligence & Compensation Law and developed from a National Institute on Environmental Litigation.

DAVID B. LARGE studied electrical engineering and business administration as an undergraduate at the University of Colorado. He continued with graduate studies in engineering at the University of California, Berkeley, followed by a year at the Lawrence Radiation Laboratory. Returning to the University of Colorado, he conducted research on radiowave propagation problems at the Environmental Science Service Administration facility in Boulder, and was awarded a Ph.D. for his research in 1968 from the University of Colorado. While engaged in applied geophysical and communications research at the Westinghouse Georesearch Laboratory, he became increasingly interested in the public policy problems arising at the interface between technology and society. Abandoning his technological research activities in 1971 to consider the broader problems of environmental degradation and social equity, he has served as an Associate with the Conservation Foundation in Washington, D.C. from 1972 until the fall of 1973, when he accepted an appointment as Program Analyst (Technology) with the recently established National Commission on Water Quality.

JAMES A. LOWDEN, Vice President, Trizec Corp., Ltd., Canada. Before joining Trizec Corp., Canada's largest public real estate investment and development company, Mr. Lowden was president of Canadian Interurban Properties, Ltd., from 1960 to 1970. He is presently vice president of the American Society of Real Estate Counselors, and has recently served as a trustee of the International Council of Shopping Centers.

ROLAND COMSTOCK is Presidential Staff Assistant at Northern States Power Co., Minneapolis, involved with legislative activities and long-range corporate planning. He was formerly director of environmental affairs for Northern States. Mr. Comstock is also a member of the Governor's Natural Resources Advisory Council in Minnesota.

MICHAEL MORRISON is a Washington D.C. journalist specializing in energy affairs. His book on energy and society will be published in spring, 1974.

Since this article was written, Atlantic Richfield has become the first oil company to call for termination of the oil depletion tax allowance and for a return of petroleum pricing to the "dynamics of the marketplace." The depletion allowance is the foundation for incremental pricing of oil which Mr. Morrison described.

IRVING LIKE is an attorney and was the draftsman for the Conservation Bill of Rights Amendment to the New York State Constitution. He has been active in litigation against the nuclear electric power-generating facility proposed by the Long Island Lighting Company at Shorehan, New York. He has published a number of articles in environmental matters and in his regular practice is extensively involved in banking and security matters.

NICHOLAS MUHLENBERG is an Associate Professor of Regional Planning in the Department of Landscape Architecture and Regional Planning of the University of Pennsylvania. His interests in natural resource management found an academic focus at the University of Michigan where he took his Master of Forestry degree, and wrote the thesis which is published in this volume.

Since then he received a Fulbright Scholarship to New Zealand, worked on tree farms in Oregon, and received a Masters degree in Economics and a Ph.D. in forest economics from Yale University.

Before joining the faculty of the University of Pennsylvania he worked for the Land and Tax Department of Crown Zellerbach Corp. in Portland, Oregon, the Food and Agriculture Organization and the Economic Commission for Europe of the United Nations, was appointed Lecturer in Forest Economics at the University of California at Berkeley, and did research for Resources for the Future at Pennsylvania State University.

He was responsible for the original formulation of the Regional Planning program of the University of Pennsylvania, in collaboration with Prof. Ian L. McHarg, an effort which led to the development of a curriculum leading towards a Masters Degree in Regional Planning.

In addition Dr. Muhlenberg has been consultant to Cape May County New Jersey, the Venezuelan Government, E.P.A. and several private consulting firms.

CONTENTS

xv

Part I

ENERGY

*

1

WORK, ENERGY, AND POWER

John W. Andrews

§1.1 Introduction. It might be said that the history of western man since the Renaissance can be interpreted in terms of his increasing understanding of energy and mastery over its use. Many may disagree with that statement especially as it seems to concentrate on the materialistic aspects of a changing society, and to ignore the spiritual and aesthetic. And surely "the truth," however it is to be interpreted, is like a solid with many faces, so that any generalization can at best describe only one of these facets at a time.

Still, energy has been a leitmotif appearing again and again at critical times during the last 400 years, both for good and for evil. The negative side is obvious. The need to develop instruments of war has always been a major spur to energy utilization; more recently, the worldwide danger of environmental degradation as a side effect of growing energy use is clear to all. On the positive side, though, it must be recognized that the increasing abundance of energy has resulted in the abolition of human slavery, the rise of the once impoverished working class to relative affluence, and the vastly expanded opportunities for creative expression in the sciences and arts which exist in our society, and for which societies less fortunate than ours are still striving.

Therefore, it cannot be said whether the net effect of this energy adventure has been for good or ill. Planet Earth is still in the dock, and the jury is still out. There probably exist, scattered throughout the universe, worlds where the inhabitants destroyed themselves, or were unable to come to terms with energy and wisely use it, so that now they eke out a primitive existence in a degraded environment. In others, perhaps, the occupants were able to overcome whatever social handicaps their particular evolutionary processes imposed upon them, and now live on high plateaus of excellence and love.

An old Chinese curse goes, "May you live in interesting times." Of course, every era thinks that it is unique, but that ours stands apart is demonstrated by Figure 1.1 which shows the extent of fossil fuel use from times long past projected well into the future. (See Chapter 2 for a detailed consideration of fossil fuels.)

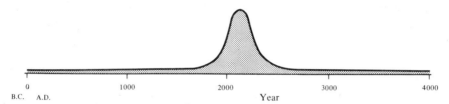

Fig 1.1.   Estimated world fossil fuel usage rate.

Until the Industrial Revolution, society was a low-energy operation, depending on human muscle and beasts of burden for most of its energy. Not too far in the future, a few hundred years at most, we will have essentially exhausted our limited supplies of fossil fuels (assuming near present rates of consumption) and we will face the choice of developing alternative forms of energy or of returning to a low-energy mode of existence which will last until the end of time. In the meantime, we face the immediate dangers of destroying our civilization with high-energy weapons of war, and of contaminating our environment as a result of intensive use of fossil fuels or the imprudent and premature deployment of new energy sources. Understanding something about energy, then, would seem a worthwhile task, not only in view of the immediate problem of short supply facing the United States and other industrialized countries, but also in light of the historical processes of our age.

§1.2. Energy, Work and Power. There are three concepts that one simply has to come to terms with if an understanding of what is happening in the field of energy is to be achieved: *energy* itself, *work* and *power*. Energy and work are on a similar footing, and one is often transformed into the other, but power is different. It is not the same as energy, even though in common parlance the two words are often used interchangeably. When people talk about "generating energy" and "producing power" they probably mean the same thing; nevertheless, it is basic to an understanding of these things to keep energy and power distinctly separate in our minds.

Each of the three terms, energy, work, and power, was borrowed from the English language, and each had a popular meaning before it was given a technical meaning. Introductory physics texts often go to great lengths to emphasize that these words do not mean the same thing when used technically as they do when used in an everyday sense, and although the texts may be correct in observing this distinction, it is well to remember that the words were not chosen at random. There is a resemblance, an analogy, between the everyday and technical meanings of these words which can be used to keep the technical meanings clear in our minds.

§1.3. Work. *The Random House Dictionary of the English Language* defines work as "exertion or effort directed to produce or accomplish something". The key words in this definition are *exertion* and *accomplish*. In the technical sense, as well as the human sense, these two ingredients must be present before work can be done. There must be an exertion, a *force* of some kind; and that force must accomplish something. It must move something through a *distance*. Impounded water serves as an example of this principle. The water backed up behind a dam exerts a considerable force on the dam, but, since it does not move the dam, it does no work. If the impounded water is allowed to flow through turbines, however, the force that it exerts on the turbine blades will cause them to move, and therefore will cause work to be done.

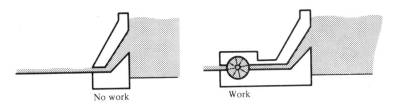

No work   Work

Fig 1.2.

Work in the product of a moving force. The amount of work done by a force is equal to the force multiplied by the distance moved in the direction of the force. Work equals force times distance.

Work in moving down slant = Slant distance × Component of Force along slant

$$= \frac{\text{Vertical distance}}{\sin \theta} \times \text{Weight} \times \sin \theta$$

$$= \text{Vertical distance} \times \text{Weight}$$

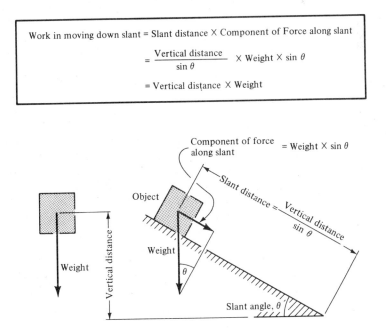

§1.4. Energy. Energy occurs in a bewildering variety of forms, but basically all forms of energy can be divided into two types. Energy is change, alteration, movement. This change can either be occurring at the present time, in which case we call it *kinetic energy,* or held in abeyance for the future, in which case it is called *potential energy.* The concept of work is important because it is a measure of the amount of energy required to do a job. Energy may be used to do work, and sometimes, as in an electric generator, we do work on a machine to convert one form of energy to a more useful form. We may also store some of the energy so converted, and this stored energy may enable us to do useful work at another time or place. In this regard, energy is like money. We do work in the human sense and thereby earn money, which we then spend to cause others to do work for us. To continue the analogy between energy and money, potential energy is like money in the bank and kinetic energy is like money in the process of being spent.

Nature has provided us with a number of energy "bank accounts". Some, like the energy of water trapped behind a dam, she is continually replenishing like a benevolent rich aunt; others, like coal, oil and natural gas have been bequeathed to us to use as we will so long as they shall last. When they are gone, they are gone. The energy resulting from the fusion of the atomic nuclei of deuterium and tritium is like money held in trust for a child—it is there in the bank, but not yet available.

Energy use typically involves a chain of transformations. When we draw on our fossil fuel bank account, we burn the fuel to produce heat. Any combustion process is like a taxable transaction. It is not possible to obtain useful work out of all the energy released during combustion. Through the use of heat engines we can recover part of the energy released as heat. This energy may be made to do useful work immediately, or it may be stored, for example, electrically in a storage battery, or mechanically, in the upper reservoir of a pumped-storage electric generating facility, but in doing useful work, heat is always generated. Eventually, all the energy of the fuel is transformed into waste heat. Even if a use were found for some of this waste heat, it would eventually become so degraded, so diffuse, so spread throughout the environment as to be of no further use, even theoretically. Figure 1.3 shows how these transformations might occur.

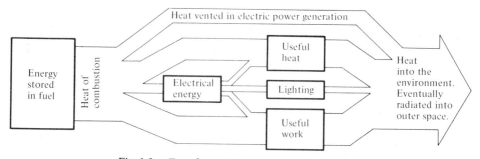

Fig 1.3. Transformation of heat obtained from fuels.

The fact that heat production always accompanies energy generation and use, and that all energy is eventually degraded into heat, gives rise to concentrations of heat which may be considered *thermal pollution* or *thermal enrichment.*

§1.5. Power. We have seen that energy can be released from storage to do useful work, as when we burn gasoline in the engine of our automobile, and work can be done to produce energy, as when steam or falling water is used to turn turbines to produce electricity. It remains for us to discuss power. *Power* refers to the ability to produce energy or do work in a given amount of time. To go back to our money analogy, earning power may be described as the amount of money a person earns per month. Two individuals may have the same earning power, but if one of them dies prematurely or is disabled, he or she will have lower total lifetime earnings than the other. A clerk with an earning power of $600 per month who works twelve months per year will earn the same gross amount in a year as a seasonal construction worker with an earning power of $1,200 per month who only works six months of the year. Similarly, a 1,000 megawatt power station which is "down" half the time will generate the same amount of electrical energy as a 500 megawatt station which runs continuously. Just as the total earnings of a person can be found by multiplying his or her earning power by the total time worked, so the energy produced by a generator (or the work done by a machine) can be determined by multiplying its power by the time it is in operation. Energy is the product of power and time. The amount of energy (a quantity produced) is equal to power (a process) times the length of time the power is applied.

§1.6. How Energy and Power are Measured. When any quantity is measured, the amount of it is compared with some standard. A board that is three times as long as a yardstick is said to be three yards long. An activity that takes a time equal to five rotations of the earth on its axis is said to be five days in duration. So too with energy and power: some standards, or units, of measurement are required.

Units of energy and power have always been a barrier to understanding. This is quite understandable for two reasons: confusion of energy units with power units, and the multiplicity of units for both energy and power. It is not necessary to know the definitions of all these units, but just as examples, here are some power units: watt, kilowatt, megawatt, gigawatt, horsepower, and Btu per hour. Energy units are even more numerous: kilowatt-hour, Btu, erg, joule, foot-pound, therm, electron volt, megaton of TNT, and even something called the Q or quad.

Probably the units that were most familiar from this list were the kilowatt and the kilowatt-hour. So let us use these as our basic units and try to develop an intuitive feeling for what they mean. Then we will relate some of them to certain uses of energy in the economy so that you will be able to sense the relative magnitudes of the quantities involved.

Let us again clearly distinguish between energy and power. Power is the rate of energy use or production, or alternatively, the rate at which work is done, that is, the amount of energy used or work done per unit time. Energy is the product of power and time or power multiplied by the length of time the power is applied. Energy may be thought of as an entity while power should be thought of as the process of generating or using energy.

One basic unit of power is the watt. The other is one thousand times larger, the kilowatt. As you probably know, the prefix kilo means one thousand. You have probably seen these units used in describing electrical power, but they can be used to describe any kind of power. How do these units of power compare with things that you can see and feel? Try the following experiment. Place this book on the floor. Now lift it onto the table. Lifting the book required you to exert force and move an object over a distance.

Lifting the book required your body to generate power of about one watt. The food you eat represents a source of potential energy for your body. Staying alive and functioning normally as a human being uses this stored energy to produce about 100 watts or a tenth of a kilowatt of power, on the average. However, only a small portion of this power is normally available for you to do useful work. The rest of the energy is used by your body to fuel its metobolic processes so that you can stay alive. During short periods of exertion, such as running up a flight of stairs, a person in good physical condition may be able to generate power at a greater rate, approaching a kilowatt or so. So, on a personal level, think of the book being raised and a football player charging full tilt up the stairs after his girlfriend when you think of a watt and a kilowatt. (Our football player does work against gravity in running upstairs, but even on the level gridiron he would do some work in accelerating from a standing start to full speed, and in overcoming the friction of air and turf (natural or synthetic). Work would be done on him by an opposing tackler in bringing him to a halt, or even by the turf if he were to slip and fall.)

Now let us suppose that the stairs up which our gridiron Lothario is charging were located in an extremely high tower, such that to get from the bottom to the top would take an hour of high-speed charging. Of course, no human being could run at his initial top speed up a steep set of stairs for anywhere near an hour, but suppose he could. How much work would he have done in lifting himself up all those stairs? Since he was developing a power of one kilowatt over a time of one hour, the answer is, of course, one kilowatt-hour. Now that's a lot of work! How much do you think you would have to pay for an equivalent amount of electrical energy? Ten dollars? A dollar? Ten cents? You're getting warm. At average 1973 prices, you would pay about two cents, for all that energy. Is it any wonder we waste it and have lost our respect for it?

Be sure to remember the kilowatt is a unit of power. The total amount of power which can be generated by power plants is listed and therefore power plants are rated in kilowatts or megawatts. (One megawatt equals one thousand kilowatts, or one million watts). Some newer plants are rated in gigawatts (One thousand megawatts or one million kilowatts or one billion watts). The unit of energy is the kilowatt-hour. They must never be used interchangeably. Remember lifting this book onto the table; since it probably took you about one second to do this, you did about one watt-second of work. The watt-second has another name, the joule (pronounced jewel). This is the basic unit of energy in the system most often used by physicists. If you feel so inclined, you may want to see if you can determine how many watt-seconds equal one kilowatt-hour.

A compact automobile crusing at 50 miles per hour on a level highway might require a power of about 10 kilowatts, although this figure would rise considerably during acceleration. Gasoline is the energy source for the internal combustion engine that powers the automobile. To release the energy stored in the gasoline, we burn the gasoline, a highly inefficient conversion process, since we must burn enough gasoline to generate approximately 70 kilowatts of power in order to obtain the 10 kilowatts of power we need to turn the wheels and move that car down the highway.

§1.6.1. Other Units of Work and Energy. At this point we could define a great many units and present a lengthy table of conversion factors, but the object of this chapter is to provide a conceptual framework which you should be able to carry around in your head without too much difficulty. We have begun to do this already by discussing the watt and kilowatt in terms of easily imagined events, and now we are going to introduce a unit which, if it is not a household word already, soon will be. This is the *Q* or *quad*. It is used to describe large amounts of energy, such as might be used by a sector of the American economy in a year. Strictly speaking, the quad is equal to 1,000,000,000,000,000 or one quadrillion British thermal units whence the name. A British thermal unit (Btu) is the amount of heat needed to add one degree Fahrenheit to the temperature of a pound of water (about a pint). Another handy unit sometimes used is the therm which equals 100,000 Btu or about 30 kilowatt-hours of heat. The heat content of fossil fuels is often quoted in therms: oil—1.5 therms per gallon or 63 therms per 42 gallon barrel; natural gas—10 therms per 1,000 cubic feet; anthracite or bituminous coal—260 therms per ton; ignite and subbituminous coal—140 to 200 therms per ton.

But rather than try to keep that Q figure in your head, think of it this way: if our nation uses one quad of energy in a year for some purpose, then on the average, each man, woman and child in the United States would use the energy contained in about a gallon of gasoline (more accurately 4/5 gallon) each and every week for that purpose. Here is a list of some common energy units, together with their definitions:

Fig 1.4.

Btu: (British Thermal Unit) the amount of heat necessary to raise the temperature of one pound of water one degree fahrenheit.

calorie or **gram-calorie:** the amount of heat necessary to raise the temperature of one gram of water one degree centigrade.

Calorie or **kilogram-Calorie** (this is the food Calorie): the amount of heat necessary to raise the temperature of one kilogram of water one degree centigrade; 1000 gram-calories.

Electron-volt: the amount of energy gained by an electron in moving through a potential difference of one volt.

Erg: the amount of work done by a force of one dyne (10^{-5} newtons) moving through a distance of one centimeter; a dyne-centimeter; 10^{-7} joules.

Foot-Pound: the amount of work done by a force of one pound moving through a distance of one foot.

Joule: the amount of work done by a force of one newton (0.225 lb.) acting through a distance of one meter; a newton-meter. Also, the amount of work done by a machine, or the amount of energy produced by a generator, with a power of one watt acting for a time of one second; a watt second.

Kilowatt-hour: the amount of work done by a machine or the amount of energy produced by a generator with a power of one kilowatt acting for a time of one hour.

Therm: 100,000 Btu.

Quad: One quadrillion (1×10^{15}) or 1,000,000,000,000,000 Btu's or ten billion (1×10^{10} or 10,000,000,000) therms.

TABLE 1.1 ENERGY AND WORK

To convert COLUMN units to ROW units, multiply COLUMN value by
the number in the box at which the ROW and COLUMN intersect.
Example: 8 ergs = ? ft-lb; multiply 8 by 7.376×10^{-8}

	joule	erg	ft-lb	kWh	eV	calorie	BTU
joule	1	10^{-7}	1.356	3.6×10^{6}	1.602×10^{-19}	4.186	1.055×10^{3}
erg	10^{7}	1	1.356×10^{7}	3.6×10^{13}	1.602×10^{-12}	4.186×10^{7}	$1.055 \times$
ft-lb	0.7376	7.376×10^{-8}	1	2.655×10^{6}	1.182×10^{-19}	3.087	7.783×10^{2}
kWh	2.778×10^{-7}	2.778×10^{-14}	3.766×10^{-7}	1	4.450×10^{-26}	1.163×10^{-6}	2.930×10^{-4}
eV	6.242×10^{18}	6.242×10^{11}	8.464×10^{18}	2.246×10^{25}	1	2.613×10^{19}	6.585×10^{21}
calorie	0.2389	2.389×10^{-8}	0.3239	8.6×10^{5}	3.827×10^{-20}	1	2.520×10^{2}
BTU	9.480×10^{-4}	9.480×10^{-11}	1.285×10^{-3}	3.413×10^{3}	1.519×10^{-22}	3.919×10^{-3}	1

§1.6.2. Energy, Power, Work and An Automobile. It was observed that wnen the car described in the text, equipped with a standard transmission, was placed in neutral while going 53 mph, it slowed to 48 mph in a tenth of a mile. The car weighed 3000 pounds (including driver). We calculated the force of friction (internal friction of moving parts beyond the transmission, plus wind friction) that the car must overcome by equating the car's loss of kinetic energy to the work done by the frictional force. Kinetic energy equals ½ mv^2, where "m" is the mass of the car (equal, because of Newton's Second Law, to the car's weight in pounds divided by 32 ft/sec², the acceleration due to gravity), and "v" is the speed of the car (measured in feet per second). The following table outlines the calculations; if your car has a transmission which can be thrown into neutral at high speed without damage, you may want to repeat the experiment for yourself.

Table 1.2. Calculations for Automobile Problem

V (mph)	V (ftsec)	½MV²(ft-lb)
53	77.7	283,000
48	70.4	232,000

The work done by wind and internal friction in slowing the car equaled the difference between these two kinetic energies—the energy lost by the car—or 51,000 foot-pounds. The distance traveled was 1/10 mile or 528 feet, so the force on the car was the work devided by the distance or about 100 pounds. The motor must provide an equal and opposite force if the car is to be kept moving rather than slowing down. At 50 m.p.h. the engine would do work in one hour equal to the 100 lb. force times the distance of 50 x 5280 feet. The work, then, during this 50 mile (or one hour) trip would be 26,000,000 ft-lb. Converting this to kilowatt hours by multiplying by 0.38 and dividing by a million (see our discussion of gravitational energy) we obtain very nearly 10 kilowatt hours. Since this is the work done in one hour the power was ten kilowatts.

At 50 m.p.h. this car gets 25 miles per gallon (it has radial tires, a standard transmission, no power options except power brakes, and is kept well-tuned). The energy equivalent of a gallon of gasoline is about 115,000 Btu or 1.15 therms (equivalent to 14 sticks of dynamite according to *Newsweek*) or 1.15 × 29.4 = 33.8 kilowatt hours (1 therm = 29.4 kwh). Since it used two gallons, its energy use was 67.6 kilowatt hours, or again, a power demand of 67.6 kilowatts, since this energy was consumed in an hour. The engine's efficiency was, therefore, about 15%.

§1.7. Gravitational Energy. Energy is a changeable beast; it appears in many forms. Considering some of the forms of energy which have been put to use by mankind, the simplest from a practical viewpoint (although by no means simple from the theoretical one) is the potential energy that can be stored in matter due to its height. This gravitational energy can be stored in water behind dams to be used by hydroelectric generating systems where the falling water releases its stored energy doing work on turbines which turn generators to produce electricity. Gravitational energy can also be used as a medium for the temporary storage of energy in pumped-storage facilities such as the one planned at Storm King Mountain on the Hudson River in New York State.

We saw that work is the product of force and distance. Work equals force multiplied by the distance over which the force moves. To lift water from the bottom of a dam to the top, we would have to exert a force on the water equal to its weight, and therefore to do work equal to this weight multiplied by the distance the water was raised, namely the height of the dam. This work we have done is stored as potential energy during the time the work is being performed, and is released when the water is allowed to flow back to its original level. If the water simply flows over the dam, the energy is released when the water crashes into the pool below the dam. If, on the other hand, the water is made to strike the blades of a turbine, it can be made to do useful work as it falls.

In our example we lifted the water to the top of the dam. In a practical hydroelectric station, the sun does the work, lifting the water vapor from the oceans and other bodies of surface water by evaporation and storing the energy resulting from this work in the air, often visible as clouds. The vapor laden air eventually releases its burden of water as precipitation — rain, snow, sleet or hail which finds its way to the reservoir of the dam in our example.

In a pumped-storage facility, electrical energy generated elsewhere is used to pump the water uphill. This does not create new energy, but does allow energy to be stored for use at times when demand is high.

Some final words about gravitational energy might be in order. First, in using the fact that work is force acting over distance (force multiplied by the distance over which the force acts), we used the vertical distance. This would be appropriate even if the water flowed down on a slant, as it would do, for example, in the Storm King Plant. In figuring work, we have to use the distance travelled in the direction of the force. Since the force of gravity is vertical, we are only concerned with the vertical distance.

Secondly, the easiest unit of energy to use here is the foot-pound, since it is natural to measure force in pounds and distance in feet. The best plan to follow when units other than our basic ones of kilowatt-hour, therm, and quad come up is to look up the conversion factor and as soon as possible convert to a familiar unit. Here, for example, if we multiply the number of foot-pounds by 0.38 and divide by a million we will obtain the number of kilowatt-hours.

In order to apply critically the techniques and concepts presented in this chapter, five case histories have been included. Each of these represents either a recent cause celebre or a banner of hope for energy hungry Americans. You should now be able to test basic assumptions and assertions in these and other public issues.

§1.7.1. Storm King.

They'll get more electricity out of a hand crank than they'll get out of Storm King in the next twenty years.

— Rod Vandivert, environmental consultant to the
Scenic Hudson Preservation Committee

More Power To The People.

Consolidated Edison Power Company advertisement

Con Ed, the big power company that supplies electricity to New York City, has been trying for over a decade to build a pumped-storage generating facility at Storm King Mountain, at a point on the Hudson River where its waters are constricted into a gorge of striking beauty. Opposing the plan has been the Scenic Hudson Preservation Committee which, beginning with an important court victory in 1965, has fought the project in the courts and in the regulatory agencies. As this is written, the legal obstacles to the project seem to have been cleared away, but Scenic Hudson has yet to throw in the towel.

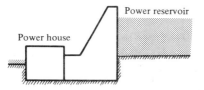

Fig 1.5. Cross section of a power dam.

Whatever one's feelings about the project, it can serve as an illustration of some of the points we have been making about energy. Con Ed's purpose is to store energy generated at night and at off-peak hours in general, when its generating facilities would otherwise be idle, and make the energy available for use at certain times of the day when the public's energy demand exceeds the rated capacity of the company's other power stations.

The facility called for by present design consists of an eight billion gallon reservoir located at an altitude of about 1000 feet above the waters of the Hudson (Figure 1.6). The reservoir is to be linked by a two-mile-long aqueduct to a plant of a type which is able to use electrical energy generated elsewhere to pump water uphill to the reservoir, or alternatively, to convert the energy of water rushing downhill from the reservoir back into electricity, with a maximum power rating of two million kilowatts. It is like a huge storage battery, which can be charged up and later used to supply power. Thus it is not a new source of energy, but a means of "smoothing out" the uneven demand on the rest of Con Ed's facilities.

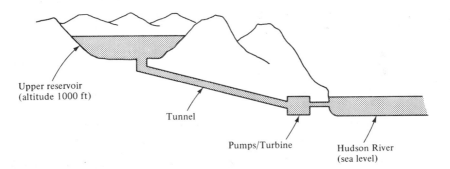

Fig 1.6. Storm King actual pumped-storage generating facility.

Opponents of the project point out that because of losses incurred in pumping the water, three kilowatt-hours of energy generated somewhere else would be used by the Storm King plant for every two kilowatt-hours given back. Proponents admit this, but claim that Storm King would replace even less efficient generating facilities now used during peak-demand hours.

The size of the reservoir—eight billion gallons—would seem to be adequate, but we are in a position to calculate for ourselves whether this is so, and in the process to check on our understanding of some of the concepts we have discussed. Let us ask the following questions. How much energy would eight billion gallons of water falling one thousand feet release? How long could a two million kilowatt plant operate on this energy?

Let us first calculate how much energy one gallon would release. We saw that the potential energy that any material has due to its height is equal to the weight of the material times its height above the level to which it falls. This is equivalent to the work the material could do in falling—the force (weight) times the distance fallen. The weight of a gallon of water is 8.3 pounds—"a pint's a pound the world around" is very nearly true—so that if it falls a distance of 1000 feet, the energy it can deliver is 8,300 foot-pounds. If you remember from the discussion of gravitational energy that the way to convert foot-pounds to kilowatt-hours is to multiply the number of foot-pounds by 0.38 and divide by a million, then our 8,300 foot-pounds is equal to 8,300 × 0.38/1,000,000 or 0.0032 kilowatt-hours.

We can estimate the amount of energy available if the eight billion gallon reservoir were drained dry by multiplying 8,000,000,000 gallons by 0.0032, the number of kilowatt-hours delivered by each gallon. We obtain close to 25 million kilowatt-hours. Of course, one does not necessarily want to drain the reservoir dry, and some energy is lost during the process. Nevertheless, it would appear that the reservoir is adequate to deliver the stated peak power of two million kilowatts over the several hours in which peak demand occurs.

§1.8 Electricity. Electricity is not the only form of energy we use. In fact, only about one-quarter of our energy goes to the generation of electricity. Still, it has been the fastest growing form of energy use, because of its versatility and cheapness.

What is electricity? In most materials, the electrons—the light, negatively charged particles which swirl like a cloud of gnats about the much heavier, positively charged nucleus of each atom of matter—are bound to their nuclei by the force of electrical attraction.

In an insulator, each electron is assigned to a given nucleus and must stay there (like a sheep which is confined to a pen). In a conductor, such as silver, copper, or aluminum however, some electrons are free to move about within the volume of the conductor (like the flying sheep in the accompanying cartoon which fly over the pens but not away from them). Intermediate materials, semiconductors, may have a few excess "orphan" electrons which jump from atom to atom looking for a home (n-type) or alternatively, may have atoms with too few electrons, like the p-type semiconductor caricatured here. In this case, a sheep from an adjacent pen may jump the fence to fill the vacancy, leaving a vacancy, or "hole" in the pen it just left. This hole may in turn be filled by a sheep from the next pen, and so on. Although no one sheep jumps more than one fence, the deficiency (the "hole") moves through many pens in the opposite direction to the sheep, acting, in the case of the "hole" in the semiconductor, like a positively charged electron.

If a conductor of electricity is stretched out into a long thin wire, the conduction electrons (the free ones) can be made to transmit energy from one end of the wire to the other. Actually, the drift speed of conduction electrons in a wire is quite slow, so they transmit energy more by bumping into each other than by moving continually toward their goal. Imagine for a moment that you are at the end of a long line of people waiting for tickets to a show. If you bump the person ahead of you, he may bump the next one, and the bump may be transmitted to the head of the line long before you ever get near the box office.

If all the bumps are transmitted in the same direction, we speak of *direct current* (DC). If we imagine the people in our queue forming a closed loop (a circuit) then we could imagine transmitting bumps alternately in each of two directions *alternating current* or (AC). In this case, instead of drifting down the wire in one direction, the electrons will oscillate back and forth about some equilibrium point. Energy can be transmitted this way just as well as by steady pushing. Typical examples of this are sawing wood and turning a hand crank.

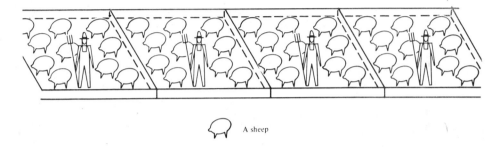

Fig 1.7. **An insulator. The sheep (electrons) must stay in their own pens (atoms).**

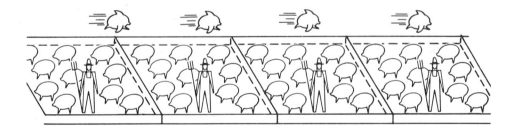

Fig 1.8. **A conductor. Some sheep can fly along the line of pens.**

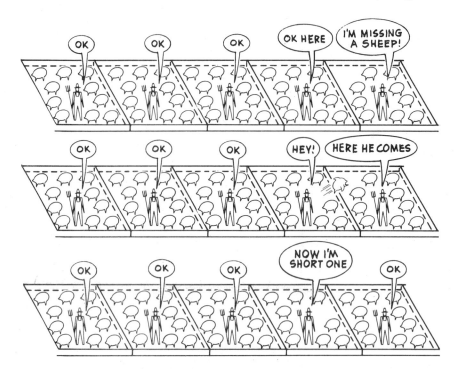

Fig 1.9. **A p-type semiconductor. The sheep (electrons) all jumped to the right, but the missing sheep (hole) moved to the left.**

Alternating current is used commercially because of the ease with which voltage can be stepped up for transmission and back down again for use in homes and factories. It is much more cumbersome to change the voltage of direct current. Furthermore, alternating current motors are simpler and more compact than their DC counterparts.

Both motors and generators operate on the principle that a changing magnetic field causes a current to flow in a wire, and conversely a current flowing generates a magnetic field. In a generator, we turn a rotor or armature around which wires are wound, causing them to move past nearby magnets. When a wire passes a positive magnetic pole, current flows in one direction; when it passes a negative pole, the direction of the current reverses. Thus, alternating current (AC) is easy to generate. If, instead, an AC current is passed through the wires, magnetic fields will be generated which alternately attract and repel the poles of the magnets, thereby causing the rotor to turn. Thus one and the same machine can be used as a motor or a generator depending on whether we turn the rotor to produce current or introduce current to obtain useful work from the rotor.

Regardless of whether the current is alternating or direct, electrons are not endlessly and mindlessly dispatched over straight lines. So as to keep a balance of posotive and negative charges at all points, the wire is formed into a loop called a cir-

cuit. The loop or circuit will require at least one energy source (in your transistor radio, the energy source will be a battery; in the circuit of which your house wiring is a part, the energy source is the central power station which may serve an entire city), and one or more loads (devices like electric motors, heaters, and light bulbs, which use electrical energy to do useful work or produce useful heat and light). In addition, the wires themselves must be considered a minor load as some heat which is not used is produced when electrons are forced to travel through the confined area of the wire.

The best way to visualize the important concepts concerning electricity is to compare an electric circuit with another kind of circuit, one in which water flows instead of electrons (Figure 1.10).

Our electric circuit is a simple one. A battery is connected by means of wires to an electric motor. The battery takes electrons from the positive terminal and "pumps" them to the negative terminal, forcing them out into the wire. (see figure 1.11 for explanation of negative and positive polarity). Some of the energy given to the electrons by the chemical push from the battery is released in the form of heat as the electrons overcome the resistance as they flow through the wire, and more as the electrons flow through the motor transforming electrical energy into mechanical energy doing useful work.

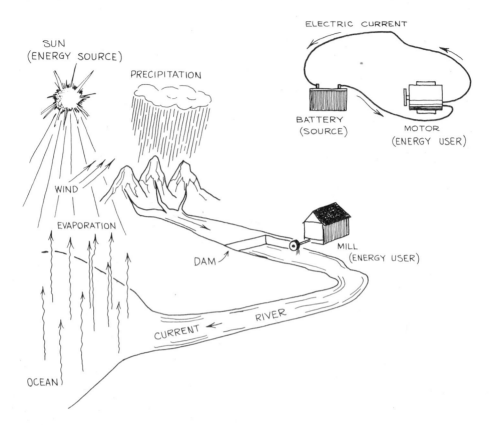

Fig 1.10.   Analogy between a river system and an electric circuit.

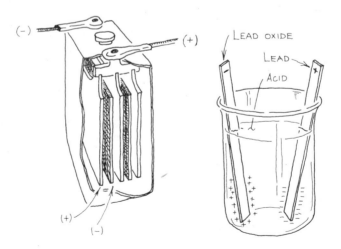

Fig 1.11.   Electrical polarity.

In the water circuit, the energy source, the sun, evaporates water from the ocean and carries it to the mountains on the winds where it falls as rain or snow. The sun acts like a gigantic pump which lifts the water from sea level to the altitude of the mountain tops. It does in the water circuit what the battery does in the electric circuit. After falling as rain, the water makes its way into mountain streams, which ultimately combine to form a river. As the river flows downhill, the potential energy it had due to its height is converted into the mechanical energy of motion. Most of this energy is lost along the way as the water overcomes the resistance to its flow represented by rapids, sunken trees, and even the bottom and banks of the river itself. This loss of energy attributed to overcoming the resistance of the channel to the flow of the water is similar to the resistance offered to the motion of electrons by the wire through which their flow is confined. When electrons are forced to move through a wire in spite of the resistance of the wire, heat is generated.

In engineering applications, current is usually considered to flow from the positive terminal of the battery to the negative, because it has the higher voltage, and it is easier to think of electrical current flowing from a high electrical height to a low one. But it is actually electrons that flow, and they travel the other way. It would have been a lot nicer if Ben Franklin, who first applied the terms "positive" and "negative" to electricity, had reversed his designations, for then electrons would have been called positive and we would be able to say that a positive charge really did flow from the positive to the negative side. But Franklin didn't know about electrons, so he had to flip a coin. (the equations and calculations, if correctly applied, work no matter which assumption you make, as long as you stick to it).

Somewhere along the river a dam is built, and the water backed up behind the dam is used to power a mill. The two factors that control the power (the rate at which energy can be extracted from the reservoir behind the dam) available from the impounded water are the height of the dam and the amount of water flowing over the dam during each unit of time.

In electrical work there is a quantity that is like the height of the dam, and another that is like the quantity of water flowing per second. The quantity that is like the height of the dam is called *voltage* or potential difference, and it is measured in *volts*. The voltage difference between the input and output terminals of a piece of electrical apparatus is the electrical "height" of the apparatus. The quantity that is like the amount of water flowing per second is called the *current* and is measured in *amperes* (commonly called *amps*). The electrical current is the quantity of electrical charge flowing per second, just as the water current is the number of pounds of water flowing per second. Just as water power is dam height multiplied by the rate of water flow or water current, so electric power is the product of voltage and electrical current. The volt and the amp are so defined that the power comes out in watts. Watts equals volts multiplied by amps.

We must be careful, then, never to speak of a power of so many volts or amps. It is the product of volts *and* amps that determines power. The voltage on the electrical outlets in your home is 110 volts whether or not you have anything plugged in. The open outlet represents a source of electrical potential energy, just as the water poised behind the dam does. The voltage available is simply the electrical height of the system. If you don't have anything plugged in, of course, your current is zero amps since no current is flowing, and the power is also zero because the product of

110 volts and zero amps is zero. (Any quantity multiplied by zero equals zero.) In order to have power, you must have both voltage *and* current. Again, power in watts equals volts times amps.

A final word on that popular unit of energy, the kilowatt-hour. You may see, in discussions involving kilowatt-hours (abbreviated kwh), a distinction between the thermal kilowatt-hour (abbreviated *kwht*) and the electrical kilowatt-hour (abbreviated *kwhe*). This is done to avoid confusion between the energy content of the fuel going into an electrical power plant and the energy available in the form of electricity coming out of the plant. Because of the inefficiency of the process of converting heat to electricity, it takes about three units of heat energy to produce one unit of electrical energy. The other two units are lost as waste heat. We might express this in the form of an equation (a mathematical Sentence).

3 kwht (the energy stored in the fuel and available for the production of heat) = 1 kwhe (electrical energy delivered by the electrical generating facility and available to do useful work) + 2 kwh (waste heat).

§1.9. Transmission of Electrical Energy: A Line On Four Corners. Recent proposals to expand the cluster of coal-fired power plants now being built and operated in the Four Corners region (the point where Arizona, New Mexico, Colorado, and Utah meet) and to build a 50,000 megawatt generating complex in Wyoming to burn the vast coal deposits of the region following strip mining, have provoked concern that the environment will be degraded in a number of different ways. One of the subjects of these discussions is the transmission lines that would be needed to carry the energy generated at these centers to urban areas in southern California and the midwest that use large amounts of electrical energy. Everyone has seen the huge towers carrying the high voltage transmission lines across the country, with their heavy insulators and sizable ground clearance.

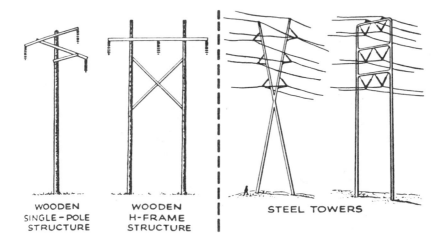

WOODEN   WOODEN
SINGLE-POLE   H-FRAME   STEEL TOWERS
STRUCTURE   STRUCTURE

Fig 1.12.

Why do the towers require such big insulators, such a great distance between conductors, and such a large ground clearance? Thinking back to our dam, it stands to reason that a high dam would have to be built out of stronger materials and have better reinforcement than a low dam. If the dam is not built well enough it may burst, allowing all the water backed up behind it to cascade downstream, wreaking much havoc. In the case of electricity the equivalent of a strong dam is a large distance between objects at different electrical heights—in this case, the conductors themselves, the towers, and the ground. Air is a good insulator—it is very good at preventing electrons from passing through it, but it has its limits. By providing enough distance (i.e., a thick enough insulating layer) between conductors and between each conductor and the towers and ground, we do the equivalent of building a strong, thick dam. If the distance is not great enough, the air may "break down" and sparking or arcing may occur. Moisture in the air is like a crack in the dam, making breakdown easier, and designers of transmission lines to have to provide for the worst possible atmospheric situation. Lightning is an example of a breakdown between two electrically charged clouds or between such a cloud and the ground.

You may also have wondered why, if electricity is delivered to your home at 110 and 220 volts, it is necessary to transmit it over lines at voltages which range up into the hundreds of thousands of volts. For a given type of power line, the amount of energy lost as heat to the resistance of the line increases as the amount of current fed through it is increased, and also as the line is lengthened. For a short line, these losses are small enough, but as the distances over which electric current must flow are increased, the heat loss becomes a significant fraction of the total energy transmitted. Some way must be found to carry the same amount of power, but reduce the heat lost, which depends on the current and the distance. Remembering that power equals voltage multiplied by current, if we increase the voltage and reduce the current, we may keep the power the same but reduce the heat losses which depend on the current. Keep the current as low as possible so the losses are low, and then raise the voltage as high as possible so that the product of voltage and current, the power, remains high.

As the distances lengthen, the need for higher and higher voltages becomes evident. Transmission lines carrying voltages of 100,000 to 300,000 volts are common, and ones of over a million volts are under consideration. Radical new approaches, such as superconducting cables, to reduce resistance or eliminate it entirely, may be developed to eliminate the need for these extremely high voltages.

§1.9.1. Transmission systems and networks. A line used solely for transmitting power from a generating plant to a distant center of load, or from one load center to another, is a transmission line. The group of interconnected transmission lines that carries the energy at a high voltage is termed the transmission system. Transmission voltages usually range from 69,000 to 500,000 volts.

Transmission lines emanating from power plants usually form a network looping through various communities and other power plants.

The *main transmission system* of bulk power supply is the system of transmission lines interconnecting a company's major power plants and the lines interconnecting with other power suppliers. Thus, the transmission system for bulk power supply is distinguished from the lower voltage transmission lines serving certain communities and industries and from the low-voltage distribution systems within the communities.

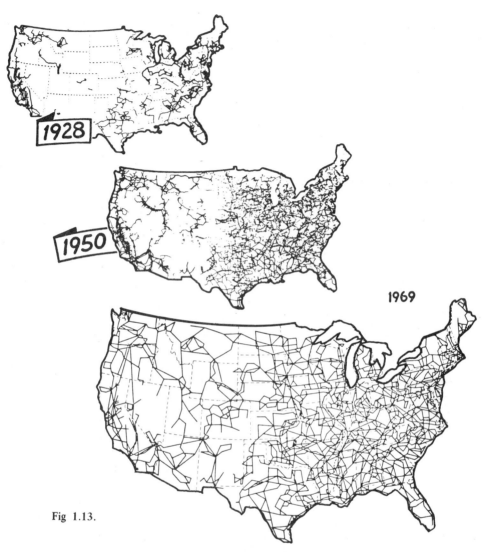

Fig 1.13.

   Much of the following information has been excerpted from Volume II of the Briefings before the Task Force on Energy of the Subcommittee on Science, Research and Development of the Committee on Science and Astronautics of the United States House of Representatives, 92nd Congress, Second Session, in March, 1972. (USGPO73-902-0) and were taken from the testimony of Professor Simpson Linke, of Cornell University, Program Manager in the Division of Advanced Technology Application of the National Science Foundation, among other witnesses.

Modern electric power systems operate in two modes: (1) the *steady-state* or *dynamic equilibrium* condition and (2) the *transient* or *unstable condition*. The condition of dynamic equilibrium is the desirable one. Unfortunately, power systems are always undergoing discrete changes in demand and supply and are subject to transmission-line interruptions and short circuits. To operate satisfactorily a system must absorb all of these disturbances and still remain in dynamic equilibrium. Whenever a disturbance of any kind causes this equilibrium condition to be lost, the system becomes unstable and the result is usually a *blackout*. Under normal conditions the system operates under several constraints, the most important one being that it must supply the entire consumer demand with minimum fluctuations in voltage and frequency. When a system cannot meet this requirement the result is a *"brownout."*

Further, the system must operate at maximum possible efficiency and with minimum overall cost; that is, for a given demand situation the most economical generators must be on-line and the energy must be transmitted over those channels that introduce minimum loss. A moment's reflection reveals that power-system operation is exactly analogous to many other enterprises: supermarket chains for example. Fortunately it is usually easier to move orange juice than it is to transmit kilowatt-hours. Finally, a relatively new constraint, but one that is certainly uppermost in the minds of consumer and supplier alike, is that power systems must now be arranged so as to operate with minimum impact on the environment.

The term "power system", as just described, may refer to a single electric utility company such as VEPCO in Virginia or Consolidated Edison in New York City. On a larger scale, the same criteria apply to groups of individual utilities that are banded together in so-called "Interconnected Systems" or "Power Pools" such as the Pennsylvania-New Jersey-Maryland Interchange (PJM) or the New York Power Pool. At present there are less than a dozen such combines in the country, each organized as a separate entity serving a specific region. Under emergency conditions it is sometimes possible for blocks of power to be moved from one region to another provided that elaborate pre-arrangements have been made. No plan or program exists at present for integrating these individual systems into a National Grid and, except for a few isolated cases, prevailing conditions do not appear to require such a network. It has been possible to interchange some 1400 megawatts between the Bonneville Power Authority and the Los Angeles, California area via the Pacific Coast d-c Intertie. (This tie has been out of service due to the recent earthquake in Southern California but is expected to be energized in the near future.) Still another restraint on the development of large-scale power-pool interconnections is the technical limitation on the practical length of overhead a-c transmission lines. Lines that are longer than 400 miles operate under severe stability and synchronization limits, and such ties are avoided in normal power system design. Similarly, underground a-c cables are even more restricted, with a maximum possible length of about 20 miles. Presumably, when the need for the National Grid emerges, such necessary technological development as long-distance direct-current transmission will be available to offset the present barriers to nationwide power distribution.

Modern electrical power systems are highly sophisticated and complicated electromechanical networks whose complexity is increasing steadily as the Nation's demand for energy expands. In spite of several untoward incidents of power disruption in the past few years, the overall record of power system development and operation in this country borders on the remarkable. The power is almost always there when we throw the switch. However, these same difficulties of the past are indicators of coming change that mark the need for new levels of technology.

§1.9.2. System synthesis and simulation in planning. An important factor in effective power planning and design is accurate computer simulation of both steady-state and transient conditions. The growing complexity of the modern power system requires that the already extensive use of computer techniques for system analysis be supplemented by application of the most advanced techniques of network synthesis and simulation.

§1.9.3. On-line optimal computer control. Present-day operation and control of power systems leans heavily upon assistance from computers, However, the systems are still very dependent upon human input for essential decision making. There is a popular misconception that power systems now operate in a completely automatic fashion. To a certain degree, automatic operation does in fact exist in that small perturbations on the system can be absorbed and proper compensation can be applied without manual assistance. On the other hand, large-scale changes in power demand cannot be inserted into the system without previous warning and elaborate pre-scheduling and pre-setting of system controls. In the event of a large transient disturbance such as the sudden loss of a major generator or transmission line, the protection of the power system relies completely on pre-set safety devices known as protective relays with their associated circuit breakers. While these mechanisms have attained a very high state-of-the-art, they are a far cry from the control functions required for completely automatic protection. As the systems grow in size and complexity, pre-scheduled decisions will be inadequate and human reaction will be too slow to maintain the proper operating criteria. The day is coming when it will be necessary for power systems to be completely automated. Realization of the completely automated power system is considered to be a major challenge to control-system engineers who see this task as an excellent application for the field of optimum-control theory. Under such control, a disturbance of potential blackout proportions would be detected and proper defensive action would be initiated before disruption of the system could occur.

§1.9.4. High-voltage direct current transmission. Parallel with the development of automatic control of individual power pools there is the necessity to provide means for future requirements of large-scale distribution of bulk power throughout the Nation. It may be necessary to construct a "super-power network" that would allow the entire country to operate as one very large integrated power pool. The key to successful development of such a grid lies in the ability to operate long-distance, high-voltage direct-current lines in this network form. Present technology does not allow feasible and economic operation of such a d-c grid. Thus the necessary research and development of direct-current transmission systems has a high order of priority in energy planning. Some of the specific tasks that must be studied are: switching and control, and methodology or operation of d-c and a-c networks in tandem. It should be pointed out that d-c transmission also holds the promise of eventually being competitive with gas pipelines for the transfer of bulk energy over trans-continental distances.

§1.9.5. Long-distance underground power transmission. Large-scale, long-distance, bulk-power transmission of the future will require use of underground cables not only to satisfy environmental concerns but to allow transmission through otherwise unavailable rights-of-way and possibly along underwater routes. The major obstacles to high-voltage underground transmission today are the very high cost of conventional cables, both in material and labor, and the large heat losses that limit the practical length of the lines. Even if the cost factor is ignored, the charging-current heat loss limitation still makes the present-day a-c cable unfit for long-distance links between major systems. As mentioned earlier, the maximum present usable length of a-c underground lines is about 20 miles. An exciting prospect for the solution of these problems is the use of superconducting and cryogenic cables. Since the former operates at liquid-helium temperatures, the technology is very difficult and costly and may not be fully developed for many years. However, high-voltage cryogenic cables (cryocables) operating at liquid-nitrogen temperatures are now feasible, and with suitable research and development, could be brought into use by the end of this decade. Use of cryocables could reduce transmission losses by a factor of from three to six and would thus have significant impact on transmission systems.

§1.10. Chemical and Nuclear Energy. Most of the energy which fuels our economy results from chemical reactions. Reactions which release energy (exothermal) are of two types. The first type involves the release of energy when complex molecules break up into smaller ones. As an analogy, imagine two billiard balls joined by a spring (Figure 1.14). If we compress the spring, and tie the two balls together with a light thread just strong enough to keep the spring compressed, we will have stored energy, since we had to do work to compress the spring. If for some reason the thread is broken, the spring will propel the two balls in opposite directions. giving them energy of motion or *kinetic energy*. If there are many such ball and

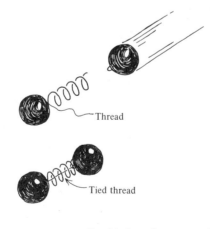

Fig 1.14. Stored energy: Billiard balls and compressed spring.

spring setups in the vicinity, the two moving balls may hit others with enough force to break their threads, sending them off to hit still others, and so on, in a *chain reaction,* with the result that within a short time most or all of the springs will have been released. The tremendous amount of energy released in such a short time would result in a tremendous explosion. On the molecular level, some kinds of molecules have excess energy stored within them that can be released very quickly. Nitroglycerin is an example of such a molecule.

The second way that chemical energy can be released is by a combination of molecules or of atoms from different molecules to form new molecules. The burning of fossil fuels is this type of reaction. Take natural gas, for example. It consists largely of methane, a molecule consisting of one carbon atom and four hydrogen atoms. Under normal conditions a mixture of methane and oxygen will not do anything unusual. To use a different type of analogy, the oxygens are like shy teenage boys at a high school dance, each waiting for the other to make the first move towards the girls. They move about in pairs, like pals, giving each other moral support (Figure 1.15). The carbon and the hydrogens stick together like the girls at the same dance, not knowing what to do.

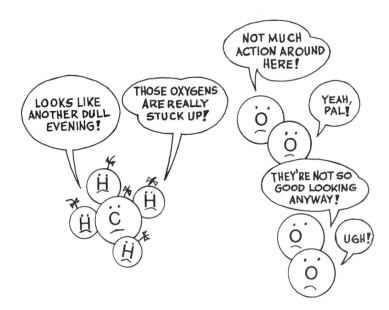

Fig 1.15.   Methane and oxygen. (Things are kind of slow.)

Now strike a spark into the mixture. With the heat turned up, the inevitable attractions begin to occur (Figure 1.16). One of the oxygen atoms splits off from his friend and begins to move in on a sexy little hydrogen atom. He likes the first one so well he grabs two, and a *menage a trois* known as a water molecule is formed. The other oxygen, not to be outdone, makes off with the other two hydrogens. The carbon, being more strong-willed than her hydrogen sisters (chemists say she has a higher valence), is not content to share an oxygen with another carbon, but latches on to two oxygen atoms all for herself. Now a different type of threesome, carbon dioxide, is formed. The entire reaction can be written as a chemical equation:

$$CH_4 + 2O_2 \rightarrow CO_2 + 2H_2O$$

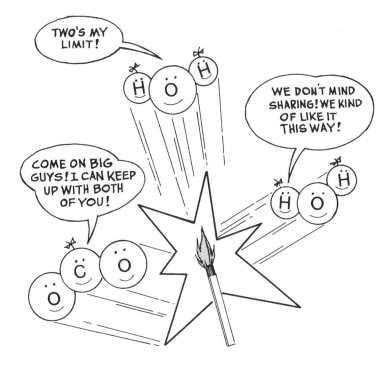

Fig 1.16. Water and carbon dioxide. (Someone lit a match.)

The attractive forces between the oxygens, on the one hand, and the carbon and hydrogens, on the other, have caused these atoms to move towards each other. Since a force has moved objects through a distance, work has been done, and the work is now available as kinetic energy of the molecules. Molecular kinetic energy, as we will discuss shortly, is nothing other than heat, and the combination of the oxygen atoms with the carbons and hydrogens is what occurs in combustion.

Chemical energy is produced by rearrangements of atoms reforming themselves into different molecular combinations. A similar thing happens within the nuclei of atoms, where, instead of whole atoms rearranging themselves, the subnuclear particles which make up the nuclei do the rearranging instead, which brings us to nuclear energy.

Most of the matter in an atom is concentrated in a small speck at the center called the *nucleus* (Figure 1.17). Most of the space of an atom is taken up by the very light, negatively charged electrons, which swirl about the nucleus like a gas or a cloud. The nucleus itself is composed of positively charged particles, *protons,* and particles with no charge, *neutrons.* These particles are held together in the small space by a force many times stronger than the force which binds the electrons to the atom.

Just as atoms can arrange themselves in many ways to form the molecules that compose the substances of the world of experience, so the protons and neutrons combine in many ways to form the nuclei of the more than 100 elements (including

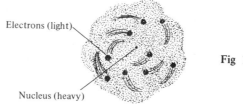

Electrons (light)

Nucleus (heavy)

Fig 1.17.   An atom.

the man-made as well as the naturally occuring elements). And just as there are two ways of obtaining chemical energy, by breakup of unstable molecules and by the combining of atoms during combustion to form new molecules, so are there two ways to obtain nuclear energy. The breakup of heavy, unstable nuclei into roughly equal parts is called *fission.* The combining of protons and neutrons in certain combinations to form new, and heavier, nuclei is called *fusion.* By analogy with chemical combustion, fusion is sometimes called thermonuclear combustion. Nuclear energy, as well as energy from the sun and from the interior of the earth (both of which are nuclear in their ultimate origin) are discussed more fully elsewhere. The object here is to establish the similarity between chemical and nuclear energy. The obvious differences between the two are due to the fact that the forces giving rise to nuclear energy are so much stronger than the forces giving rise to chemical energy. The amount of energy given off in nuclear reactions is so great that Einstein's prediction of the quantitative relation between matter and energy ($E = mc^2$) can be tested. It has been found that the energy released results in a loss of just the amount of mass expected from Einstein's equation. However, even in chemical reactions, there is every reason to believe that the energy given off is accompanied by a loss of mass as well. The amount of mass lost in a chemical reaction, however, is so small that no instrument yet devised has been able to measure it.

§1.10.1. Solar Power. In an interview conducted for the cassette tape series, *Energy, A Dialogue,* published in 1973 by the American Association for the Advancement of Science, Dr. Aden and Marjorie Meinel of the University of Arizona discussed the prospects for harnessing energy from the sun to produce useful power. They propose to cover desert land with solar energy collectors, which will capture sunlight and pass it to a fluid which will carry the heat to a molten salt storage reservoir; this in turn will heat water to generate steam to turn turbines to produce electricity. In the course of their conversation, the Meinels said that a square mile of land covered with collectors would be needed to generate 100 megawatts of power. Let us see whether this claim can be substantiated. (See also §2.22.)

    Solar energy is, of course, a form of nuclear energy where the sun itself is the thermonuclear reactor. The amount of solar energy reaching the vicinity of the earth's orbit is described in terms of the *solar constant,* which is the power delivered, at the average distance of the earth's orbit from the sun, to a unit area perpendicular to the sun's rays (Figure 1.18). Of course, the power per unit area decreases with the square of the distance from the sun, so its value at earth's distance is what is important to us. The value of the solar constant is 3,300,000 kilowatts per square mile. This is much greater than the 100,000 kilowatts (100 megawatts) per square mile quoted by the Meinels, but we must remember that a kilowatt of power at the earth's orbit, but outside the atmosphere, is a far cry from a kilowatt of electricity generated at the earth's surface and available as power for use.

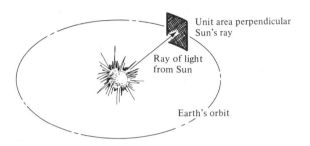

Fig 1.18.   Solar constant = power delivered through shaded area.

There are in fact four factors, each of which tends to reduce the available power: the rotation of the earth, the effect of latitude, losses in the atmosphere, and inefficiency in the collection and generation apparatus. Because the collectors are earthbound, they are not always directly facing the sun. Half of the time it is night, and even during the day the sun's rays do not strike perpendicularly on the collectors (except possibly at noon on one or two days of the year). The collectors move in a path represented by the circumference of a circle, whereas the sunlight intercepted is represented by a swath whose width is the diameter (Figure 1.19). Thus the effective power collected must be reduced, because of the earth's rotation, by a factor equal to the ratio of the diameter of the circular path to the circumference. This factor is equal to $1/\pi$ or about 0.3.

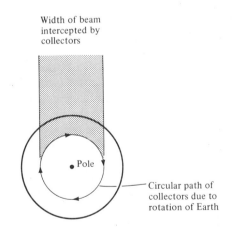

Fig 1.19.   Reduction in solar power due to Earth's rotation.

The second reduction of power is determined by the latitude of the area where the collectors are placed. The higher the latitude, the less energy is absorbed by a given collector (Figure 1.20). In the extreme case of a collector at the pole, no energy is collected at all. This is, of course, strictly true at the equinoxes, the beginning of spring or fall, but in temperate latitudes the seasonal variations— the fact that in summer the sun's rays come in closer to the perpendicular and in winter at a greater slant—should average out. At the latitude of Arizona, the collectors will intercept about 80% (the cosine of the latitude) as much sunlight as they would if the sun were overhead at noon each day, so for the effect of latitude we need to multiply the available power by a factor of 0.8.

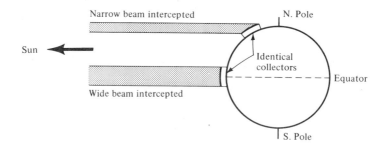

Fig 1.20.   Reduction in solar power due to latitude.

The atmosphere is not perfectly transparent. On the average, slightly less than half of the solar radiation which enters the atmosphere reaches the ground. Radiation losses occur because of absorption and reflection by the various gases which make up the atmosphere and by clouds and dust. In the desert, where clouds are few, the fraction of undepleted solar radiation would be somewhat greater than average, about 60% or so. So for this effect we have to multiply the available power by 0.6.

Finally, the process of converting sunlight to electricity is not yet very efficient. Present solar cells, of the type used in space satellites, have efficiencies of up to 12% and there are hopes that 20% might be achieved. As yet, solar cells are very expensive. If a steam-electric turbine-generator system is used, it may have to operate at lower temperatures than present fossil-fueled or nuclear plants, and as we will see this would mean lower efficiency. Also, the solar collectors are not perfectly efficient. Present power plants operate at efficiencies ranging between 30% and 40%, so we may have to settle for about 20% for solar power. The Meinels themselves estimate a conversion efficiency of 20 to 25%. This is not necessarily a disaster, for the incident energy is, after all, free. But because of this inefficiency, it is necessary to multiply the available power by a further factor of about 0.2.

If we take all these factors together, we obtain available power of 3,300,000 kw (the solar constant) × 0.3 × 0.8 × 0.6 × 0.2, or just over 100,000 kilowatts, which is in agreement with the Meinels' claim. In spite of the limitations imposed by these four factors, the potential power available from solar energy is still highly significant, and should be explored more thoroughly.

§1.11. Heat and Temperature. For a long time it was thought that heat was a substance distinct from ordinary matter. A science, thermodynamics, arose to explain the behavior of heat without knowing what heat was, just as Sir Isaac Newton explained the behavior of particles of matter without knowing what matter was. While thermodynamics as it then existed was quite useful, and indeed made possible the Industrial Revolution, it remained for physicists in the last century to make one of the great steps toward the unification of knowledge—the realization that heat is nothing more or less than the random motion of the molecules of which matter is composed.

To better understand heat and temperature, let us imagine two billiard tables. Suppose that instead of felt the tables are covered with a hard, bare slate, so that with little or no friction a ball, once hit, will continue to bounce around for a long time. One of the tables is set up with a standard rack of fifteen balls, while the other is set up with only two balls (Figure 1.21). The two tables, of course, are supposed to

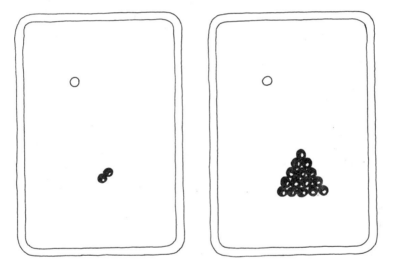

Fig 1.21.   Two set-ups at the billiard table.

represent two systems, one with many molecules and the other with only a few. In each case the player hits the cue ball with equal force, sending it into the balls on the table. On each table the energy of the cue ball will, after a short time, be more or less equally distributed among all the balls on that table (Figure 1.22).

On both tables the balls are moving randomly. Before, when the cue ball was hurtling forward, the energy was anything but random. It was concentrated in a single ball, and had some mechanism been set up to trap this ball, it could have been made to do useful work. But now the motion is random. The kinetic energy of the cue ball has been distributed in a random fashion among all the balls on the table. If these balls had been molecules, we would have said that the kinetic energy had been transformed into heat energy. Since the cue ball came in with the same speed on each table, the same amount of energy was transformed into heat in each case. But on the first table it had to be parcelled out among many more balls (molecules) than

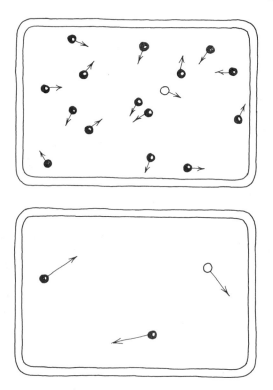

Fig 1.22.   After the "break".

in the second, so that the average energy of each molecule was much less. So in addition to talking about the total amount of heat which we added to the system, we need a different measure, one which will tell us the average amount of energy of each individual molecule. This measure is called *temperature*.

Before we began our billiard ball experiment the balls were stationary. There was no movement at all. We define the temperature of a condition like this, where the molecules have no energy at all, as *absolute zero*. This is the lowest possible temperature. Its value is 460° below zero on the Fahrenheit temperature scale. Even the coldest temperatures on earth, in the neighborhood of 100° below zero, do not begin to approach this depth of cold. Absolute zero has been approached quite closely in laboratory experiments, and temperatures on the outermost planets, on which the sun must seem little more than a distant star, are not much above absolute zero. We can assume that the molecules on the planet Pluto move about quite cautiously.

§1.12. Heat Conduction. Suppose we push our two billiard tables together and replace the sides that touch with a barrier of stretched rubber, so that if two balls hit the barrier simultaneously in roughly the same spot, each one will "feel" the impact of the other (Figure 1.23). What would you expect to happen? If one of the balls were moving slowly when it hit, and the one on the other side were moving very fast, wouldn't you expect the fast one to give a "kick" to the slow one? The slow one

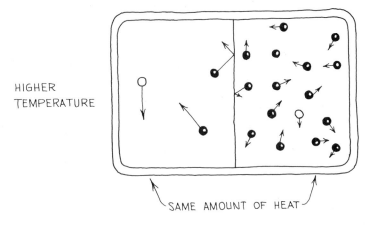

Fig 1.23.   Special table—two halves separated by stretched rubber band. As in Fig 1.22, situation immediately after the "break." Both halves possess the same amount of heat. Sparsely populated half is at higher temperature. (The balls are moving faster there.)

would come out of the collision moving a little faster, and the fast one, then, would have to come out moving a little slower since it had to give up some of its energy to the slow one. If we started out, as in our case, with fast molecules on one side and slow molecules on the other, then after many such collisions the speeds on both sides would eventually even themselves out (Figure 1.24), with the final speeds, and therefore the final temperature, being some kind of compromise between the two sides. Energy, heat energy since it is randomized motion on both sides, was transferred from the hotter (faster) side to the colder (slower) side, and after a while the

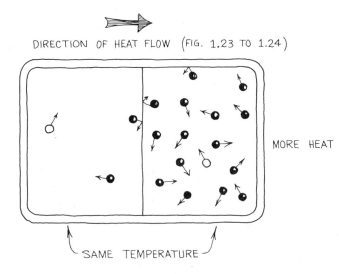

Fig   1.24.   Sometime later. Speeds of balls on both sides have had a chance to equalize. Both sides are at the same temperature. The more densely populated half has more heat.

temperatures of the two sides became equal. But be careful to note well that although the temperatures of the two sides may be equal, the crowded side has much more energy. Before, both sides had the same energy, but the sparsely populated side was at a much higher temperature.

§1.12.1. Geothermal energy. In September of 1972, in a conference on geothermal energy, former Secretary of the Interior Walter Hickel proposed a $680 million research effort over ten years to develop the geothermal resources of the United States. Geothermal energy is heat from the earth's interior (See also §2.19), thought to result largely from the decay of radioactive nuclei in the earth's core.

In the report it was stated that the heat extracted from a cubic mile of hot rock in the process of lowering its temperature 200° Centigrade (by pumping water through fissures in the rock) would be equivalent to 300 million barrels of oil. Let us try to verify this statement.

Although rocks vary, a nice rule of thumb that is roughly true is that a given volume of rock will give up the same amount of heat in cooling one degree Centigrade as the same volume of water will give up in cooling one degree Fahrenheit. (See Figure 1.25 for the relationships among the several temperature scales). Thus to do a calculation with rock, simply replace the volume of rock with an equal

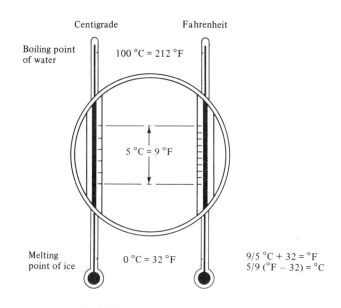

Fig 1.25. Temperature scales.

volume of water and change degrees Centigrade to degrees Fahrenheit. Now with water and degrees Fahrenheit, we can talk about Btu's.

Remember that a Btu is the amount of heat required when a pound of water is heated one degree Fahrenheit, or given up when the same water is cooled by the same amount. Using the above rule of thumb, let us replace the cubic mile of rock with a cubic mile of water, and the 200° Centigrade with 200° Fahrenheit. Then from one cubic foot of water we will gain about 12,000 Btu, since a cubic foot of water weighs approximately 60 pounds. The volume of a cubic mile is nearly 150,000,000,000 cubic feet (5,280 × 5,280 × 5,280) so our total heat gain is 1,800,000,000,000,000 Btu or 1.8 quads.

Now let us try relating this figure to oil. A barrel of oil has an energy content of about 63 therms or 6,300,000 Btu. Multiplying this by 300,000,000 barrels, we obtain 1,890,000,000,000,000 Btu, or 1.89 quads, which is certainly quite close to our above figure in light of the rough rule of thumb concerning rock and water which we have used.

Here again, as with solar power, there exists a huge untapped energy resource that demands considerably more attention than it has received.

§1.14. Energy Degradation. In our billiard ball experiment we saw the cue ball distribute its energy among all the balls in the rack, resulting in random motion. Did you ever see sixteen randomly moving balls suddenly concentrate all their energy in one ball, the rest of them coming to rest in a clump? Of course not, and you never will! Similarly, on the molecular level, it is easy to transform mechanical or electrical energy into heat, but you do not see the reverse happen spontaneously. Left to its own devices, energy eventually degrades into heat. Man, however, has constructed devices which can at least temporarily reverse this trend and convert some heat into useful work. Such devices are called *heat engines* and are the major users of energy in our economy.

§1.13. Energy Bookkeeping. In everything we have done, we have seen energy changing from one form to another. The potential energy of water in a high reservoir becomes kinetic energy of falling water, which becomes kinetic energy of whirling turbine blades and generator rotors, which becomes, perhaps, mechanical energy, or light, or heat. Sometimes it seems that energy is lost, but if we are good enough accountants, and look for all the different forms of energy (mechanical, chemical, electrical, gravitational, heat-—expecially heat) then we will find that we always have the same amount. It is not destroyed, and neither is it created. (Einstein's equation $E = mc^2$ merely includes mass as a form of energy.) There is no counterpart to the Federal Reserve System (See Chapter 11) to funnel new energy into the market. The United States Government can print dollars, but it can't print kilowatt-hours. This constancy of the total amount of energy is the substance of a basic principle, the *First Law of Thermodynamics*. It is because of this law that we can forget about such things as perpetual motion machines. If we want useful work to be done, we have to provide an energy source to do that work.

§1.15. Heat Engines. The purpose of an engine whether it is the internal comustion engine in your automobile, a jet engine in a 747, the steam engine in an old time "iron horse," the diesel engine in its modern counterpart, or the huge main engine of an Apollo moon rocket, is to convert the random motion of molecules which we experience as heat into a united, useful force which can do work in a single direction. In a human organization, if each of the members pulls in his or her own direction, no work—in the human sense—gets done. But if they work together, much may be accomplished. So too, in a organization of molecules, useful work—in the physical sense—can be performed if we can get the molecules to work together. Of course, in a human organization, it is seldom possible to achieve perfect harmony among the members; there is always a certain amount of wasted effort. In an organization of molecules, it is never possible to achieve perfect harmony either. A certain amount of heat is always wasted. This fact, certainly understandable to anyone who has worked in an organization of human beings, has been elevated in the realm of molecules to the status of a law—the *Second Law of Thermodynamics*. The Second Law is often stated in terms of the concept of *entropy* which will be considered in §1.17.

§1.16. Generating Electricity With Steam. Let us examine the type of steam cycle used in most electric generating plants, whether nuclear or fossil fueled (Figure 1.26). Heat energy is transmitted to water from a heat source—a coal, oil or natural gas fired boiler, or perhaps the core of a boiling water nuclear reactor—and

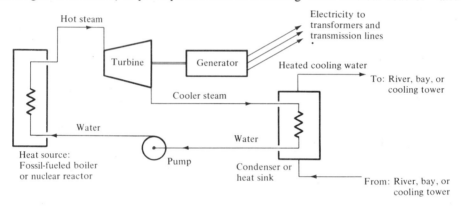

Fig   1.26.

the water is converted to high-pressure, high-temperature steam, which is then passed through a turbine. The blades of the turbine are forced to turn by the onrushing steam, which does work on the turbine blades as they turn, and this work is converted to electrical energy by the generator to which the turbine is connected.

After passing through the turbine, the steam is cooler than when it went in, for in doing work on the turbine it lost some of the heat which was the form of the energy which it carried. This steam could be pumped directly back to the heat source, but if this were done, the pump would have to work harder and harder to prevent a serious back pressure from developing.

Since we have used the word back-pressure, perhaps we ought to say a bit about it. Very simply, a turbine spins because the pressure at the inlet is greater than the pressure at the outlet. If the two pressures were the same, there would be no reason for gas to flow through the turbine. Automobile buffs know that anything in a car's exhaust system, such as a muffler, which impedes the flow of exhaust gases causes a loss of power. This is due to the rise in pressure at the exhaust manifold due to the back-up of these gases. Some of the energy which might have been used to move the car has to be used just to force the exhaust gases out of the cylinder and into the exhaust manifold. In the turbine, too, high pressure at the exhaust outlet reduces the power which the turbine can transmit to the generator.

If the pump were not used, the pressure everywhere in the system would soon equalize, and the turbine, which requires a higher pressure at its inlet than at its outlet if it is to do useful work, would stop. Alternatively, if the pump were used to reduce this back-pressure, it would soon be using as much power to do its work as the turbine was producing, and none would be left over to transmit to the generator.

This problem is solved by passing the steam which leaves the turbine through a condenser. The steam is cooled and turned back into water, which lowers its pressure and volume considerably, thereby reducing the turbine back-pressure and the work load on the pump.

The condenser is sometime called a *heat sink* because heat from the power plant "sinks" into it. It is cooled, typically, by water from a river, lake or bay. The cooling water is returned to its source after it is used, at a higher temperature than drawn, because it is now carrying the excess heat from the steam it has cooled. This, of course, is the source of thermal effluent from power plants. If more heat must be vented than the river, lake or bay can take without serious ecological damage, cooling ponds or towers must be used to direct the heat to another part of the environment, usually the atmosphere, which may be able to absorb it without damage, although in some areas the release of such waste heat to the atmosphere can result in microclimatological changes which include increasing the occurence of fog and increasing the frequency and extent of rain and snowfalls.

The amount of heat absorbed from the heat source tends to be directly proportional to the temperature of the source. As the temperature of the source increases, so does the amount of heat which may be absorbed from the source. The amount of heat which must be vented to the heat sink tends to be directly proportional to the temperature of the sink. Thus we would prefer, in the interests of thermal efficiency, that the source be as hot as possible and the sink be as cold as possible, so that we absorb as much heat as we can and waste as little as possible. Common sense might say that the hotter a power plant runs, the more thermal effluent and potential thermal pollution it might produce, but, in fact, the exact opposite is true. For safety reasons, nuclear power plants are not permitted to run as hot as conventional fossil-fueled plants, and this is why nuclear plants tend to generate more waste heat. This does not, by itself, mean that nuclear plants should not be built; all of the effects of all types of electric power generating facilities need to be taken into account before public decisions are made. Also, newer types of nuclear reactors are expected to operate at increased levels of thermal efficiency.

§1.16.1. Thermal pollution. Hearings before the Joint Committee on Atomic Energy of the United States Congress on the Environmental Effects of Electric Power, November 4, 1969, produced the following testimony which shall be used as the basis for considering a problem in thermal efficiency of electric power generation.

Mr. Perry: (Research Advisor to the Assistant Secretary of the Interior for Mineral Resources)	The contribution to the total thermal pollution load in nuclear plants in 1969 is very small. However, with to-day's nuclear electric generating technology, about 70% more thermal pollution per kilowatt can be expected than from the most efficient fossil-fuel plant.
Representative Hosmer: (United States Representative Craig Hosmer, Member of the JCAE)	I don't understand that figure. About 70% more than from a nuclear plant?
Mr. Perry:	Various figures have been used ranging from the 40% to 70% depending on what efficiency you assign to the average nuclear plant and what efficiency you assign to the fossil-fuels plant. For the purpose of leaving the record uncluttered I gave you [AEC] Commissioner Tape's numbers out of a recent speech.
Representative Hosmer:	Well, he was talking about the discharge into the water.
Mr. Perry:	Yes, sir, that is what I am talking about.
Representative Hosmer:	And not the total discharge.
Mr. Perry:	I am talking about thermal pollution to water at this point.
Representative Hosmer:	It is because essentially all of the nuclear thermal effluent will enter the water, whereas, that is not the case with the conventional plant which discharges considerable heat into the atmosphere.
Mr. Perry:	Yes, sir.
Chairman Holifield: (United States Representative Chet Holifield, Chairman of the JCAE)	. . . In order to have it balanced, and we are very anxious that there be a balance based on the facts, I think this is an important thing to understand, that practically all of the waste heat goes into the water in the case of a nuclear plant.
Representative Hosmer:	Your figure is 75% of thermal goes into the water and 25% around the powerhouse up the stack and elsewhere.
Mr. Perry:	That is right.

The thermal efficiency of light-water nuclear reactors now being built and operated in the United States is around 33%. The best fossil-fueled plants now under construction may achieve an efficiency of 40%. This 7% difference does not at first seem very significant, but the figure quoted by Mr. Perry that nuclear plants gener-ate 70% more waterborne thermal effluent than fossil-fueled plants looms larger. Is it possible to reconcile these facts?

If a plant is 33% efficient it means that one-third of its energy is converted into electricity and that two-thirds is converted into waste heat. The ratio of waste heat to useful energy is then 2 to 1. For a 40% efficient plant, two-fifths of its energy is con-verted into electricity and three-fifths is wasted. The ratio of waste heat to useful energy here is 3 to 2 or 1.5 to 1. If both plants generate 100 units of energy, the nuclear plant wastes 200 units of heat, and the fossil plant wastes 150 units of heat. Furthermore, all or nearly all of the waste heat from the nuclear plant, according to Mr. Perry, is delivered to receiving waters, whereas, as Representative Hosmer emphasized, only 75% of the waste heat from the fossil-fueled plant goes to the waters, the other 25% leaving via the stack. So in our example, 75% of 150 units or 112 units of waste heat are released to the water from the fossil-fueled plant, and 200 units from the nuclear. 200 units is 78% larger than 112, not so very far from Mr. Perry's figure. The calculations are summarized in the following table.

Table 1.3. Calculations for Thermal Pollution Problem

Efficiency (%) & Type	Ratio, Waste to Useful Energy	Useful Energy (Units)	Waste Heat (Units)	Waste Up Stack	Waste Heat Into Water
33% Nuclear	2 to 1	100	200	0	200
40% Fossil-Fueled	1½ to 1	100	150	38	112

§1.17. Entropy. The concept of entropy is used to express the amount of disorder in a system. A well-shuffled pack of cards is more disorderly than a deck which has been arranged in suits. The billiard balls in random motion are certainly more disorderly than the motionless "rack" of 15 arranged in a neat triangle. If a pack of cards were shuffled, you would not expect the suits to sort themselves out, any more than you would expect the randomly moving balls to return to their former ordered state. Systems tend, if left to their own devices, to become more disordered. This statement, that the disorder or entropy of systems tends to increase as time passes, is another statement of the Second Law of Thermodynamics.

Some systems, however, seem to maintain their orderly state in spite of the Second Law. Some, of course, do not change at all. A perfect crystal floating in intergalactic space might maintain its high degree of order over eons, but this is simply to say that it is not changing at all. Other systems certainly do change, and in the direction of greater order. A single cell, growing within the womb, being born, and developing into an adult human being is an example of such a system. Ideally, human society itself, growing from a loosely knit group of independent rustics to a highly interdependent network of people interacting constantly both in person and through artificial information processes, is another. Such systems, however, maintain their order at the price of increased disorder in their environment. That this is true for human societies is obvious; however, at present human societies produce vastly more disorder in their environments than is required by the Second Law of Thermodynamics.

*

Part II

ENERGY
AND
RESOURCES

*

2

ENERGY RESOURCES

M. King Hubbert

§2.0. **Energy and Power.** Inasmuch as this chapter deals with energy and power, both of which are precisely defined physical quantities, it will not be possible to discuss these quantities intelligibly unless their meaning and the units in which they are measured are understood.

In physics, the Energy of a system is measured in terms of its capacity to perform work. Work, in turn, is measured by the product of an applied force and the distance the object acted upon by the force is moved. Hence,

$$\text{Work} = \text{force} \times \text{distance}$$

There are as many units of work as there are units of force and of distance, but here, for brevity, we shall limit ourselves to the most nearly universal system of measurement, the metric system. In this system, the unit of length is the meter, the unit of time, the mean solar second (the second of ordinary clock time), and the unit of mass, the kilogram. The unit of force is the newton, which is defined as that force which, when applied to a mass of 1 kilogram free to move without friction, will cause it to accelerate its velocity at a rate of 1 meter per second for each second the force is applied. In free fall, the force of gravity acting upon a 1-kilogram mass causes it to accelerate at a rate of approximately 9.8 (m/sec)/sec. The force must therefore be approximately 9.8 newtons. This is expressed by saying that a 1-kilogram mass weighs 9.8 newtons. A mass that weighs 1 newton would accordingly be

1 kilogram/9.8 = 0.102 kg = 102 grams

The unit of work in the metric system is the joule, defined by

1 joule = 1 newton × 1 meter

Hence, approximately, 1 joule is the amount of work required to lift a mass of 0.102 kilogram a height of 1 meter. The word "approximately" is emphasized because the force of gravity on a 1-kilogram mass is not exactly 9.8 newtons nor is it constant. In fact, this force varies somewhat both with elevation above sea level and with latitude. The magnitude of the joule, however, is defined independently of gravity.

Bodies may possess two kinds of mechanical energy, *potential* and *kinetic*. Potential energy is an energy of position, or of configuration. Thus, if a mass of 1 kilogram rests on a shelf 1 meter above the floor, it is said to possess 9.8 joules of potential energy with respect to floor level, because this is the amount of work that it could perform in being lowered to the floor. A moving body possesses kinetic energy with respect to a state of rest. This is the amount of work required to change the body from a state of rest to that of motion, and it is also the amount of work that could be done by the body in having its motion arrested.

A different manifestation of energy is that of thermal energy, or *heat*. When work is done on a mechanical system having friction, it is a common observation that heat is produced. Careful measurements have shown that when a given amount of work is performed on such a system, the same amount of heat is produced. This makes it possible to measure heat in units of work. The heat produced by the expenditure of 1 joule of mechanical work is defined to be 1 thermal joule.

In a similar manner, the work or heat involved in all kinds of physical processes—mechanical, thermal, electrical, or chemical—is measureable in joules. In fact, it has been found that in all such systems, when energy in a given form disappears, an equivalent amount of energy in some other form appears, and the total amount of energy remains constant. This is known as the principle of *conservation of energy*, or the *First Law of Thermodynamics*.

A very important extension of this principle is due to Albert Einstein, who, in a short paper published in 1905, deduced theoretically that there must also be an equivalence between the energy of a body and its mass. Einstein showed that when a body of mass m loses an amount of energy $\triangle E$, the mass is also diminished by an amount $\triangle m$, given by

$$\triangle m = \triangle E/c^2$$

where c (equal to 3×10^8 meters per second) is the velocity of light.

The validity of this equation has subsequently been amply confirmed experimentally. If both sides of this equation are divided by the mass m, the equation can also be written in the form

$$\triangle m/m = (1/c^2)\,(\triangle E/m)$$

For ordinary chemical reactions, such as combustion, the ratio of the energy $\triangle E$ generated per unit of mass m is the order of 10^7 joules per kilogram. When this is multiplied by $1/c^2$, the value of $\triangle m/m$ is found to be about 10^{-10}. Hence the mass reduction is only about 1 part in 10 billion, which is negligible for chemical purposes. For nuclear phenomena, on the other hand, the ratio of the energy released per unit of mass, $\triangle E/m$, is very much larger—on the order of 10^{13} joules per kilogram, or about a million times larger than the energy per unit mass released by chemical reactions. In this case, the ratio of the mass reduction to the original mass, $\triangle m/m$, is about 10^{-4}, or 1 part of 10,000. Such a mass reduction accompanying the release of energy is of fundamental importance when dealing with nuclear phenomena.

There are many processes involving work or energy to which a definite quantity of energy can be assigned. A waterfall is such a process. This is capable of doing a given amount of work per unit of time, but the total amount of work that can be done increases without limit as time increases. For such a process, we require a different measure, namely the rate at which work is done, or *power*. Accordingly, we define the power of such a system by

$$Power = work/time$$

When the work is measured in joules and the time in seconds, the power is expressed in joules per second, or *watts*. For larger units of power, we have the *kilowatt* (1,000 watts) and the *megawatt* (1 million, or 10^6 watts.)

Power and time can also be used to express *quantities* of *energy*. Since, by definition, 1 watt is equal to 1 joule/second, then it follows that

$$1 \text{ joule} = 1 \text{ watt-second}$$

Other similar units of energy are the kilowatt-hour and the megawatt-day. The kilowatt-hour is the familiar unit for which one pays his monthly electric bill. This represents 1,000 watts for 3,600 seconds, or 3,600,000 joules. A megawatt-day represents 1 million watts for 86,400 seconds, or 8.64×10^{10} joules.

While energy can be transformed from one form to another without diminution of its amount, energy changes, in general, have a unidirectional and irreversible character. For example, when an object has a higher temperature than its surroundings, heat spontaneously flows from the warmer body to the cooler surroundings. The reverse, however, never occurs, and the system cannot be restored to its original state except by some external means. In mechanical systems, such as a free-swinging pendulum, the mechanical motion is retarded by friction and the system eventually comes to rest. The potential and kinetic energy possessed initially are completely converted by friction into heat. The reverse process, however, never occurs.

Yet, as the steam engine demonstrates, heat can be converted into work. This is only possible provided a difference of temperature exists. In the case of the steam engine, we have a higher temperature heat reservoir, the furnace, and a lower temperature reservoir, a supply of cooling water. Heat at higher temperature is absorbed by the boiler and steam is produced at higher pressure. This, upon expansion, drives an engine which does mechanical or electrical work. Finally, the exapnded steam is passed through the condenser, from which the heat is discharged to a stream of cooling water, which is increased somewhat in temperature. The steam is condensed to a liquid which is then pumped back into the boiler, and the cycle is repeated.

During this cycle, a quantity of heat Q_1 is taken from the furnace; a fraction of this is converted into work W, and the remaining heat Q_2 is discharged to the cooling water through the condenser. From the principle of the conservation of energy, the sum of the work W plus the heat Q_2 discharged by the condenser must equal the amount of high-temperature heat Q_1 taken from the furnace. Hence,

$$W + Q_2 = \mathbf{Q}_1$$

or

$$W = Q_1 - Q_2$$

The ratio

$$W/Q_1 = (Q_1 - Q_2)/Q_1$$

which expresses the fraction of the thermal energy taken from the furnace that is converted into work, is known as the *thermal efficiency* of the engine. By a somewhat more elaborate analysis than we can develop here, it can be shown that the maximum possible efficiency of a steam engine is given by

$$(W/Q_1)_{max} = (T_1 - T_2)/T_1$$

where T_1 is the temperature of the steam and T_2 that of the condenser, both given in degrees on the absolute temperature scale (degrees Celsius plus 273.15°).

From this it is seen that the fraction of the high-temperature heat that can be converted into work is proportional to the difference between the two temperatures. If the difference is zero, no work at all can be obtained. For this reason, in designing semi-electric power plants, an effort is made to obtain as high a steam temperature as possible. For steam-electric power plants in the United States in 1970, the average thermal efficiency is about 0.33 or 33%. The largest and most modern plants have somewhat higher efficiencies — up to as high as 38%. Present nuclear-electric plants are somewhat less efficient (thermally) than those deriving their energy from fossil fuels.

In steam-electric power plants, some energy in the case of fuel-fired plants is discharged through the smokestack. Of the thermal energy absorbed by the boiler, about one-third is converted into electrical energy and the remaining two-thirds is discharged through the condenser to the environment at low temperature. The electrical energy, through lighting, direct heating, or frictional dissipation, is also converted into waste heat. Hence, the end product of the entire operation is the conversion of the initial chemical or nuclear energy of the fuel into waste heat at the mean temperature of the environment.

§2.1 Introduction. If we are to appreciate the significance of energy resources in the evolution of our contemporary society is will be necessary not only for us to understand the principal physical aspects of the conversion of energy in the complex of activities transpiring on the earth, but also to view these activities in a somewhat longer historical perspective than is customary. For those of us who live in the more industrialized areas of the world—particularly in the United States—it is difficult to appreciate the unique character of the industrial and social evolution in which we are participating. During our own lifetimes, and during the immediately preceding period of history with which we are most familiar, the pattern of activity we have observed most consistently has been one of continuous change, usually continuous growth or increase. We have seen a population begun by a small number of European immigrants to North America expand within a few centuries to over 200 million, while still maintaining such a growth-rate, even now, as to double within the next 40 years. We have seen villages grow into large cities. We have seen primeval forests and prairies transformed into widespread agricultural developments. We have seen a transition from a handicraft and agrarian culture to one of complex industrialization. Within a few generations we have witnessed the transition from human and animal power to continent-wide electrical power supernetworks; from the horse and buggy to the airplane.

Out of this experience it is not surprising that we have come to regard continual growth and increase as being the normal order of things.

However, if we are to appraise more accurately what our present position is in our social and industrial evolution, and what limitations may be placed upon our future, it is necessary that we consider, not only for the present but in historical perspective, certain fundamental relationships which underlie all our activities. Of these the most general are the properties of matter and those of energy.

From such a viewpoint the earth may be regarded as a material system whose gain or loss of matter over the period of our interest is negligible. Into and out of this system, however, there occurs a continuous flux of energy in consequence of which the material constituents of the outer part of the earth undergo continuous or intermittent circulation. The material constituents of the earth comprise the familiar chemical elements. These, with the exception of a small number of radioactive elements, may be regarded as being nontransmutable and constant in amount in processes occurring naturally on the earth.

For the present discussion our attention will be directed primarily to the flux and degradation of a supply of energy, and secondarily to the corresponding circulation of the earth's material components.

§ 2.2 Flux of Energy on the Earth.

The overall flux of energy on the earth is shown qualitatively and diagrammatically in the flow-sheet of Figure 2.1.

The energy inputs into the earth's surface environment are principally from three sources.

(1) the energy derived from the sun by means of solar radiation,

(2) the energy derived from the mechanical kinetic and potential energy of the earth-sun-moon system which is manifested principally in the oceanic tides and tidal currents, and

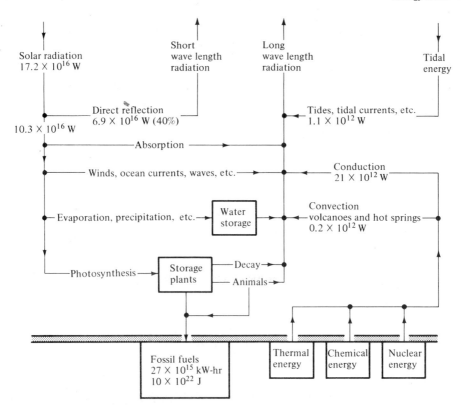

Fig 2.1. Energy Flow Sheet for the Earth

(3) The energy derived from the interior of the earth itself in the form of outward heat conduction, and heat convected to the surface by volcanos and hot springs.

Secondary sources of energy of much smaller magnitude than those cited are the energy received by radiation from the stars, the planets, and the moon, and the energy released from the interior of the earth in the process of erecting and eroding mountain ranges.

No definite quantity can be assigned to the energy from any of the foregoing sources because we are confronted not with a fixed quantity of energy but a continuous flux of energy from the various sources, at nearly constant rates. The rate of energy flux is measurable in terms of power, defined by the equation

$$\text{power} = \frac{\text{energy}}{\text{time}},$$

and if the energy is measured in terms of the work unit, the *joule,* and the time in seconds, the power is then in joules per second, or *watts*.

§ 2.2.1 Energy from Solar Radiation. The rate of energy flux from the sun, or the solar power, intercepted by the earth is readily obtainable from the solar constant, and the area of the earth's diametral plane. The solar constant is the quantity of energy which crosses a unit area normal to the sun's rays in unit time in free space outside the earth's atmosphere, at a distance from the sun equal to the mean distance to the earth. It is, accordingly, the power transmitted by the sun's rays per unit cross-sectional area at the mean distance of the earth, 17.2×10^{16} watts.

In heat units, the value of the solar constant, I, has been found to be 1.94 calories per minute per square centimeter.[3] This can be converted explicitly to power units by noting that 1 calorie of heat is equal to 4.19 joules of work, and 1 minute is 60 seconds. The solar constant in watts/cm^2 is, accordingly, given by

$$I = \frac{1.94 \times 4.19 \text{ joules/cm}^2}{60 \text{ seconds}}$$

$$= 0.135 \text{ watts/cm}^2.$$

The total solar power intercepted by the earth is then

$$P = IA = I\pi r^2,$$

where A is the diametral area of the earth and r, equal to 6.37×10^8 cm, is the mean radius of the earth. Supplying the numerical values of I and r, we then obtain for the total solar power incident upon the earth

$$P = 17.2 \times 10^{16} \text{ watts.}$$

For comparison, the installed generating capacity of all the electric utilities in the United States in 1959 amounted to 15.7×10^{10} watts.[1] Hence, the power of the solar radiation intercepted by the earth is about a million times the power capacity of all the electric utilities in the United States in 1959.

§ 2.2.2 Energy From the Earth's Interior. The second largest input of energy into the earth's surface environment is that which escapes from the interior of the earth, which is estimated to be at a rate of about 21×10^{12} watts. Of this, about 99 per cent is by thermal conduction, and only about 1 per cent by convection in volcanos and hot springs.

§ 2.2.3 Tidal Energy. The tidal energy is derived from the combined potential and kinetic energy of the earth-moon-sun system. The total rate of dissipation of this energy, as indicated by the rates of change of the earth's period of rotation and the moon's period of revolution, is estimated[2] to be about 1.4×10^{19} ergs/sec, or 1.4×10^{12} watts. Of this, about 1.1×10^{12} watts, or about 80 per cent, is estimated to be accounted for by oceanic tidal friction in bays and esturaries around the world.

Thus, tidal power is about an order of magnitude (te times) smaller than that of the heat escaping from the earth's interior, and both together are one thousand times less than the power impinging upon the earth from solar radiation.

§ 2.2.4 Energy Flow-Sheet.

In view of its predominance, our principal concern is in tracing the flow of the 17.2×10^{16} watts of solar power that is being shed continuously on the earth. About 40 per cent of this, or 6.9×10^{16} watts,[3] known as the *albedo*, is directly reflected back into space. This leaves about 10.3×10^{16} watts which are effective in propelling the various material circulations occurring on the earth.

No further quantitative breakdown will be attempted. However, a part of the remaining solar power is absorbed directly by the atmosphere, the oceans, and the lithosphere, and is converted into heat. A large part of this heat is immediately reradiated back into space as long-wavelength thermal radiation. Another part, however, sets up differences of temperature in the atmosphere and the oceans, in such a manner that convective currents of both water and air are generated, producing the winds, ocean currents, and waves. The oceans and the atmosphere serve in this manner as the working fluids of a world-girdling heat engine whereby a fraction of the thermal energy from sunshine is converted into mechanical energy. The mechanical energy of the wind, waves, and currents is again dissipated by friction into heat at the lowest temperature of the surroundings.

Still another part of the solar energy follows the evaporation, precipitation, and surface run-off channel of the hydrologic cycle. Heat energy is absorbed during the evaporation of water, but it is again released when the water is precipitated. However, the water vapor, being a part of the atmosphere, is convected to high elevations by means of the convective energy already discussed; and, when precipitation occurs at these elevations, the water possesses potential energy, which again is dissipated back to low-temperature heat on the descent to sea level. It is this energy, however, that is responsible for all precipitation on the land, and for the potential and kinetic energy of surface lakes and streams.

A final fraction of incident solar radiation is that which is captured by the leaves of plants by the process of photosynthesis. Although enormously complex in detail, this is the driving mechanism for the synthesis of common inorganic chemicals, such as H_2O, and CO_2, into the chemical compounds of living plants. Schematically this process is represented by the reaction

$$\text{Energy} + CO_2 + H_2O \rightarrow \text{Carbohydrates} + O_2,$$

during which solar energy becomes captured and stored as chemical energy. By the reverse reaction, as in the burning of wood,

$$O_2 + \text{Carbohydrates} \rightarrow CO_2 + H_2O + \text{Heat},$$

and the stored energy is released as thermal energy.

The energy-flow channel whose first step is photosynthesis is that which sustains the entire complex of organisms on the earth. We have the familiar food chain:

$$Plants \rightarrow Herbivores \rightarrow Carnivores \rightarrow Parasites \rightarrow \ldots.$$

in which the energy of each link is a small fraction of that of the preceding, the remainder being dissipated as heat. The end-product of this chain is the complete degradation of the photosynthetic energy to heat at the ambient temperature, and the conversion of the material constituents back to their initial inorganic state.

§ 2.2.5 The Fossil Fuels. If the energy stored in plants by photosynthesis could be systematically retained, as for example in the form of firewood, it is clear that the aggregate amount would increase without limit, and could, in a few decades or centuries, become very large indeed. Actually, in the natural state, the rate of decay of organic compounds and the release of their stored energy as low-temperature heat is very nearly equal to the contemporary rate of photosynthesis. However, in a few favored places such as swamps and peat bogs, vegetable material becomes submerged in a reducing environment so that the rate of decay is greatly retarded and a storage of a small fraction of the photosynthesized energy becomes possible.

This, in principle, is what has been happening during the last 500 million years of geologic history. During that time a minute fraction of the existing organisms have become buried in sedimentary muds under conditions preventing their complete decay. These accumulated organic remains comprise our present stores of the fossil fuels: coal, petroleum, natural gas, and related products, the energy content of these fuels being derived from the solar energy of this 500 million-year period which was stored chemically by contemporary photosynthesis.

§ 2.2.6 Summary. The energy flow-diagram, which we have just reviewed, represents, in broad outline, all the major channels of energy flux into and out of the earth's surface environment. By the First Law of Thermodynamics, the quantity of energy in any particular channel, although repeatedly transformed in transit, remains constant in amount. It follows, therefore, that, with the exception of an insignificant amount of energy storage, the energy which leaves the earth by long-wavelength thermal radiation into space must be equal to the combined energy inputs from solar and stellar radiation, from tidal forces, and from the earth's interior.

By the Second Law of Thermodynamics, however, this flux of energy is unidirectional and irreversible. It arrives as short-wavelength electromagnetic radiation, corresponding to the temperature of the sun; or as mechanical energy of the tides; or as thermal energy from a temperature higher than that of the earth's surface environment. By a series of irreversible degradations it ultimately is reduced to thermal energy at the lowest temperature of its environment, after which it is radiated from the earth in the form of spent, long-wavelength, low-temperature radiation.

During this energy flux and degradation the material constituents of the earth's surface, while remaining essentially constant in amount, are circulated. The wind blows; oceanic currents, tides, and waves are formed; rain falls and rivers flow; volcanos erupt and geysers spew; and plants grow and animals eat, move about, procreate, and die.

But for this energy flux none of these things would or could happen and the matter of the earth's surface would be as dead or inactive as that of the moon.

Biologically, the human species is simply a member of the energy-consuming chain which begins with the energy capture and storage of plants by photosynthesis. Man is both an herbivore and a carnivore, and, as such, is merely another member of the biological complex, depending for his essential energy supply—his food—upon other members of the complex, and ultimately on the energy from the sun captured and stored in plants by photosynthesis.

In addition, however, man has been able to do what no other animal has ever achieved; he has learned to tap other channels of the energy flow-sheet, and he has managed to divert the energy flow from its customary path into other channels appropriate to his own uses.

An understanding of these processes is essential if we are to appreciate the significance of energy resources in determining what is possible and what is impossible in human affairs.

§ 2.3 Evolution of Man's Ability to Control Energy.   Since energy is an essential ingredient in all terrestrial activity, organic and inorganic, it follows that the history of the evolution of human culture must also be a history of man's increasing ability to control and manipulate energy.

Consider the earliest stages of this evolution. From geological and archeological evidence, organic evolution had proceeded far enough that by about a million years ago one of the ape-like species had reached the stage where his few skeletal remains are now classed as those of early man. How many of this species there may have been at that time can only be conjectured, but from the scarcity of the remains it may be surmised that the numbers were not large—possibly comparable to those of gorillas or chimpanzees at the present time.

This species must have coexisted in some sort of ecological adjustment with the other members of the biologic complex of which it was a member, and upon which it depended for a share of the solar energy essential to its existence. At this hypothetical stage its sole capacity for the utilization of energy was limited to the food it was able to eat—the order of 2,000 kilocalories per capita per day.

Between that stage and the dawn of recorded history, this species distinguished itself from all others in its inventiveness of means for the capture of a larger and larger fraction of the available flux of energy. The invention of clothing, the use of tools and weapons, the control of fire, the domestication of animals and plants, and other similar developments all had this in common: Each increased the fraction of the contemporary flux of solar energy which was available for the use of the human species, and each upset the ecological balance in such a manner as to favor the increase in the human population, with corresponding adjustments in all other populations of the biologic complex.

Although little is known about the time when many of these developments first occurred, tool making and the use of fire date back at least as far as Peking man (estimated at about 500,000 years ago), but from the length of time involved the rate of change must have been extremely slow.[8] The pace quickened, however, at about 10,000 to 12,000 years ago when, with the domestication of animals and the cultivation of plants, man began to change from a food-gathering to a food-producing species.[4]

After a few thousand years of cultural incubation, there followed almost simultaneously at about 300 B.C. in each of three localities, (the Tigris-Eupharates delta and the Indus and the Nile valleys) the rise of cities with populations estimated at 8,000–10,000 supported by an intensive agriculture.

At least as early as about 1900 B.C. the use of oxen for ploughing is depicted in paintings in Egyptian tombs.[8] Similarly, pictures of sailing ships of advanced design occur in Egypt as early as 1500 B.C.[4]

This quickening of pace continued for the next few thousand years, but the energy supply available was dominantly that which was tapped from the biological channel of solar energy. It permitted a very large increase in the population density in favorable agricultural areas, and a corresponding increase of the total human population as the new culture spread geographically, but throughout this period the energy available per capita was still not much more—possibly only two or three times greater —than that of the food consumed.

Energy from a nonbiological source was first obtained when the energy of the winds and the hydrologic cycle was tapped for human uses. This apparently occurred first with the use of sails for the propulsion of boats and ships. Then followed water mills and windmills.

According to Forbes[6] both the water mill and the windmill are thought to have originated in the Middle East, the water mill during the last century or so B.C., but the windmill not until about 900—1000 A.D. The first water mills were small affairs, with a horizontal wheel and vertical shaft requiring a continuous stream of water and capable of turning small family-size grain miills. This type of mill was improved by the Roman, Vitruvius, during the first century B.C. by making the wheel vertical and gearing the horizontal shaft to a vertical shaft turning the millstone.

However, water mills were not extensively used by the Romans until near the end of the Roman Empire. From this time forward, even during the Dark Ages, the use of the water mill spread throughout Western Europe, until by the sixteenth century it had been adapted to every kind of industrial use requiring stationary power. This use has continued subsequently in both Europe and North America.

However, it has only been since about the beginning of the twentieth century that advancing technology, particularly the transmission of power by electricity, has made it practical to build water mills larger than the tens-to-hundreds of kilowatts range of power capacity. This new technology made the small mills obsolete at the same time that it rendered practical the building of water-power plants in the hundreds-of-megawatts range.

Windmills appear to have been first developed in the Persian province of Seistan about the tenth century A.D. Windmills began to be built in the Low Countries and elsewhere in Western Europe about the thirteenth century, but whether as an independent invention, or introduced by the Muslims by way of Morocco and Spain, is uncertain. In any case, since the thirteenth century, windmills have been used in Western Europe and later in North America and the West Indies for such uses as grinding grain, pumping water, and operating mills for crushing sugar cane.

Escape from this dependence upon contemporary solar energy with its inherent limitations in the quantity utilizable per person was not possible until a new and hitherto unknown source of energy should become available. Such a source was represented by the fossil fuels. Although Marco Polo reported that the Chinese used "black rocks" for fuel,[10] and recent studies indicate that the Chinese may have used coal in small amounts for two or three milenia previously, the use of coal as a major source of energy did not begin until about the twelfth century, when the inhabitants of the northeast coast of England discovered that certain black rocks found along the seashore, and thereafter known as "sea coles," would burn.

Since this initial discovery coal has been mined continuously, first in England and shortly thereafter in what are now Belgium, France and Western Germany, and finally in all coal-bearing areas of the world, in ever-increasing amounts. Then, about a century ago, first in Romania in 1857 and then in the United States in 1859, petroleum in commercial quantities began to be produced, thus tapping the second of the great stores of energy preserved in the fossil fuels. Other fossil fuels, large in amount, are the tar sands and the oil shales. Although oil has been obtained in limited quantities from oil shale for more than a century, the period of large-scale exploitation of the tar sands and the oil shales is still in the future.

Finally, only within the last few decades a way has been found to tap a still larger and more concentrated reservoir of potential energy, that of nuclear energy.

While the evolution of the means of controlling energy had been proceeding at a gradually accelerating rate for many millenia, it did not reach its crescendo stage until after the exploitation of the fossil fuels had begun. Once it was learned that "sea coles" would burn, it did not take long to discover that these loose chunks found along the shore had been derived from the outcropping strata in the sea cliffs above, which were gradually being undercut by the waves. The digging of these strata, first along the cliffs, and then by means of holes sunk to the beds from above, initiated the mining of coal in Western Europe.

So superior was this fuel to wood and peat that the digging proceeded apace. It is recorded that in 1234 King Henry III confirmed a privilege for the mining of coal granted to Newcastle-upon-Tyne by King John.[7] At this time coal was already being transported by barge to London, where by 1273 the smoke from coal burning had become so obnoxious as to provoke complaints from the gentry. In addition to its use as a domestic fuel, coal was promptly adopted as a fuel for lime burning, and was used by blacksmiths and for other post-smelting metallurgical purposes, and for glass making.

Statistics of early production are few, but it is recorded that coal shipments from Newcastle-upon-Tyne in the year 1563–1564 amounted to 32, 951 tons. By 1658–1659, nearly a century later, the yearly production had increased to 529,032 tons—more than 16-fold. Between 1580 and 1660 the imports of coal to London increased 20–25-fold. In the meantime, coal mining in Britain had become general in England, Scotland, and Wales, and the annual production for the whole country by 1660, or shortly thereafter, had reached about 2 million tons per year, which is estimated to have been five times as much as the production of the rest of the world.[10] By 1750 the annual production had reached 7 million tons.[13]

This rapid increase in the mining of coal immediately created grave technological problems. The influx of water into the mines forced the development of continually larger and better pumps. First, water was removed by bailing, then by pumps powered by human labor, and finally by animal power, with pumps driven by as many as 100 horses on treadmills.

Ultimately, so desperate had this problem become that attention was directed to the powers of steam and the newly discovered properties of a vacuum.[5] This led in 1698 to the development by Thomas Savery of the first successful water pump powered by steam. Water was lifted through a vertical pipe to fill the vacuum induced by the condensation of steam in an otherwise closed chamber. By the repetition of this cycle, with the opening and closing of appropriate valves, water could be pumped indefinitely.

This was followed shortly by the "atmospheric engine" of Thomas Newcomen in 1712, which was the first practical steam engine to be developed. This consisted of a walking beam, to one end of which was attached the plunger of a pump, and to the other the piston rod from a vertical steam cylinder. Steam at atmospheric pressure filled the cylinder during the nonworking stroke, and the work was done by atmospheric pressure on the piston when a vacuum was created in the cylinder by the injection of a jet of water.

The use of this engine for pumping water spread rapidly throughout Britain and also to the Continent. However, as it had no rotary motion it did not meet the needs of mills driven by water, except as a means for pumping water from the tailrace to the mill pond, permitting rotary power to be extracted by the water wheel.

Fundamental modification of the Newcomen engine did not occur until more than 50 years later when James Watt introduced a succession of radical improvements, including a separate condenser, a double-acting cylinder and piston, a governor, and, most important of all, a rotary shaft and fly wheel, making the engine suitable for the driving of all types of rotary machinery. It was only after this, late in the eighteenth century, that the steam engine was able to compete with, and eventually to displace, water as a principal source of industrial power.

A second problem that was made critical by the mining of coal was the land transportation of heavily laden wagons of coal. While the principal transportation of coal was by water, that from the collieries to the docksides was by horse-drawn wagons. This led to the development of railroads with longitudinal wooden rails, but with the wagons drawn by horses. Finally, the idea of putting the steam engine on wheels and making it self-propelling was successfully accomplished by Richard Trevithick in 1802. Shortly thereafter, the use of the steam engine for the propulsion of boats was successfully accomplished by Robert Fulton and others.

Thus, by the second decade of the nineteenth century the steam engine had been adapted to supply all contemporary needs for mechanical power: the pumping of water, the driving of stationary industrial machinery, and transportation by water and land. However, the transmission of power was still limited to mechanical means, and hence to short distances.

The end of this era was foreshadowed when, in November 1831, Michael Faraday announced his epoch-making discovery of electromagnetic induction. Within a year Hippolyte Pixii publicly exhibited in Paris the first magnetoelectric machine, a hand-cranked magnetoelectric generator. After this, the further development of magnetoelectric generators and equipment proceeded in England, France, and Germany at a rapid rate. By 1857 a successful experiment of powering an arc light with a steam-driven generator of about 1-1/2 kilowatts capacity was demonstrated by Holmes in London. In 1858 illumination of a lighthouse in this manner was accomplished. By 1875 in France, and by 1878 in London, whole buildings were being illuminated.[9] Finally, in 1881 the generation and public distribution of electric power by a central-power station was initiated when Thomas A. Edison installed the Pearl Street power station and its associated distribution network in New York. From that time forward, the steady advance of the technology of electrical-power generation, distribution, and utilization has advanced to the extent that it has rendered obsolete most other forms of stationary power.

Another major use of the energy from coal, of which scant mention has been made, has been in the smelting and processing of metals. The working of metals has been one of the major uses of coal since the beginning, but it was not until about the middle of the eighteenth century that coal supplanted charcoal for smelting. This was made possible only after it had been discovered how to rid coal of its injurious sulfur and gases by coking—a procedure quite analogous to the manufacture of charcoal from wood. Since the eighteenth century the metallurgical industries, principally iron and steel, have become almost solely dependent upon coal as a source of fuel, and thus among it s largest consumers.

Oil and natural gas, as was mentioned earlier, came into commercial production during the last half of the nineteenth century, first for heat and light, and then as fuel for steam-power plants and to some extent in metallurgical industries. A dominant new use for petroleum was generated with the development in the 1880's of the high-speed, internal-combustion engine. This led immediately to the development of motorized vehicles for travel by land, water, and eventually by air. Gradually, oil and natural gas have succeeded in large measure in displacing coal as the traditional fuel for steamships, railroad locomotives, and even for central electric-power plants.

Finally, progress is well underway in the controlled use of the last and largest known source of potential energy, the atomic nucleus. During the brief period since the attainment of the first controlled fission of uranium at Chicago on December 2, 1942, central power plants in the hundred-megawatt range have been built and are already in operation in the United States, Great Britain, and the U.S.S.R.; nuclear-propelled ships and submarines are also in operation.

§2.4 Growth of Human Population.

As was pointed out earlier, the human proclivity for capturing an ever larger fraction of the total flux of the energy on the earth, and eventually for tapping the large supplies of stored energy, has had the effect of continuously upsetting the ecological equilibrium in the direction of an increase in the human population. The magnitude of the upset and the rates at which it has occurred are best seen by plotting the estimates of the human population graphically as a function of time.

This has been done in Figure 2.2 for the period from 1000–2000 A.D., inclusive, using Putnam's[12] graphs of the means of various demographic estimates for the period 1000–1890, inclusive, United Nations[14] estimates for the period 1900–1950, inclusive, and estimates of Frank W. Notestein[11] for the period from 1960 to 2000.

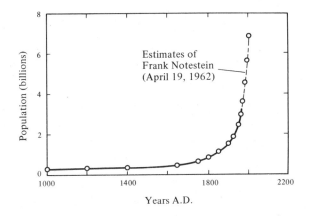

Fig 2.2. Growth of World Population

For the period earlier than that shown in Figure 2.2, Putnam[12] estimates the world population for the year 10,000 B.C. to have been about 1 million and that at 1 A.D. at about 275 million, with a maximum of around 290 million about 225 A.D. and a minimum of 270 million at about 700 A.D.

For the period earlier than 10,000 B.C. about all we have to go on is the archeological evidence that the culture was Paleolithic and the subsistence was by hunting and food-gathering rather than food-producing, and the ecological evidence that the gradual evolution of Paleolithic culture should have been, on the whole, in the direction of a population increase with time.

We infer, therefore, that the human population at 1 million B.C. must have been less than that at 10,000 B.C., but equal to or greater than 2, the least number of biologically possible. However, since this population probably arose by continuous evolution from its immediate forebears, and since for very small numbers of a population the chances of extinction are very high, it is doubtful that a population as small as two individuals ever existed. What the minimum number may have been is unknown, but it is improbable that it was ever as small as 1,000.

Assuming 1,000 as a minimum number at 1 million B.C., we have a basis for judgment concerning the rates of growth of the population during the principal divisions of subsequent history. An initial population of 1,000 would only have to be doubled 21.5 times to reach 3.0 billion, which is the estimate of the world population in 1960. Accepting Putnam's estimates of populations of 1 million at 10,000 B.C. and 275 million at 0 B.C., then the first 10 of these 21.5 doublings would have occurred by 10,000 B.C., and 18 by 0 B.C. The remaining 3.5 doublings have occurred between the beginning of the Christian Era and the present time.

Consider, however, the lengths of time required for the successive doublings. If only ten doublings occurred during the million years prior to 10,000 B.C., the the *average* length of time required for each must have been 100,000 years. A change in a population increasing at such a rate would probably not be detectable by two censuses taken a thousand years apart. We do not assume that the population during this period actually grew in this manner. It probably fluctuated up and down with famines, plagues, and climatic changes, but its *average* growth rate over the whole time must have been not very different from this.

From the period from 10,000 B.C. to 0 B.C. about 8 doublings occurred with an average length of time for each of 1,250 years. This plainly shows the quickening of the growth rate over that of the preceding period—an increase of about 8-fold.

Then, during the Christian Era, 3.5 more doublings have occurred with an *average* length of about 560 years. This, however, fails to tell the whole story, because the time for each successive doubling has been shorter than for the one before. Thus, the first doubling after 0 A.D. occurred at about 1690, the second at about 1845, and the third at about 1937. Thus, during the interval since 0 A.D. the first doubling required 1,690 years, the second 155, and the third only 92.

That this reduction of the doubling period, or increase in the rate of growth, is still continuing may be seen by the population increase for the decade 1950–1960. The United Nations (1958) estimate of the population in 1950 is 2.497 billion. By 1960 this had increased to an estimated 2.996, or roughly 3.0 billion. This corresponds to a rate

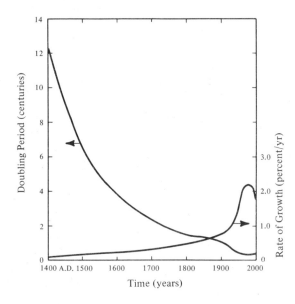

Fig 2.3. Decrease of Doubling Period and Rate of Increase of World Population

of increase of 1.82 per cent per year, at which rate the population would double in only 38.2 years. The instantaneous rates of growth and the corresponding lengths of time which would be required for the population to double are plotted graphically in Figure 2.3 for the world population data represented in Figure 2.2.

What emerges from this examination is the very great contrast between the population growth during the last 1,000 years, particularly during the last few decades, and all preceding history. If we may define the term "normal" as describing a state of affairs which exists most of the time, then we must recognize that the normal state of the human population, and of biologic populations in general, is a state of extremely slow secular change. We must, accordingly, regard the rate of growth of the human population and the concurrent disturbances of all other biologic populations during the last few centuries as being extremely abnormal. It represents, in fact, one of the greatest biological upheavals known in geological as well as in human history.

§ 2.5   **Production Data and Fossil Fuel Reserves.**   The historical background in the use of energy from the fossil fuels has already been given in outline form in § 2.3. From here on it will be more informative if we consider the rates of growth of energy consumption from these sources, with the data presented in graphical form.

§ 2.5.1 **Production Data and Coal Reserves.**   The historical background in the use of energy from the fossil fuels has already been given in outline form in §2.3. From here on it will be more informative if we consider the rates of growth of energy consumption from these sources, with the data presented in graphical form.

§ 2.5.2 **World Production of Coal.**   World production statistics before 1860 are not available, but, as we noted (§2.3) in the principal production during the first few centuries was in Britain, where the production began in the twelfth century and increased steadily over the next seven centuries. The British production rate reached 2 million tons per year by 1660 and 7 million by 1750, and world production reached 134 million metric tons by 1860.

Statistical data on annual production are available from 1860 to 1960, and the rate of production as a function of time is shown graphically in Figure 2.4 for that period. At a glance it will be seen that the growth in the rate of production during this period falls into three distinct phases.

(1) a period of steady growth extending from 1860 until 1913, during which the production rate increased from 134×10^6 to $1,257 \times 10^6$ metric tons per year,

(2) a period of unsettled growth and oscillation extending from 1913 to 1954, during which the production rate increased from $1,257 \times 10^6$ to $1,631 \times 10^6$ metric tons, and, finally,

(3) a period from 1954 to 1960 when the production rate assumed a spurt of renewed growth from $1,631 \times 10^6$ to $2,414 \times 10^6$ metric tons per year.

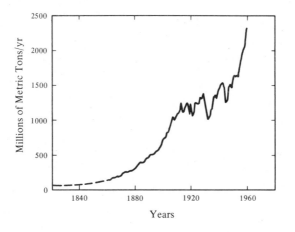

Fig 2.4. World Production of Coal

During the intermediate period from 1913 to 1954 production increased much more slowly, averaging only about 0.64 per cent per year, while during the last period from 1954 to 1960 the rate of growth has been very nearly the same as for the earlier period—about 4.2 per cent per year.

The nature of this growth is brought out more clearly in Figure 2.5 where the same data are plotted logarithmically against time. Here the growth from 1860 to 1913 is seen to plot as an essentially straight line. This indicates that during that period the rate of coal production increased with time at an exponential, or compound-interest, rate of 4.2 per cent per year, or at such a rate of growth that the production rate doubled every 16.8 years.

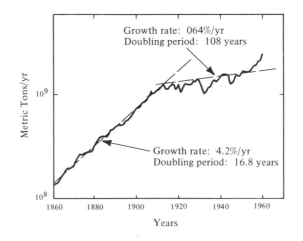

Fig 2.5. World Production of Coal (Logarithmic Scale)

Because Figures 2.4 and 2.5 present data for only the last century of approximately 900 years during which coal has been mined, they do not properly convey an appreciation of the relative importance of coal mining during this period compared with that of earlier history. A better sense of this may be gained if we consider the cumulative production of coal during the total period. For the period prior to 1860, from the few production statistics and the knowledge that coal mining increased continually, it can be estimated that the total coal mined from the twelfth century until 1860 could only have about 5.4 billion (5.4×10^9) metric tons. That mined during the 100 years from 1860 to 1960 amounted to 93.6×10^9 metric tons, giving a total of 99.0×10^9 metric tons for all coal mined from the beginning until 1960. However, the first half of this required the seven centuries up to 1927, whereas the second half required only the 33 years from 1927 to 1960. Only 20 per cent of the coal mined by 1960 was produced before 1900, and the remaining 80 per cent has been produced since that time.

§ 2.5.3 World Production of Crude Oil. Figures 2.6 and 2.7 show graphs of the world production of crude oil, in which the production rates are plotted arithmetically and logarithmically, respectively, against time. Production actually began in 1857, but the rates before 1885 were too small to plot on Figure 2.6. In this case, except for minor setbacks during the depression of the 1930's and during World War II, the production rate has been characterized by steady growth.

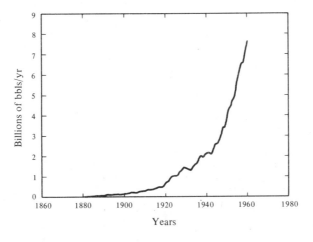

Fig 2.6. World Production of Crude Oil

On the logarithmic scale of Figure 2.7 it will be seen that for the 50-year period from 1880 to 1930, the production rate increased linearly with time. Before 1880, although the production rate was very small, the rate of increase was even greater than that after 1880. Subsequent to 1880 the rate of increase has slackened. From 1880 to 1930 the production rate increased at an exponential rate of 7.4 per cent per year, with a doubling period of only 9.7 years.

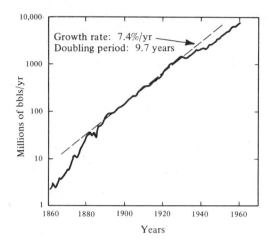

Fig 2.7. World Production of Crude Oil (Logarithmic Scale)

§2.5.4 Energy from Coal and Crude Oil Combined. Finally, in Figure 2.8 there has been plotted on an arithmetic scale the production of energy from both coal and crude oil, in which the thermal energy of each fuel is expressed in the common unit, the kilowatt-hour of heat, obtained by combustion. Until about 1900, it will be noted, crude oil contributed a negligible amount of the total energy. From that time on, the fraction contributed by crude oil has steadily increased, by 1960, it amounted to almost one-half of the total.

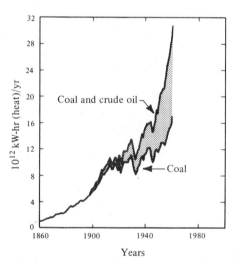

Fig 2.8. World Production of Energy from Coal and Crude Oil

§2.5.5 United States Production of Coal. The production of coal in the United States started about 1820, when 14 tons are reported to have been mined. Since that time the production of coal increased steadily until about 1907, after which the rate has fluctuated between the extremes of about 400 and 700 million short tons per year as shown in Figure 2.9.

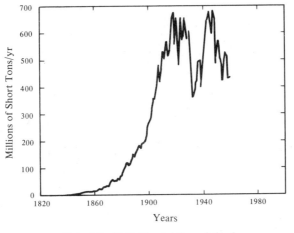

Fig   2.9. U.S. Production of Coal

In Figure 2.10 the same data are shown plotted on a logarithmic scale. Here again, after an initial more rapid rate, the growth settled down to a linear plot on semilogarithmic paper indicating a steady exponential rate of increase. This persisted from about 1850 to 1907, during which period the production rate increased 6.6 per cent per year, with a doubling period of 10.5 years. After 1907 the growth practically ceased, due in large measure to the increasing displacement of coal by the complementary fuels, oil and gas.

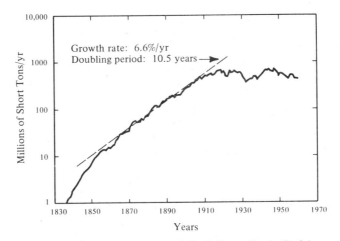

Fig   2.10. U.S. Production of Coal (Logarithmic Sacle)

§ 2.5.6 U.S. Production of Crude Oil. The production of crude oil in the United States since 1860 is shown graphically on arithmetic and logarithmic scales, respectively, in Figures 2.11 and 2.12. Oil was first discovered in the United States by the Drake well drilled at Titusville, Pennsylvania, in 1859. Since that time, with only an occasional setback, the production rate has continually increased. On the semi-logarithmic plotting of Figure 2.12, the production rate increased exponentially from about 1875 to 1929 at 7.9 per cent per year, doubling every 8.7 years. Since 1929 the growth has continued, but at a decreasing rate.

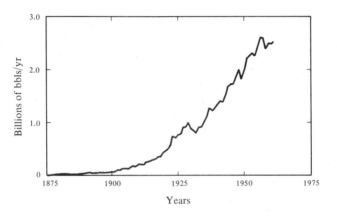

Fig   2.11. Figure 2.11. U.S. Production of Crude Oil

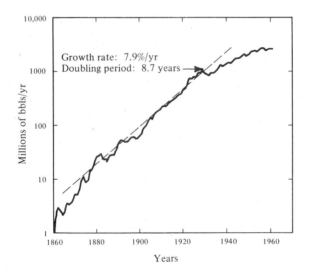

Fig   2.12. U.S. Production of Crude Oil (Logarithmic Scale)

§2.5.7 United States Production of Natural Gas. Figure 2.13 shows the U.S. production of marketed natural gas since about 1905. In the early days of the petroleum industry only a small amount of gas could be utilized, and the rest was disposed of by flaring. Gradually, however, gathering and distributing pipelines have been built, and facilities for using gas a fuel made available, so that very little gas is now wasted. As gas is genetically related to oil, the rate of growth of gas production and consumption is similar, except for a time delay, to that of oil.

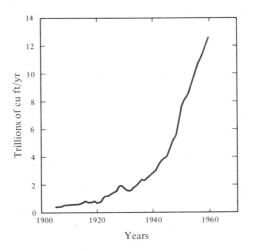

Fig 2.13. U.S. Marketed Production of Natural Gas

§2.5.8 Energy from Coal, Oil, Gas, and Water Power. Finally, the total energy produced in the United States from coal, oil, gas, and water power combined is shown in Figure 2.14[17,18,1] plotted on an arithmetic scale, and in Figure 2.15, plotted logarithmically. In the latter plot it will be seen that the straight-line section of the curve, or the period of exponential growth, persisted from 1845 until 1907, after which the growth rate abruptly dropped to a much smaller value. During the 60-year period of exponential growth the rate of increase was 7.4 per cent per year, with a doubling period of 9.7 years.

From 1907 to 1960 the consumption of energy from the fossil fuels and water power increased from 14.6×10^{15} Btu per year to 44.9×10^{15}. The mean exponential rate of growth for the 53-year period dropped to only 2.04 per cent, and the mean doubling period increased to 34 years. The amounts of energy contributed from the separate sources for the period 1920–1960 are shown in Figure 2.16. During this period the percentage contribution by water power increased only from 3.1 to 3.9 per cent. The dramatic transition, however, has been the displacement of coal by oil and gas. In 1920, 89 per cent of the energy consumed was supplied by coal and only 8 per

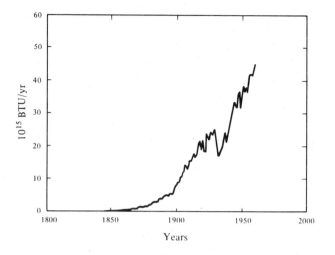

Fig 2.14. U.S. Consumption of Energy (Coal, Oil, Gas, and Water Power)

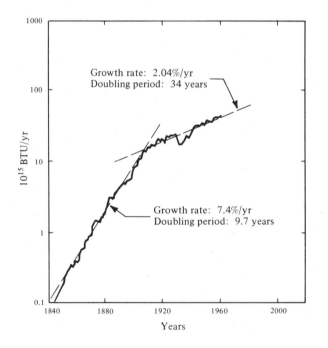

Fig 2.15. U.S. Consumption of Energy (Logarithmic Scale)

cent by oil and gas; by 1960 the contribution of coal had dropped to only 23 per cent, while that by oil and gas had increased to 73 per cent, or about three-quarters of the total.

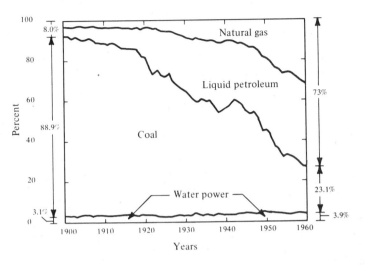

Fig 2.16. U.S. Consumption of Energy since 1900. Percentages Contributed by Coal, Oil, Gas and Water Power

The importance of the information on the U.S. consumption of energy from coal, oil, gas, and water power, with respect to the industrial rate of growth can hardly be overemphasized, since, with the exception of energy dervied from biologic sources, and a small amount of wind power, almost every wheel that turns, every industrial process that is in operation, and a predominant amount of space heating are made possible by the energy from these sources. Furthermore, it is this energy consumption which distinguishes the activities in the United States from those of other major areas of the world whose energy supplies are limited principally to biological sources. Hence, the curve of the consumption of nonbiological energy in the United States is, in effect, a physical integration of all the industrial activity in the country.

While it is true that the industrial output per unit of energy consumed is also increasing with time, because of physical limitations this tends asymptotically to a maximum. Hence, if the total rate of energy consumption were to be maintained constant, the industrial output would continue to rise, but at a decreasing rate of growth, until it also leveled off to an essentially constant rate. The curves of Figures 2.14 and 2.15, therefore, may be considered to represent minimum rates of the industrial growth of the United States. During most of the nineteenth century the industrial rate of growth was somewhat greater than 7 per cent per year, and the rate of output doubled in somewhat less than 10 years. During most of the twentieth century there has been a drastic reduction in this rate of growth.

§2.6 Future Production of Fossil Fuels. The history of the production and consumption of energy from the fossil fuels for both the world and the United States is graphically and accurately summarized in Figures 2.4 to 2.16. Beginning from zero, it is seen how the consumption of these fuels has gradually increased until during the last century the rates of consumption have reached magnitudes many times greater than the energy derived from all other sources in the industrialized areas of the world. Furthermore, as we have noted, most of this has occured within the last 30 years.

It is difficult to contemplate these curves without wondering: How long can we keep this up?

That it cannot continue indefinitely can be seen very simply. The supply of fossil fuels initially in the ground before human exploitation began was some fixed finite amount. As was observed earlier, these fuels are the residues of organisms which became buried in the sedimentary muds and sands over a period of some 500 million years of geological history. Their energy content represents solar energy, stored by photosynthesis as chemical energy, from that same span of time. Geologically, this process is still continuing but probably at a rate not greatly different from that of the past. Hence, the new fossil fuels to be generated during the next million years will probably not differ greatly from 1/500th of that of the last 500 million years, and that for the next 1,000 years correspondingly less.

Hence, we may regard the initial supply of fossil fuels as constituting a nonrenewable resource which is exhaustible. When we burn oil or coal, the energy content, after various degradations during use, degenerates to unusable heat at the lowest ambient temperature, and then leaves the earth as long-wavelength radiation. The material content is reduced to common inorganic chemicals such as water and carbon dioxide, and a residue of mineral ash.

This fact provides us with the most powerful means we have available for anticipating the future history of the consumption of these sources of energy. If we plot a curve of the production rate P against time t on arithmetic paper, as we have done in Figures 2.4, 2.6, 2.8, 2.9, 2.11, and 2.13, for any nonrenewable resource, this curve must have the following properties:

   1.   It must begin with $P=0$, and, after passing through one or more maxima, it must ultimately decline to zero. This last state would be due either to the exhaustion of the resource or to the abandonment of its production for other reasons.
   2.   The cumulative production Q up to any given time is given by the equation

$$Q = \int_0^t (dQ/dt)\, dt = \int_0^t P\, dt, \qquad (2.1)$$

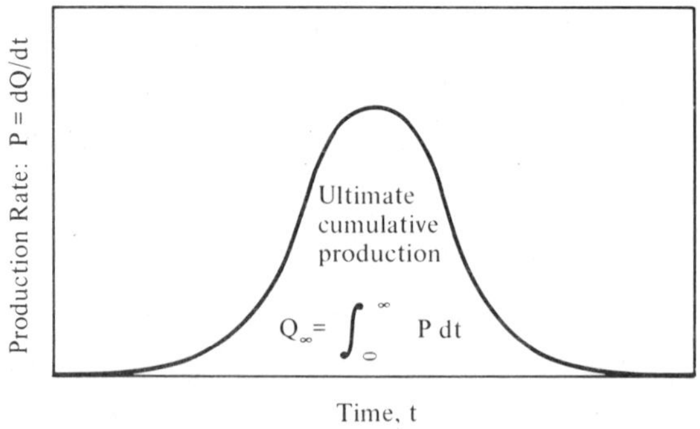

Fig   2.17. **Production of an Exhaustible Resource**

and this, on the graphical plot, is proportional to the area A between the rate-of-production curve and the time axis. This principle is illustrated in Figure 2.17, where the ultimate cumulative production Q_∞ at very large time is proportional to the total area under the curve.

The fundamental fact with which we here must deal is this: *The quantity ultimately produced is equal to or less than the quantity initially present*, or

$$Q_\infty \leqslant Q_i \qquad\qquad (2.2)$$

Hence, if we can estimate Q_i, the amount of the quantity initially present, the curve of production rate P versus time t must begin at zero and end at zero, and it must not encompass an area greater than that corresponding to Q_i.

§2.6.1 Application to Coal. The production of coal lends itself readily to this type of treatment because coal occurs in stratified deposits which frequently extend over large areas, or whole sedimentary basins, and hence are amenable to comparatively accurate estimates of their amount. Coal beds frequently crop out on the

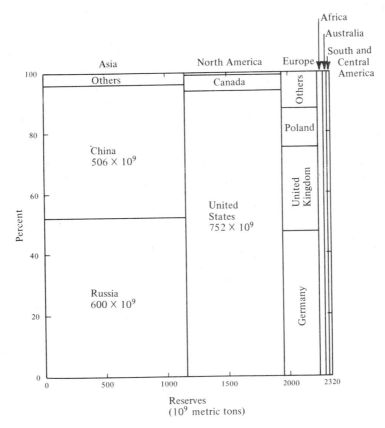

Fig 2.18. Recoverable World Coal Reserves

surface of the ground, and in the subsurface they can be mapped by comparatively few widely spaced drill holes.

The first world inventory of coal resources was made during the Twelfth International Geological Congress at Toronto in the year 1913. Any estimates that could be made at that time were necessarily very provisional. Nevertheless, the estimate of minable coal resources for the whole world amounted to about 8×10^{12} metric tons, which is only about 50 per cent higher than the present estimates.

Since that time, extensive and intensive geological exploration has been extended to all parts of the world. Also during the last decade the Fuels Branch of the United States Geological Survey has been engaged in a detailed re-examination of the coal resources of the United States, and also has maintained currently the estimates being made of the coal resources of the rest of the world.

The latest (1962) such world summary is that prepared by Paul Averitt[15] from the preliminary reports of the Geological Survey. These estimates are shown graphically in an abbreviated form in Figure 2.18.

It is to be emphasized that these are the *remaining* coal reserves, and to obtain the initial reserves we shall have to add the quantity already produced, which for the world was about 99×10^9 metric tons by the end of 1960. That for the United States amounted to 32×10^9 metric tons. Then, from the data in Figure 2.18 and the above figures on cumulative production, the estimates of initial minable coal reserves of the world as of 1961 are $2,419 \times 10^9$ metric tons; the corresponding figure for the United States is 785×10^9 metric tons.

The figures are of *recoverable reserves* defined as " $\cdots$ reserves in the ground, as of the date of the estimate, that past experience suggests can actually be produced in the future"[16]. Elsewhere it is explained that these include all seams 14 inches or more thick, occurring at depths of 3,000 feet or less, with an allowance for nonrecovery of 50 per cent of the coal in place.

Before proceeding further it is worthy of note that the coal reserves of the world are far from equitably distributed among the world's people. The continent of Asia, for example, has 49.4 %, or almost exactly half, of the world's coal reserves, nearly all of which are in the U.S.S.R. and China. North America has 34.4 %, or about one-third; Europe has 13.0 per cent; and the remaining 3.2 % is divided between the three whole continents: Africa, South America, and Australia.

By countries, the United States has approximately one-third, Russia one-fourth, and China one-fifth of the world's coal reserves. Of the 13.0% in Europe, Germany has about one-half, the United Kingdom one-fourth, and Poland one-eighth.

A fairly widespread delusion among the citizens of the United States is that the country owes its phenomenal industrial development, as contrasted with the lack of development of regions such as Africa, South and Central America, and India, to the superiority of American personal and institutional characteristics. It may be well to remind ourselves that, but for a fortuitous combination of a large fraction of the world's resources of coal and iron in the eastern United States, the growth of which we are justly proud could never have occurred.

Returning now to the problem of predicting the future of coal production, let us apply the technique illustrated in Figure 2.17. For the world the results are shown in Figure 2.19 and for the United States in Figure 2.20. In Figure 2.19 the world production of coal through 1960 is first plotted. From this point the graph must continue with time until it passes through one or more maxima, and then the production of coal must ultimately decline to zero. The area under the curve, however, must not exceed that corresponding to the estimated initial reserves, $2,400 \times 10^9$ metric tons.

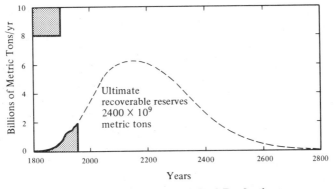

Fig 2.19. Ultimate World Coal Production

A scale for the conversion of area to tons of coal is shown in the upper left-hand corner of the chart. Here one square in the coordinate grid is seen to have the dimensions

$$2 \times 10^9 \text{ metric tons/yr} \times 100 \text{ yrs},$$

and so represents 200×10^9 metric tons. Hence, the area under the production curve between the beginning of coal mining and its end cannot exceed 12 grid rectangles, representing $2,400 \times 10^9$ metric tons of coal.

The curve is drawn subject to these conditions. The shape of the curve, of course, is not known, but if the world should continue to be heavily dependent on coal, and if the peak of production should reach as much as 6×10^9 tons/yr—about three times the present production rate— this peak would occur about the year 2150, or 200 years hence. If the production rate went higher than this the peak would occur sooner; if less high the date of the peak would be postponed.

In Figure 2.20 the coal production of the United States is treated in a similar manner, except that the coal is measured in short tons instead of metric tons.

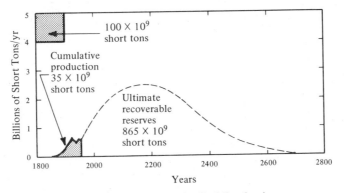

Fig 2.20. Ultimate U.S. Coal Production

By January 1, 1961, the cumulative coal production of the United States amounted to 35.2×10^9 short tons. The remaining reserves by that date were 830×10^9 short tons. The initial reserves, which are the sum of these two figures, are thus 865×10^9 short tons.

The grid rectangle in Figure 2.20 represents 100×10^9 short tons, so the coal-production curve must be drawn in such a manner as to enclose 8.65 grid rectangles. Again, assuming that we continue to require coal, and assuming a production peak of 2.5×10^9 tons per year —two more doublings of the present rate– the peak again would occur about 200 years hence.

§2.7 Future Production of Petroleum and Natural Gas.

The technique described in §2.6 is also applicable to petroleum and natural gas, only in this case it is much more difficult to estimate the producible amounts of these fuels initially present. Because of this difficulty, as indicated by the wide disparity among recent estimates by different investigators, it will be necessary to consider petroleum and natural gas in much more detail than was the case for coal.

Such an extended examination not only is justified, but also is becoming increasingly urgent, in view of the fact that oil and gas are approaching equality with coal as a source of energy on a world scale, whereas, in the United States, the energy consumption from these fuels is already three times as large as that from coal. Yet, preliminary evidence indicates that the total energy reserves from oil and gas are much smaller than those from coal. Thus, because of the relative smallness of the reserves and their rapid rate of depletion, critical problems are due to arise with respect to supplies of oil and gas much sooner than with any other source of energy.

§2.7.1 Petroleum Classification.

Since a great deal of unnecessary confusion in discussing petroleumreserve problems arises from the failure to distinguish between the different classes of petroleum fluids, let us first define what these classes are.

The first of these fluids is *crude oil*, which is the liquid petroleum obtainable from an oil reservoir after the gaseous constituents have been removed or have escaped. Next comes *natural gas*, consisting principally of methane (CH_4), the constituent of petroleum fluids which remains gaseous at standard conditions of temperature and pressure. Finally, there are the *natural-gas liquids*, which are the liquid constituents obtained from wells which otherwise produce natural gas.

The sum of the liquid phases, crude oil and natural-gas liquids, is frequently combined statistically and classed as *petroleum liquids*, or *liquid hydrocarbons*.

The extraction of natural-gas liquids became significant only after about 1920. Since that time its production rate has risen in the United States until by 1961 it represented about 15 per cent of the total production of liquid hydrocarbons. Thus, originally crude oil was the sole liquid hydrocarbon, but more recently natural-gas liquids have achieved a significant fraction of total production.

§2.8 Estimation of Crude Oil Reserves of the United States.

A great deal of confusion has been introduced into discussions of petroleum reserves by the failure to distinguish between crude oil and total petroleum liquids. In what follows this distinction will clearly be made. We shall first deal with crude oil, for which our data are the most complete, and then use the results obtained as a basis for estimating the reserves of natural gas and natural-gas liquids. Also, since the petroleum industry in the United States is more advanced in its evolution toward total depletion than that of any comparable area of the rest of the world, we shall use the data of the United States as a yardstick for estimating the reserves of other areas.

§ 2.8.1 **Geological Background.** Before proceeding with this problem in detail, let us first consider a few of the fundamental facts concerning the manner of occurrence of oil and gas underground. If a well is drilled deep enough at any place on the earth it will eventually encounter some form of dense, crystalline rock such as granite, or gneiss, or schist, of either igneous or metamorphic origin, whose grains are so tightly packed that the pore space is practically zero.

We shall refer to this system of crystalline rocks, which is continuous over the whole surface of the earth, as the *basement* or *basement complex*.

In many parts of the world, such as eastern Canada, Scandinavia, and a large part of Africa, the rocks of the basement complex occur at the surface of the ground. In other areas these rocks are covered with a veneer of unmetamorphosed rocks such as sandstones, shales, and limestones, which are sedimentary in origin. The thicknesses of these deposits of sedimentary rocks vary from zero to possibly 10 miles or more. The average thickness is probably not more than about a mile. The sediments having thicknesses of one to several miles occupy basin-like depressions in the upper surface of the basement complex.

These unmetamorphosed sediments comprise the habitat of the fossil fuels. They are the sands and muds in which the organic remains of the geologic past were buried and preserved. These rocks, or contiguous fractured basement rocks, are therefore the only rocks in which commercial quantities of fossil fuels have ever been found, or are ever expected to be found.

The unmetamorphosed sedimentary rocks are mostly porous, with the pore volume comprising about 20 per cent of the total volume. This pore space forms a three-dimensional interconnected network which normally is completely filled with water. Exceptionally, in very local regions of space whose horizontal dimensions may range from a few hundred feet to some tens of miles, oil and gas may have displaced the water in certain strata of the sedimentary deposit. These local concentrations of oil and gas in the sedimentary rocks are the sources of our commercial production of these fluids.

This knowledge provides us with a powerful geological basis against unbridled speculation as to the occurrence of oil and gas. The initial supply is finite; the rate of renewal is negligible; and the occurrences are limited to those areas of the earth where the basement rocks are covered by thick sedimentary deposits.

The geographical distribution of all of such basins on earth is reasonably well known. If we can estimate about how much oil and gas is contained per unit volume in the sediments in the better-known areas, such as the United States, then, by assuming comparable oil and gas contents in similar sedimentary basins in the rest of the world, an estimate in advance of extensive development can be made of the possible oil and gas that other areas may eventually produce.

This, in essence, is the geological basis for estimating the ultimate petroleum reserves. It is an essential method, but as we shall see, it has inherent limitations of accuracy. The sedimentary rocks of the United States and its continental shelves to a depth of two miles have a volume of about 3×10^6 cubic miles, or about 14×10^6 km^3. With an average porosity of 20 per cent the pore volume of these rocks would be about 2.8×10^6 km^3. Now suppose that these rocks contain 1,000 billion barrels of crude oil in commercially producible concentrations. The volume of this amount of oil would be 159 km^3, which would represent a fraction of 5.7×10^{-5} of the entire pore volume, or about 6 parts per 100,000.

There is no geological information in existence that will permit us to know whether this is a high figure or a low figure. We have no *a priori* way of knowing whether the average content of oil occurring in commercial quantities in sedimentary rocks should be a few parts in 100,000, or ten times or one-tenth this amount.

If the oil production of the United States is to be used as a primary standard for estimating the petroleum potentialities of the rest of the world, then the only possible way we have of determining how much oil the United States will produce is by pure empiricism, based on our actual experience in the exploration and production of petroleum. The United States experience can then be used to estimate what may be expected from other comparable regions.

§ 2.8.2 **Reserve Estimates.** According to Wallace E. Pratt,[36, 37, 38] then Vice President for Exploration and Production of the Standard Oil Company of New Jersey, the world's largest oil company, one of the Jersey Standard geologists, L. G. Weeks, had made an extensive world-wide study along the general lines sketched above. This report has never been published, but in 1948 Weeks[41] published a summary of the results that had been obtained. This consisted of estimates of the *ultimate potential reserves* of various areas, defined as the total amount of crude oil that could reasonably be expected to be produced by productive methods, and under economic conditions, prevailing in 1947. For our purposes the two principal results were:

Land area of the United States
  (excluding Alaska) 110×10^9 barrels
Land areas of entire world 610×10^9 barrels

Two years later, during a discussion on petroleum reserves at the United Nations Scientific Conference on the Conservation and Utilization of Resources, held at Lake Success, New York, Weeks[42] amplified his earlier estimate by adding about 400×10^9 barrels for the continental-shelf areas of the world, and arrived at a round estimate of $1,000 \times 10^9$ barrels for the whole world. This was in criticism of an estimate of $1,500 \times 10^9$ barrels by A. I. Levorsen.[28]

Weeks gave his own appraisal of the reliability of these figures in the following words

I look upon my estimates for the United States as reasonable at this time. Furthermore, I now know of no good reason for considering that the incidence of oil occurrence in the United States should be much, if any, above that of the average for the world. As previously stated, I feel that the actual measure of oil recoverable by conventional methods and under present economies is more likely to be 50 per cent larger than 10 per cent smaller than my estimate of same. However, again I must warn that these are not proved reserves. The actual figure of ultimate reserves may very easily vary from my figure by considerably more than the percentages I have just cited (see Ref. 42, p. 109).

It should be emphasized that the foregoing estimates were for crude oil only.

In March 1956 Hubbert[26] added 20 billion barrels to Weeks' estimate for the land area of the United States (excluding Alaska) and 20 billion barrels for the U. S. offshore areas, and arrived at a figure of 150×10^9 barrels for the ultimate potential reserves of crude oil in the United States, and $1,250 \times 10^9$ barrels for the whole world. Almost simultaneously, Pratt[39] published an estimate for the United States of 170×10^9 barrels of liquid hydrocarbons (which implies about 145×10^9 barrels of crude oil); and Pogue and Hill[35] of the Chase Manhattan Bank published a figure of 165×10^9 barrels of crude oil for the ultimate potential reserves of the United States.

At the meeting of the American Association of Petroleum Geologists in Dallas in March 1959, G. Moses Knebel, Chief Geologist of the Standard Oil Company of New Jersey, stated that he and his staff had a few years previously made a comprehensive review of the oil potentialities of both North and South America, and that their estimates for the United States were in substantial agreement with the 150 billion-barrel figure of Hubbert, an estimate which was still regarded as valid. He later disclosed privately that their estimate for the United States was 203 billion barrels of liquid hydrocarbons. Of this, crude oil would comprise about 85 per cent, or about 173 billion barrels.

These figures are cited because they represent a very good cross-section of informed petroleum-industry opinion at that time. Pratt's estimate was based, in part, on twenty-two returns to a questionnaire he had sent to a selected group of well-informed people in the petroleum industry. The high figure in these returns was an estimate of 200×10^9 barrels of crude oil by the consulting firm DeGolyer and MacNaughton.

The only discordant figure of this series was an estimate of 300 billion barrels from an anonymous source in the Department of the Interior.[27]

Shortly after 1956, however, all consistency in the estimates of petroleum reserves vanished. Within a year after the Pogue and Hill estimate of 165 billion barrels, Hill, Hammar and Winger,[25] also of the Chase Manhattan Bank, raised the Pogue and Hill estimate to 250 billion barrels. In 1958, published estimates ranged from a low figure of 165 billion barrels by Davis[24] of Gulf Oil Corporation to a high of about 372 billion barrels by Netschert[32] of Resources for the Future.

In 1958 L. G. Weeks[43] raised his earlier estimate of 110 billion barrels for the ultimate potential reserves of crude oil for the land area of the United States to 240 billion barrels of liquid petroleum for both the land and offshore areas. This quantity was said to represent "...the ultimate potential liquid petroleum resources, recoverable by conventional primary methods in terms of current economics..." Of this, about 85 per cent, or 204 billion barrels, would be represented by crude oil.

What Weeks meant by "conventional primary methods" is not entirely clear since his 240 billion-barrel figure was stated to include both cumulative production and proved reserves, each of which is a composite of oil already produced, or producible, by both primary and secondary methods. He did mention, however, that a means might ultimately be found to recover by secondary methods an additional quantity as large as the one cited. A year later[44] this ambiguity was resolved. In a new estimate Weeks raised the figure of 240 billion barrels of liquid petroleum recoverable by conventional primary methods to 270 billion and then added 190 billion barrels producible by "secondary recovery," giving a total of 460 billion barrels. Again, about 85 per cent of this, or about 391 billion barrels, would be represented by crude oil.

This last estimate was still adhered to by Weeks as recently as May 1961.[45]

The 1958 and 1959 estimates of Weeks were used by Paul Averitt[16] of the United States Geological Survey as the basis for his figure of 470 billion barrels of liquid petroleum (or 400 billion barrels of crude oil) for the United States exclusive of Alaska. However, what appears to be the "official" estimate of the U.S.G.S. is that prepared by A. D. Zapp[46] for presentation by V. E. McKelvey to the Natural Resources Subcommittee of the Federal Science Council, November 28, 1961. Zapp's estimate of the ultimate U.S. resources of crude oil (including past production) was 590 billion barrels. Concerning this estimate, V. E. McKelvey, in the same report, remarked:

> Those who have studied Zapp's method are much impressed with it and we in the Geological Survey have much confidence in his estimates (see Ref. 29, p. 12).

A published exposition of Zapp's method[46] has subsequently appeared in the U. S. Geological Survey Bulletin 1142-H, entitled "Future Petroleum Producing Capacity

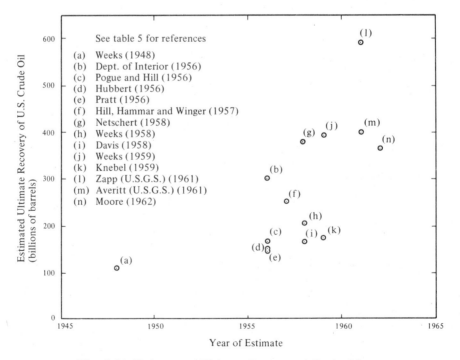

Fig 2.21. Estimates of Ultimate Recovery of Crude Oil

of the United States." In this, no estimate is given explicitly of the ultimate amount of crude oil the United States may be expected to produce, but such an estimate is implied in two statements on page H-24:

1. But this much is certain: it cannot be safely assumed that even the 20-percent mark has been reached in exploration for petroleum in the United States, excluding Alaska and excluding rocks deeper than 20,000 feet.
2. With the crude yardstick of at least 100 billion barrels of oil found so far, and a rough appraisal of the extent of exploration so far, an objective estimate of the approximate minimum ultimate "reserves" appears to be in sight.

As an aside, petroleum-exploration people are intimately familiar with the initial 20 per cent of the petroleum exploration postulated by Zapp, but many are at a loss as to how to proceed with respect to his postulated remaining 80 per cent.

From a study of petroleum-industry statistics, C. L. Moore of the U.S. Department of the Interior, Office of Oil and Gas in 1962 arrived at an estimate of 364 billion barrels for the ultimate U.S. recovery of crude oil.

These various estimates are shown graphically in Figure 2.21. To review the often lengthy arguments whereby they were derived would be time-consuming and profitless, as the extent of their unreliability is attested by the range of disagreement exhibited among the estimates themselves. There exists some definite quantity of crude oil, Q_∞ (at the moment unknown) which will ultimately be produced in the United States. The estimates plotted in Figure 2.21 are each intended to represent this quantity. Suppose that the correct value happened to be 590 billion barrels, the highest figure cited. Then the lowest figure since 1955, 145 billion barrels, would be in error by 445 billion barrels; and the errors of the other estimates, except the correct one, would range between 190 and 445 billion barrels.

If the smallest figure happened to be the correct one, then all the others would be erroneously high, with the errors again ranging from 5 to 445 billion barrels.

If the correct figure happened to fall about mid-range, say at 370 billion barrels, then the errors on either side would range between zero and about 200 billion barrels.

It is thus demonstrable, without making any hypothesis whatever of what the true value of Q_∞ should be, that the preponderance of recent attempts to determine this quantity are grossly in error. This raises the question of whether the desired quantity is intrinsically indeterminate, except within these wide limits, or whether from data now available it should be possible to determine this quantity within a much narrower range of uncertainty. It is the thesis of the present report (Dr. Hubbert is now referring to his 1962 report to the National Academy of Sciences–National Research Council) that such data do exist, and that from them a much more reliable estimate can be made.

§ 2.9 New Method for Estimating the Ultimate Crude-Oil Production of the United States. The method we shall now employ makes explicit use only of two of the most reliable series of statistics of the petroleum industry: (1) the quantity of crude oil produced in the United States per year, for which data are available annually since 1860, and (2) the estimates of proved reserves of crude oil in the United States made annually since 1937 by the Committee on Petroleum Reserves of the American Petroleum Institute.

The data on the annual production of crude oil requires no comment. The meaning of the term "proved reserves," as defined by the Reserve Committee, however, needs to be clearly understood, because the Reserve Committee operates on the basis of this definition, and their reserve figures are not susceptible to any other interpretation. The following is a partial quotation from the definition of the term "proved reserves of crude oil" taken from the Report of Committee on Petroleum Reserves of the American Petroleum Institute (API) of March 9, 1962 (p. 3):

Proved Reserves of Crude Oil—Definition

The reserves listed in this Report, as in all previous Annual Reports, refer solely to "proved" reserves. These are the volumes of crude oil which geological and engineering information indicate, beyond reasonable doubt, to be recoverable in the future from an oil reservoir under existing economic and operating conditions. They represent strictly technical judgments, and are not knowingly influenced by policies of conservatism or optimism. They are listed only by the definition of the term "proved." They do not include what are commonly referred to as "probable" or "possible" reserves.

* * *

Both drilled and undrilled acreage are considered in the estimates of the proved reserves. However, the undrilled proved reserves are limited to those drilling units immediately adjacent to the developed areas which are virtually certain of productive development, except where the geological information on the producing horizons insures continuity across the undrilled acreage.

The report adds that the estimates do not include oil that may become available by fluid injection or other methods from fields in which such operations have not yet been applied.

Each year's report presents data in each of the following classifications:

1. Estimate of proved reserves at the end of the preceding year.
2. Changes in proved reserves due to extensions and revisions during the subject year.
3. Proved reserves discovered in new fields and in new pools in old fields during the subject year.
4. Production during the subject year.
5. Proved reserves as of December 31 of the subject year. (Items $1 + 2 + 3 - 4$)
6. Changes in reserves during the subject year. (Items $5 - 1$)

Added reserves due to extensions and revisions (Item 2) each year are on the order of 6 to 7 times the reserves due to new discoveries (Item 3).

The significance of the A. P. I. estimates of proved reserves can perhaps best be understood by considering a hypothetical field discovered in a given year. Suppose the field is destined, ultimately, to produce a total of 100 million barrels. Suppose that during the year of discovery only five wells were drilled. The proved reserve estimate would perhaps show:

Reserves added by extensions and revisions: None
Reserves due to new discovery: 150,000 barrels

For a number of years each successive year would then show sizeable reserve additions due to extensions and revisions, but none by new discoveries. Then, as the field approaches complete development, the changes due to extensions and revisions would diminish from one year to the next, ultimately to zero.

The sum of the reserves added, year by year, in this manner would ultimately equal the total amount of oil which the field will produce. This process might continue, however, for thirty or forty years after the date of initial discovery. The reason is that, although the field may eventually produce 100 million barrels of oil, this amount of oil was not discovered at the date of discovery of the field; it was discovered only gradually as the field was developed.

The estimates of proved reserves for the whole United States have exactly the same significance. In fact, all the oil we can claim to have discovered in the United States up to the end of any given year is the total amount of oil already taken from the ground up to that date, the cumulative production, plus the proved reserves. We may call this quantity the "cumulative discoveries" up to that date; or, if one prefers, the "cumulative proved discoveries."

If we represent the cumulative production by the symbol Q_P, the cumulative proved discoveries by Q_D, and the proved reserves by Q_R, then for each year,

$$Q_D = Q_P + Q_R \tag{2.3}$$

The relation between rates of change of these quantities with time is obtained by taking the derivative with respect to time of equation (3), giving

$$\frac{dQ_D}{dt} = \frac{dQ_P}{dt} + \frac{dQ_R}{dt}, \tag{2.4}$$

in which dQ_D/dt is the rate of discovery, dQ_P/dt is the rate of production, and dQ_R/dt is the rate of increase of the proved reserves.

The manner in which the three quantities cumulative proved discoveries, Q_D, cumulative production Q_P, and proved reserves Q_R must vary with time during the entire history of petroleum production from start to finish must be approximately as follows: The cumulative production Q_P, when plotted as a function of time, will increase slowly during the early stages of petroleum exploitation, increase more and

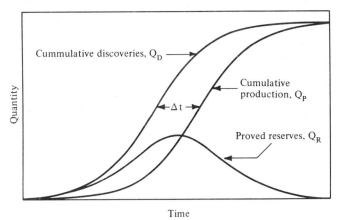

Fig 2.22. Cumulative Discoveries and Production and Proved Reserves

more rapidly with time to about the halfway point, and then continue its ascent by rising more and more slowly, finally leveling off to the ultimate figure Q_∞ as production ceases.

The curve of proved reserves Q_R will start at zero, rise gradually until a maximum is reached at about the halfway point, and then gradually decline to zero.

As oil must be found before it can be produced, the curve of cumulative proved discoveries must closely resemble that of cumulative production, except that it must plot ahead of the production curve by some time interval Δt, which itself may vary during the cycle.

A plot of the family of the three curves Q_D, Q_P, and Q_R is shown in Figure 2.22 as they may be expected to appear in the case of cumulative production of crude oil in the United States. All present evidence indicates that the U. S. discovery and production is following a single growth cycle, rather than a multiple cycle like the State of Illinois which has two production peaks 30 years apart. One- and two-cycle growths are illustrated in Figure 2.23.

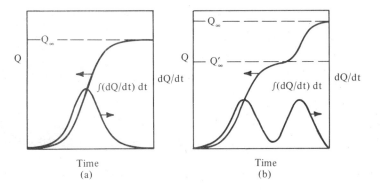

Fig. 2.23. Growth Curves and Production Rates for Single- and Multiple-Cycle Developments

Because of the close similarity between the curve of cumulative proved discoveries and that of cumulative production, it follows that the study of the discovery curve must give one a preview of what production will do at a time of approximately Δt in the future.

Taking the time derivatives of the three curves shown in Figure 2.22 gives us the rate of discovery, rate of production, and rate of increase of proved reserves, which are plotted as a function of time in Figure 2.24. It will be noted that the rate of discovery will reach a peak at about mid-range and, thereafter, gradually decline to zero. The rate of production will reach a peak at a time about Δt after that of discovery, and the increase of proved reserves will change from positive to negative about halfway between the discovery and production peaks. The reserves themselves, Q_R, will reach a maximum at this same time.

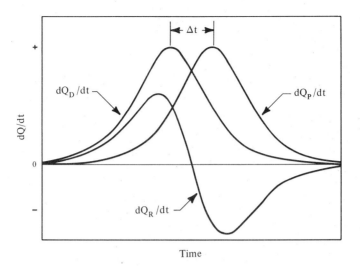

Fig 2.24. Rates of Discovery, Production and Change of Proved Reserves

The relations between the three curves at this mid-point can be seen by noting that when reserves reach their maximum value, their derivative

$$\frac{dQ_R}{dt} = 0 \tag{2.5}$$

which, when inserted into equation (4), gives

$$\frac{dQ_D}{dt} = \frac{dQ_P}{dt}. \tag{2.6}$$

This tells us that when reserves reach their maximum value the curves of discovery rate and production rate will cross, production going up and discovery going down. This is shown in Figure 2.24.

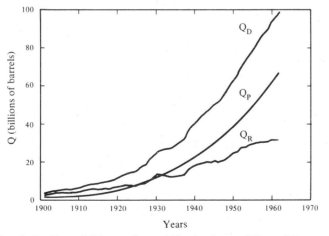

Fig 2.25. Cumulative Proved Discoveries and Production and Proved Reserves of U.S. Crude Oil

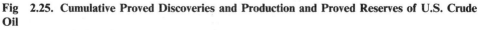

§2.9.1 Observations.   This is the theoretical framework in which we now propose to examine the crude-oil production data and proved-reserve data of the United
States (excluding Alaska). (In all subsequent discussions the petroleum data for the
United States are to be understood to refer to the conterminous part of the United
States, and to exclude Hawaii and Alaska, unless stated otherwise.) Graphs of
cumulative production Q_P, proved reserves Q_R, and cumulative proved discoveries Q_D
for the United States are shown in Figure 2.25[21] The curve for Q_D is the sum of the
first two.

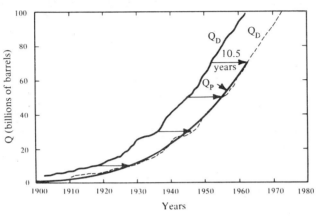

**Fig   2.26. Time Lag Between Cumulative Proved Discoveries and Cumulative Production of U.
S. Crude Oil**

     In order to obtain the approximate magnitude of Δt, we trace the Q_D curve on tracing
paper and then translate it parallel to the time axis until the closest fit with Q_P is obtained.
This is shown in Figure 2.26. Ten years is too small and 11 years is too large; the best fit is at
about 10.5 years. Thus, since 1925 cumulative production in the United States has lagged
discovery by the nearly constant interval of 10–11 years.

     Growth phenomena such as those represented by the Q_D and Q_P curves, which start
slowly, gradually accelerate, and finally level off to a maximum, are said to follow a logistic
growth curve and are describable by an empirical equation of the form....

$$y = \frac{h}{1 + ae^{-bx}}, \tag{2.7}$$

in which h, a, and b are parameters whose magnitudes are to be determined by the data, and
e the base of natural logarithms. Adapting this equation to the data of Figure 2.25, we have
for the curve of cumulative proved discoveries

$$Q_D = \frac{Q_\infty}{1 + ae^{-bt}} \tag{2.8}$$

in which Q_∞ is the asymptotic value to which Q_D will tend as the time t becomes unlimitedly
large.

     The best values of the parameters for the Q_D data can be determined by converting
equation (8) to a linear form. By transposition

$$\frac{Q_\infty}{Q_D} - 1 = ae^{-bt}.$$

Then, by taking the logarithms of both sides, we obtain

$$\log[(Q_\infty/Q_D)-1]=\log a - bt \log e, \tag{2.9}$$

which is a linear equation between $\log[(Q_\infty/Q_D)-1]$ and t.

The quantity $[Q_\infty/Q_D-1]$ is then plotted as a function of time on semilogarithmic paper, using an assumed value of Q_∞. If the correct value is used for Q_∞, and if the data otherwise satisfy equation (8), the curve will be a straight line. By repeating this procedure, using several different values for Q_∞, it is possible to find the best value for this quantity. Then the other two parameters a and b can be obtained from the linear graph.

As determined in this manner, the increase of cumulative discoveries Q_D with time has been found to be approximated very closely by the equation

$$Q_D = \frac{170\times 10^9 \text{ barrels}}{1+46.8\,e^{-0.0687\,(t-1900)}}; \tag{2.10}$$

and cumulative production Q_P by

$$Q_P = \frac{170\times 10^9 \text{ barrels}}{1+46.8\,e^{-0.0687\,(t-1910.5)}}. \tag{2.11}$$

Analytically, the curve for Q_R is given by the difference between equations (2.10) and (2.11).

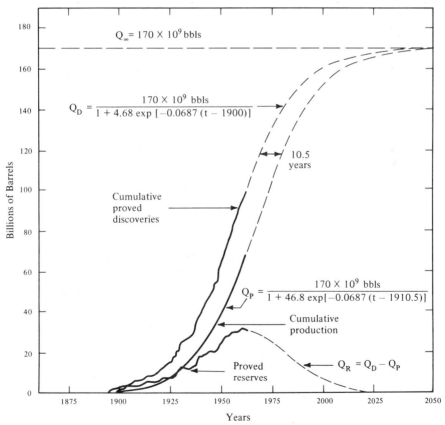

Fig 2.27. Cumulative Proved Discoveries and Production and Proved Reserves of U.S. Crude Oil

The results of these calculations and the closeness of the fit between the actual data for Q_D, Q_P and Q_R (shown in solid curves) and the computed curves (shown dashed) are presented graphically in Figure 2.27.

The discovery curve has plainly passed its inflection point at about 85 billion barrels, and this should be about the halfway point. This agrees with the asymptote of $Q_\infty = 170 \times 10^9$ barrels as given by the curves.

The significance of the cumulative production curve needs no particular discussion. It will simply level off to the maximum Q_∞ when production is finished. The discovery curve Q_D, however, merits further attention, because this curve is the embodiment of the results of all the improvements which have been made in discovery techniques, in drilling techniques, in recovery techniques, and all the oil added by geographical extensions within the United States and its offshore areas, since the beginning of the industry.

We thus do not have to worry about how much oil may be contained in known oil fields over and above the American Petroleum Institute (API) estimates of proved reserves, or how much improvement may be effected in the future in both exploration and productive techniques, for these will all be added in the future, as they have been in the past, by revisions and extensions in addition to new discoveries. And there is as

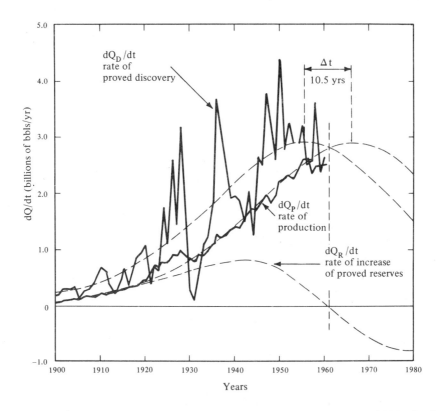

Fig 2.28. Rates of Discovery, Production and Increase of Proved Reserves of U.S. Crude Oil

yet no evidence of an impending departure in the future from the orderly progression which has characterized the evolution of the petroleum industry during the last hundred years.

In Figure 2.28 is shown the actual year-by-year plotting of the rates of discovery and of crude-oil production in the United States since 1900, on which have been superposed the analytically determined rates from equations (10) and (11). The rate of discovery, as is to be expected, oscillates rather widely from year to year, yet the data plainly indicate that the peak of the discovery rate occurred in the early or middle 1950's. The analytical-derivative curve places the date of this peak at 1956 and that for the production rate at about 1966–67. The analytical derivative of the curve of proved reserves crosses the zero point from increasing to decreasing reserves at about 1961–62, which should be the date of the peak of the proved reserves.

The rate of increase of proved reserves is shown in detail in Figure 2.29. Here, superposed on the actual data is the rate curve (shown dashed) as determined analytically. Here again, although the reserve additions oscillate widely from year to year, it will be seen that the analytical curve follows faithfully the trend of the actual data.

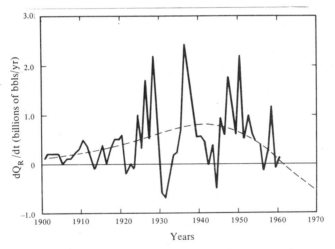

Fig 2.29 Rate of Increase of Proved Reserves of U.S. Crude Oil

A composite view on a longer time scale of the rates of discovery and production and the increase of proved reserves is given in Figure 2.30. From the evolution which has occurred up to now it is difficult to escape the conclusion that the petroleum industry in the United States is somewhere near the halfway point in its exploration for and production of crude oil. By the end of 1961 the cumulative production of crude oil had reached 67.37 billion barrels, and proved reserves were estimated at 31.76 billion barrels, from which the cumulative proved discoveries amounted to 99.1 billion barrels.

However, the peak rate of discovery occurred about 6 to 7 years previously, at about 1956; proved reserves appear to be very nearly at their maximum in 1962; and the peak of production is expected to occur by about 1967 or earlier. Unless the

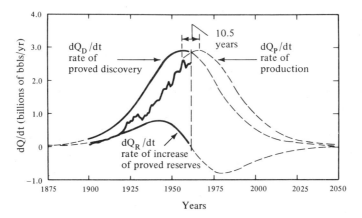

Fig   2.30. Rates of Discovery, Production and Increase of Proved Reserves of U.S. Crude Oil

evolution of the industry departs radically in the future from the orderly progression
it has followed for the last hundred years, the most probable estimate that can now be
derived from past experience for the ultimate cumulative production of crude oil is
about 170 billion barrels.

With regard to the date of the peak of crude-oil production, mention should be
made of a minor qualifying circumstance. Due in large measure to petroleum imports,
which have been building up since World War II and now amount to approximately
20 per cent of domestic production, the present rate of production is somewhat less
than full capacity. According to a recent report of the National Petroleum Council,[31]
the total crude-oil production capacity of the United States, excluding Elk Hills
shut-in capacity, was 10.422 million barrels per day, or 3.8 billion barrels per year, in
1960. This figure for capacity assumes that all wells are operating at capacity,
independently of whether pipelines and storage facilities could handle the production
at this rate. Actual production for the year 1960 was 2.47 billion barrels.

This descrepancy between the actual rate of production and a hypothetical
maximum productive capacity, therefore, allows some latitude in the exact year at
which the peak of production could occur. Conceivably, if, for some reason comparable to the Suez Crisis, the production were to be at the maximum capacity for some
given year, then in whatever year this may have occurred between 1962 and possibly
1975 the peak of production could occur. This possibility, however, is largely
irrelevant with respect to the present analysis, in which only long-term secular
changes, rather than the fluctuations which occur from year to year, are the subject of
concern.

The real significance of the curtailment of U.S. production is that it conserves the
domestic reserves of crude oil and thus tends to postpone the date of the production
decline due to diminishing reserves of oil. Had there been no imports and had the
domestic industry been operating at capacity ever since World War II, the oil that has
been imported would have had to be replaced by oil from domestic reserves. This
would have advanced the peak date of production with respect to that which is now
anticipated.

§ 2.9.2 Verification by Means of Data on Large and Small Fields. An independent check on whether Q_∞, the ultimate cumulative production of crude oil in the United States, is of the order of 170 billion barrels is afforded by a study of the large and small fields separately. Since 1943 the *Oil and Gas Journal*, in its Review-Forecast number which is issued annually about the last week in January, has been publishing statistics on the oil fields in the United States in which the large or so-called "giant" fields have been segregated for special attention. These are defined as those fields whose ultimate production is estimated to exceed 100 million barrels. All other fields are classed as small fields.

In the January 29, 1962 issue of this Journal, on page 135, the following data are given for all the oil fields in the United States:

Number of giant fields	240
Estimated ultimate production:	
All fields	103.26×10^9 barrels
Giant fields	59.22×10^9 barrels
Percent by giant fields	57.4

From this information, the average size of the large fields is found to be 0.247×10^9 barrels.

The number of small fields was not given, but a few years ago an independent estimate was made of the cumulative number of such fields which had been discov-

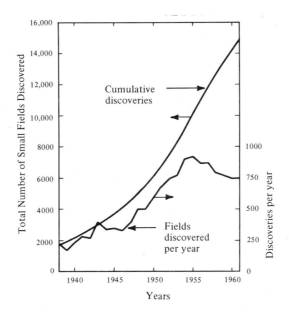

Fig 2.31. Cumulative Discoveries and Rate of Discovery of U.S. Small Fields, 1938–1961

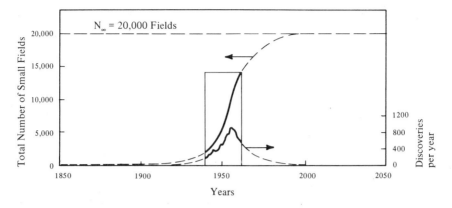

Fig   2.32. Long-Time Outlook for Discovery of Small Oil Fields in the U.S.

ered by the end of 1957. This was about 12,000, and about 3,000 more have been discovered subsequently, giving a total by the end of 1961 of about 15,000.

A table on the discovery rate of all fields up to 1959 is given by B. W. Blanpied.[22] Of these, all but an insignificant fraction are small fields. These two results, the cumulative number of small fields discovered, and the number discovered per year, are shown graphically in Figure 2.31, The same data with longer time scale are shown in Figure 2.32. The peak in the discovery rate occurred in 1955, when the total number of small fields was 10,000. Assuming that this peak represents about the halfway point in small-field discovery, then the ultimate number of small fields is estimated to be about 20,000.

The ultimate liquid hydrocarbons credited by the *Oil and Gas Journal* to the then discovered small fields by the end of 1955 was 36.5 billion barrels,[33] of which the crude-oil content would be about 85 per cent, or 31.0×10^9 billion barrels. Then assuming that this represents about half the ultimate for all the small fields we obtain an estimate of 62.0 billion barrels of crude oil as the ultimate production of all the small fields including those still to be discovered.

A corresponding ultimate figure for the large fields could be obtained if we could estimate how many big fields are likely ultimately to be discovered. This should be a particularly significant figure since, despite their small number, the large fields account for nearly 60 per cent of all the oil so far discovered in the United States. The ultimate number of large fields, N_∞, which can hardly be larger than a few hundred, is the quantity we now seek to determine.

The obvious way to do this is simply to enumerate the fields, giving them the serial numbers 1 to 240 in the order of their dates of discovery, and then plot the curve of the number N as a function of the dates of discovery, to see whether evidence of the approach to an ultimate number N_∞ can be detected. This has been done in Figure 2.33, and the curve for the fields listed in the January 1962 issue appears to be approaching an ultimate number of about 250.

That this is a false conclusion can be seen by the curve of analogous data as of December 31, 1951, also shown in Figure 2.33. This curve appears to have an

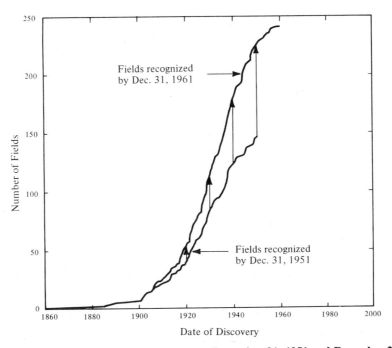

Fig 2.33. Large U.S. Oil Fields Recognized by December 31, 1951 and December 31, 1961

asymptote, or limiting value, at a number of about 175 fields. Thus, it will be noted that these curves increase not only longitudinally as ordinary growth curves do, but they also skid sidewise.

While this may be unexpected, the reason for it is simple. For an ordinary growth curve, such as that of cumulative production, the data for each successsive year are added to the curve only at its extremity. For the large fields, however, each field has two separate dates, a date of discovery (when the field is classified as small) and a date of recognition as a large field. In other words every field which ultimately becomes a large field must go through an embryo, or incubation stage, as a small field before it ultimately hatches out as a large field. The first date is the date of discovery; the second is the date of recognition as a large field.

The fallacy involved in plotting the fields by dates of discovery lies in the fact that they cannot be plotted until after recognition, which may be years or decades after discovery. Thus, when a field discovered in 1945 is not recognized as a big field until 1961, then inserting it into the curve at 1945 displaces the whole curve up by 1 point from 1945 onward. The repetition of this process for each field added produces the behavior shown in Figure 2.33.

However, if the fields are plotted by dates of recognition only, as is shown in Figure 2.34, the curve behaves as any growth-curve should. The false asymptote is missing and the curve has the appearance of being about halfway to its true asymptote. The logistic equation for this curve is

$$N = \frac{460 \text{ fields}}{1 + 110 e^{-0.078 (t-1900)}}. \tag{2.12}$$

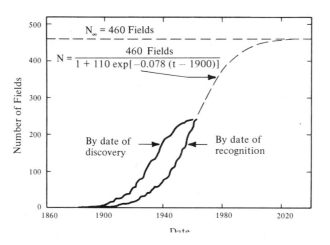

Fig 2.34. Large U.S. Oil Fields Plotted by Date of Discovery and by Date of Recognition

We may, accordingly, either limit our analysis to the curve of the number of large fields plotted by their dates of recognition, or we may attempt, in addition, to estimate the limiting position of the curve plotted by dates of discovery *after all the large fields have been discovered and recognized.*

This can be done approximately by investigating the statistical nature of the time delay, τ, defined as

$$\tau = t_r - t_d \tag{2.13}$$

where t_r is the time of recognition and t_d the time of discovery of a given large field. A curve of the number of fields plotted cumulatively against increase time-delay τ for a sample of 186 fields, excluding fields discovered after 1940, is shown in Figure 2.35. It is clearly seen that the cumulative number of fields as a function of τ is represented very closely by the equation

$$n = n_\infty (1 - e^{-0.046\tau}), \tag{2.14}$$

where n_∞ is the asymptotic number for the sample. Expressing the number of fields as a fraction of this asymptotic number we have

$$\frac{n}{n_\infty} = \frac{n}{206} = (1 - e^{-0.046\tau}), \tag{2.15}$$

which is shown graphically in Figure 2.36. From this it will be seen that about half the large fields have time delays between discovery and recognition of more than 15 years.

The advantage of this time-delay curve is that it permits us to apply a correction to the number of fields discovered in any given year that have been recognized by some definite later date. For example, suppose that by December 1961 ΔN fields discovered in 1946 have been recognized. As the time delay from 1946 to 1961 is 15 years, then according to Figure 2.36 only about half the fields discovered in 1946 should have been recognized by 1961. We therefore estimate that of all the large fields discovered in 1946 which will ultimately be recognized as large fields, only about half were recognized by 1961. We accordingly apply the correction

$$\Delta N' = 2\Delta N.$$

An analogous procedure is followed for each year prior to 1961 with τ equal successively to 1, 2, 3,..., n years. The $\Delta N'$ are then integrated into a new curve which should represent, approximately, the cumulative number of large fields ultimately to be recognized, plotted by dates of discovery.

The results of such a computation applied to the data of December 31, 1961, are shown in Figure 2.37. The lower solid-line curve shows the fields already recognized by dates of discovery, and the upper solid-line curve shows the fields probably already discovered by the end of 1961 which will ultimately be recognized as large fields. The difference between the two curves represents the embryo large fields probably already discovered, but not yet recognized.

The logistic equation for the revised number of fields by dates of discovery shown in Figure 2.37 is

$$N = \frac{460 \text{ fields}}{1 + 5.0 e^{-0.0856 (t - 1920)}}. \qquad (2.16)$$

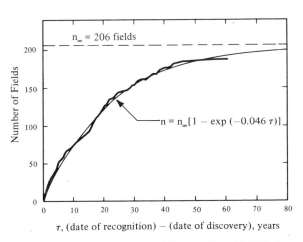

Fig 2.35. Time Lag Between Discovery and Recognition of U.S. Large Fields (Fields Discovered Since 1940 Excluded)

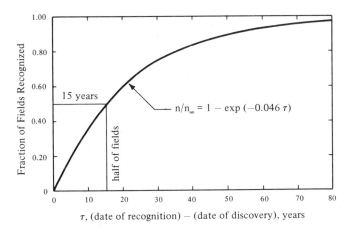

Fig 2.36. Fraction of U.S. Large Fields Recognized Within Time-Delay τ After Discovery

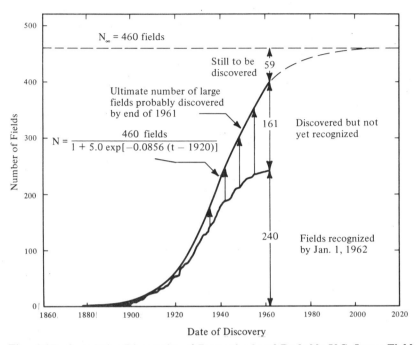

Number of Fields (vertical axis)

$N_{\infty} = 460$ fields

Still to be discovered 59

Ultimate number of large fields probably discovered by end of 1961

$$N = \frac{460 \text{ fields}}{1 + 5.0 \exp[-0.0856 (t - 1920)]}$$

161   Discovered but not yet recognized

240   Fields recognized by Jan. 1, 1962

Date of Discovery

Fig   2.37. Cumulative Discoveries of Recognized and Probable U.S. Large Fields

It therefore appears from the data on both the number of large fields plotted by date of recognition, and the revised curve on the probable fields by date of discovery, that the ultimate number of large fields is about 460. Of these, as it will be seen from Figure 2.37, about 401 have probably already been discovered, leaving about 59 still to be discovered. Of the 401 fields already discovered 240 have already been recognized by the end of 1961, and about 161 are in the incubation stage as small fields which with further development will eventually become large fields.

§ 2.9.3 Average Size of Large Fields. It has already been pointed out that according to the estimate of the *Oil and Gas Journal* the average size of the 240 large fields of December 31, 1961, is 0.247×10^9 barrels, or about one-quarter of a billion barrels.

Figure 2.38 shows the average size of successive groups of 25 large fields each in the order of discovery. This indicates that there is little ground to expect the average size of the large fields in the future to be very different from that of the past. Assuming, then, a constant average size, the ultimate amount of crude oil expectable from 460 large fields should be about 113×10^9 barrels.

If we now add the 62×10^9 barrels for the small fields to the 113×10^9 barrels for the large fields, we obtain an estimate for the total ultimate production of crude oil in the United States of 175×10^9, or 175 billion, barrels.

The method of estimation based on the use of the *Oil and Gas Journal* data for the large and small fields is not considered to have as high a reliability as that using

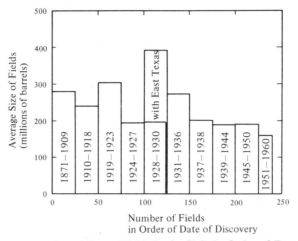

Fig 2.38. **Average Size of Large Fields in the U.S. in Order of Dates of Discovery**

the growth curves. It is nevertheless considered to be valid as to order of magnitude, and to this extent it corroborates the estimate of 170 billion barrels obtained previously. As a contingency, however, we shall adapt the higher figure of 175 billion barrels as representing our present estimate of the ultimate potential reserve of crude oil in the United States. Of this, 67 billion barrels have already been produced, and 99 billion barrels (including that already produced) have already been discovered, leaving about 76 billion barrels still to be discovered.

If a contingency allowance were to be made of how much the actual figure of Q_∞ might exceed the present estimate of 175 billion barrels, a figure higher than an additional 50 billion would be hard to justify.

Should the future of 175 billion barrels be approximately correct, the future crude-oil production of the United States would have to follow a curve closely resembling that shown in Figure 2.39.

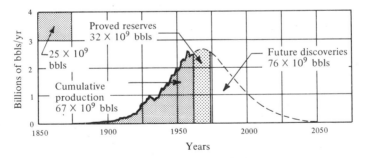

Fig 2.39. **Estimate of Ultimate U.S. Production of Crude Oil.** One grid rectangle represents 25 billion barrels of oil. The total area under the curve from start to finish could, therefore, comprise only 7 rectangles, and the culmination in the rate of production should occur in the late 1960's. If the figure of 225 billion barrels (including the 50 billion-barrel contingency allowance) should be more nearly correct, then the curve would encompass an area of 9 grid rectangles and the culmination would occur in the early 1970's.

§2.10 Ultimate Potential Crude-Oil Reserves of the World. Using the United States estimate as a yardstick, we may now give an approximate estimate of the ultimate potential crude-oil reserves of the world. This is obtained by using Weeks'[41] estimate as a base and then applying modifications which subsequent developments indicate to be necessary. This is shown graphically for the major geographical and political subdivisions of the world in Figure 2.40. The total world estimate comes to 1,250 billion barrels, of which 850 is for land areas and 400 is allowed for the offshore areas.

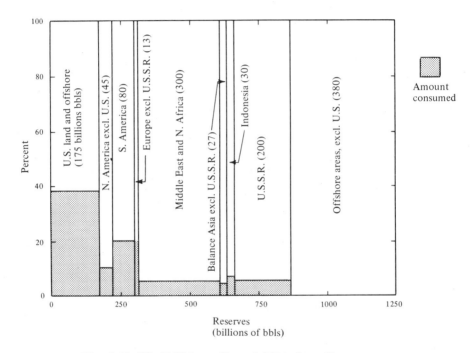

Fig 2.40. World Ultimate Potential Petroleum Reserves

Of particular interest is the preponderance of the reserves of the Middle East and North Africa (300 billion barrels) over those of any other geographical region of comparable area. Also, it is to be noted that because the United States was the world's largest producer of crude oil for nearly a century, it is also the farthest advanced toward ultimate depletion of any of the major oil-producing areas.

A curve of the ultimate world production is shown in Figure 2.41. Using 1,250 billion barrels as the ultimate potential reserve, and assuming a peak rate of production of 12.5 billion barrels per year—about twice the present production rate—the culmination of world production should occur about the year 2000 A.D.

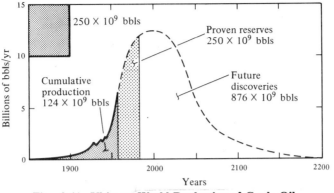

Fig 2.41. Ultimate World Production of Crude Oil

§2.11 United States Production and Ultimate Reserves of Natural Gas. The production rate of marketed natural gas in the United States has already been shown in Figure 2.13. Because pipelines for long-distance transmission of natural gas have only become available since World War II, the consumption of gas in the United States has reached a relatively less advanced state toward ultimate depletion than crude oil. It is, accordingly, not yet possible to estimate the ultimate asymptote of the curves of cumulative production and cumulative proved discoveries for natural gas, as was done for crude oil.

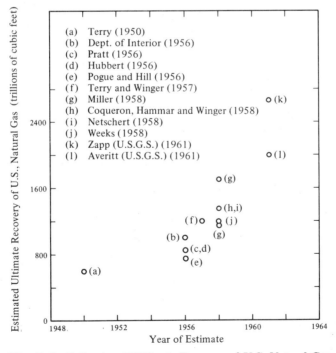

Fig 2.42. Estimates of Ultimate Recovery of U.S. Natural Gas

The next best procedure is to make use of the fact that natural gas and crude oil are genetically related, and then to base the estimate of the ultimate amount of natural gas on that of the ultimate amount of crude oil, using the observed ratio of gas to oil. It follows that the estimates of the ultimate potential reserves of gas obtained in this manner will vary, percentagewise, about as widely as the estimates of the reserves of crude oil. This is borne out by Figure 2.42 in which the principal published estimates since 1950 are presented. These estimates range from a low value of 600 trillion cubic feet by Terry[40] to a high value of 2,650 trillion cubic feet by Zapp[46] of the United States Geological survey.

The remarks made earlier with regard to published estimates of the ultimate reserves of crude oil apply equally to those for natural gas. Regardless of what the correct value for the ultimate gas reserves may be, most of the published estimates are seriously erroneous.

For our own estimate we shall take our figure of 175 billion barrels of crude oil as a base, and then apply the ratios of gas to oil obtained from petroleum-industry experience. One aspect of this experience is shown in Figure 2.43. Here a graph is shown of the ratio of cumulative proved discoveries of natural gas to the cumulative discoveries of crude oil in the United States for each year from 1925 to 1961. It will be seen that the gas-oil ratio gradually increased during this period from about 2,200 ft^3/bbl in 1925 to about 4,900 in 1961. Although this ratio is still increasing, it is probably too low, because of the large volumes of gas dissipated without any record during the history of the petroleum industry prior to World War II. A more reliable ratio should, therefore, be obtained from the gas and oil discovered during recent decades. In Figure 2.44 is shown a five-year running average of the ratio of the gas discovered per year to the oil discovered per year for each year from 1941 to 1961. The curve fluctuates between a low value of 4,000 ft^3/bbl and a high value of 10,000 ft^3/bbl, but without any pronounced secular trend. The average value for the 20-year period is 6,250 ft^3/bbl.

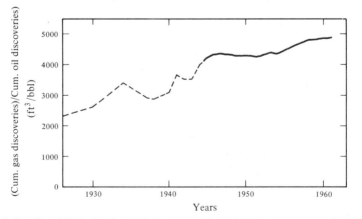

Fig   2.43. Gas-Oil Ratios for U.S. Based on Cumulative Discoveries of Oil and Gas

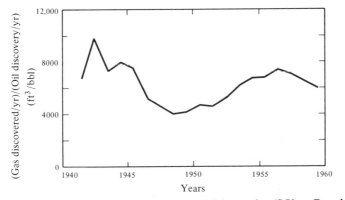

Fig 2.44. Gas-Oil Ratios for U.S. Based on Annual Discoveries (5-Year Running Average)

The figure of 6,250 ft^3/bbl represents the average gas-oil ratio for very large samples of gas and oil taken near the mid-range of the industry's history at a time when particular stress has been placed on exploration for gas. It does not appear likely, therefore, that this ratio will increase by a great deal in the future. However, as a contingency, let us assume that the ratio of gas to oil for future discoveries may be as high as 7,500 ft^3/bbl.

These two gas-oil ratios, a low of 6,250 ft^3/bbl and a high 7,500 ft^3/bbl, will accordingly be used for a low and a high estimate of the ultimate reserves of natural gas in the United States.

Using these two ratios, and the estimate of the ultimate crude-oil production of 175×10^9 barrels, we obtain the corresponding estimates of the ultimate reserves of natural gas in the following manner:

Estimated ultimate production of crude oil	175.0×10^9 bbls
Cumulative discoveries of crude oil to 12-31-61	-99.1×10^9 bbls
Undiscovered reserves of crude oil, 12-31-61	75.9×10^9 bbls

Ultimate Reserves of Natural Gas

	Minimum Estimate	Maximum Estimate
Undiscovered crude oil	75.9×10^9 bbls	75.9×10^9 bbls
Gas-oil ratio	$\times 6,250$ ft^3/bbl	$\times 7,500$ ft^3/bbl
Undiscovered nat. gas	474×10^{12} ft^3	569×10^{12} ft^3
Cum. disc. nat. gas	$+484 \times 10^{12}$ ft^3	$+484 \times 10^{12}$ ft^3
Ultimate potential reserve nat. gas	958×10^{12} ft^3	1.053×10^{12} ft^3
Average	$\approx 1,000 \times 10^{12}$ ft^3	

We accordingly adapt the round figure of $1,000 \times 10^{12}$ ft³, which is very nearly the arithmetical mean between our low and high estimates, as our present best estimate of the ultimate reserves of natural gas in the United States.

Using the asymptote of $1,000 \times 10^{12}$ ft³ for the curves of cumulative proved discoveries and cumulative production of natural gas, we are then able to evaluate the logistic equations for these curves. The curve of cumulative production is obtained for the period from 1859 to 1917 from that of crude oil by assuming the production of 2,000 ft³ of gas per barrel of oil. From 1917 to 1961 actual gas-production statistics are used (Dept. of Commerce, 1949, p. 146; 1954, p. 20; 1953–1961). Estimates of proved reserves of natural gas have been made annually by the Reserves Committee of the American Gas Association since 1945 (American Gas Association, 1945–1961). The addition of cumulative production and proved reserves from 1945 to 1961 then gives that portion of the curve of cumulative proved discoveries.

The logistic equations for cumulative discoveries and cumulative production, respectively, are then found to be

$$Q_D = \frac{1,000 \times 10^{12} \text{ ft}^3}{1 + 465 e^{-0.0793(t - 1884)}}, \tag{2.17}$$

and

$$Q_P = \frac{1,000 \times 10^{12} \text{ ft}^3}{1 + 465 e^{-0.0793(t - 1900)}}. \tag{2.18}$$

As heretofore, the equation for the proved reserves Q_R is the difference between those for Q_D and Q_P.

This family of curves is shown in Figure 2.45. It will be noted that in this case the time lag Δt between the curve of cumulative discoveries and cumulative production is about 16 years as compared with the 10–11-year lag for crude oil. This is due principally to the fact that a large backlog of proved reserves of natural gas was being accumulated before the present large pipelines for gas distribution were put into operation.

The time derivatives of the family of curves in Figure 2.45 are shown in Figure 2.46. These represent, respectively, the rate of discovery, the rate of production, and the rate of increase of proved reserves. The date of the peak in the rate of discovery of natural gas should be about when the curve of cumulated proved discoveries reaches

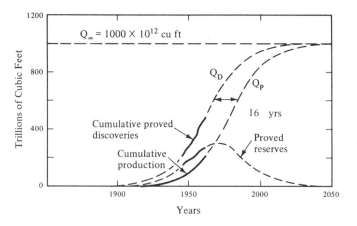

Fig   2.45. Cumulative Discovery and Production and Proved Reserves of U.S. Natural Gas

one-half the ultimate, Q_∞, or about 500×10^{12} ft^3. To the end of 1961 cumulated proved discoveries amounted to 484×10^{12} ft^3, and the rate of discovery during recent years has been about 18×10^{12} ft^3/yr. Accordingly, the halfway point, or the inflection point of the curve, should be reached by about the end of 1962. This should also be about the date of the peak of natural-gas discoveries. The peak in production should occur about 16 years later, or about 1978, and the peak of proved reserves near the mid-point between these two dates, or about 1970.

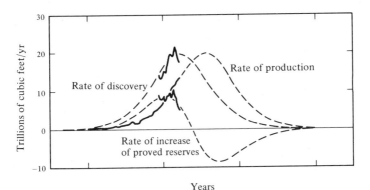

Fig 2.46. Rates of Discovery, Production and Increase of Proved Reserves of U.S. Natural Gas

Figure 2.47 shows the future production of natural gas as derived from the data of Figures 2.45 and 2.46, for both the low and the high estimates of ultimate reserves.

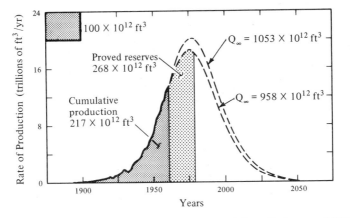

Fig 2.47. U.S. Production of Natural Gas for High and Low Estimates of Ultimate Reserves

§2.12 United States Production and Ultimate Reserves of Natural-Gas Liquids. The annual production of natural-gas liquids in the United States is shown graphically in Figure 2.48.[19, 21] Because natural-gas liquids are a by-product of the production of natural gas, an estimate of the ultimate potential reserves of natural-gas liquids may be made very simply from the estimate of the ultimate reserves of natural gas, and the ratio of natural gas to natural-gas liquids in past production experience.

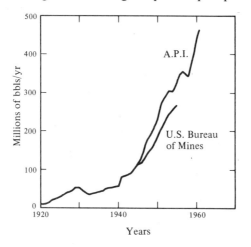

Fig 2.48. Rate of Production of U.S. Natural-Gas Liquids

From the statistics of the American Gas Association on production rates and proved reserves of both natural gas and natural-gas liquids (American Gas Association, 1948–1962), the cumulative proved discoveries of natural gas during the period 1947 to 1961, inclusive, increased by 250.2×10^{12} ft^3. During the same period the increase of the cumulative proved discoveries of natural-gas liquids increased by 8.45×10^9 bbls. The ratio of the gas to the natural-gas liquids discovered during this period amounts to 29.6×10^3 ft^3/bbl. If we assume that this ratio will remain approximately unchanged for the undiscovered gas reserves we can use it to estimate the undiscovered natural-gas liquids.

By December 31, 1961, the cumulative proved discoveries of natural gas amounted to 484×10^{12} ft^3. Subtracting this from the estimated ultimate natural-gas reserves of $1,000 \times 10^{12}$ ft^3 gives an estimate of 516×10^{12} ft^3 of natural gas still to be discovered. Then, by dividing the undiscovered gas by the ratio of gas to natural-gas liquids, we get

$$\frac{516 \times 10^{12} \text{ ft}^3}{29.6 \times 10^3 \text{ ft}^3/\text{bbl}} = 17.4 \times 10^9 \text{ bbls}$$

as the estimate of undiscovered natural-gas liquids. Adding to this the estimated cumulative discoveries of natural-gas liquids we obtain:

Cum. disc. nat.-gas liq. through 12-31-61	13.0×10^9 bbls
Nat.-gas liq. to be discovered as of 12-31-61	17.4×10^9 bbls
Est. ultimate potential res. nat.-gas liq.	30.4×10^9 bbls

as the estimated ultimate potential reserves of natural-gas liquids for the United States. Rounding this off to 30×10^9 bbls and adding it to the 175×10^9 bbls for crude oil, we then obtain 205×10^9 bbls as our present estimate of the ultimate potential reserves of liquid hydrocarbons of the United States.

The curves for cumulative proved discoveries, cumulative production, and proved reserves for natural-gas liquids are shown in Figure 2.49.

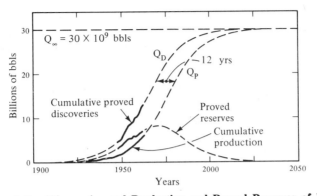

Fig 2.49. Cumulative Discoveries and Production and Proved Reserves of U.S. Natural-Gas Liquids

§2.13 United States Production and Ultimate Reserves of Liquid Hydrocarbons. By combining the U.S. data for crude oil with those for natural-gas liquids, we obtain

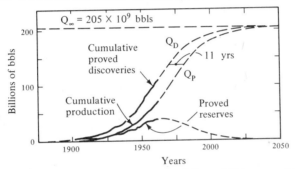

Fig 2.50. Cumulative Discovery and Production and Proved Reserves of U.S. Liquid Hydrocarbons

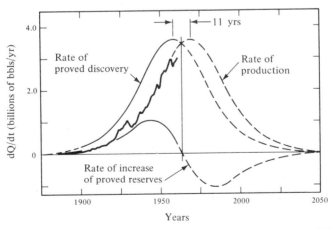

Fig 2.51. Rates of Proved Discovery, Production and Increase of Proved Reserves of U.S. Liquid Hydrocarbons

composite U.S. data for total liquid petroleum. These data for cumulative production, proved reserves, and cumulative proved discoveries are plotted graphically in Figure 2.50. The logistic curves for production and discovery, as obtained graphically from the data and independently of earlier considerations, still give an asymptotic value of 205 billion barrels for Q_∞, the ultimate expectable cumulative production.

The time derivatives of the curves are shown in Figure 2.51.

From the composite data on total petroleum liquids, the lead time, Δt, of discovery with respect to production is about 11 years. The peak discovery rate appears to have occurred about 1958; the peak of proved reserves is expected to occur at about 1964, and that of the rate of production at about 1969.

§ 2.14 Oil Shales and Tar Sands.   The world reserves of oil shales and tar sands are much less well known than those of the United States and Canada. The United States has the largest known reserve of oil shales in the world, and Canada the largest reserve of tar sands. The principal oil shale in the United States is the Green River shale in western Colorado, southwestern Wyoming, and eastern Utah. The tar sands of Canada occur in four known localities in Alberta, with reserves possibly as large as 600×10^9 bbls of crude-oil equivalent.

The reserves for the United States used here are those prepared by the United States Geological Survey and presented by V. E. McKelvey to the Natural Resources Subcommittee of the Federal Council, on November 28, 1961. According to this report, the estimates of the reserves of shale oil in the United States in the categories of *known, potential,* and *known marginal* reserves amount to 850×10^9 bbls. The reserves in the corresponding categories for oil in bituminous rocks, or tar sands, amount to only about 2.6×10^9 bbls.

The corresponding world figures in the same report are:

Shale oil	$1,297 \times 10^9$ bbls
Oil in bituminous rocks	$> 490 \times 10^9$ bbls

Potential marginal reserves in each of these categories could be much larger. The foregoing figures are those used here, although it is recognized that they are minimal figures.

§ 2.15 Ultimate World Reserves of Natural Gas and Natural-Gas Liquids. Although markets do not as yet exist for the natural gas and natural-gas liquids of the oil and gas fields in the parts of the world remote from centers of industrialization, there is promise that such markets soon will exist. Recent developments in the transportation of natural gas in a liquified form by means of insulated and refrigerated tankers make it possible to transport natural gas from any region of production to remote centers of consumption.

Although statistical data do not exist for natural gas and natural-gas liquids on a world scale, the approximate amounts potentially available can be estimated from the estimated reserves of crude oil and the amount of natural gas and natural-gas liquids produced per bbl of crude oil in he United States. For a world estimate we assume 6,000 ft^3 of natural gas per bbl of crude oil, and that natural-gas liquids and crude oil comprise, resepctively, 15 and 85 per cent of the total liquid hydrocarbons. This gives 0.1765 of a bbl of natural-gas liquids per bbl of crude oil.

Then, on the basis of our estimate of $1,250 \times 10^9$ bbls of crude oil as the ultimate reserves of the world, the ultimate reserves of natural gas and of natural-gas liquids will be $7,500 \times 10^{12}$ ft^3 and 220×10^9 bbls, respectively.

This would give a world estimate of $1,470 \times 10^9$, or roughly, $1,500 \times 10^9$ bbls for liquid hydrocarbons.

§2.15.1 Total Energy of the Fossil Fuels. Having reviewed the ultimate potential reserves of the various classes of the fossil fuels, we need now to compare them with respect to their total energy contents. For this purpose we adopt the heat of combustion expressed in the energy unit, the kilowatt-hour. A kilowatt-hour represents the work done at a rate of 10^3 joules/second during a time of 1 hour of 3,600 seconds. It, therefore, represents 3.60×10^6 joules. A kilowatt-hour of heat is the heat produced by a kilowatt-hour of work. For the world reserves of energy from the fossil fuels a convenient larger unit is 10^{15} kilowatt-hours.

§2.15.2 Ultimate World Reserves. In Figure 2.52 are shown the present estimates of the ultimate reserves of energy for the different classes of fossil fuels, and the fraction of each which has been consumed already. The total ultimate energy for all the fossil fuels is approximately 27.4×10^{15} kilowatt-hours of heat. Of this 71.6 per cent is represented by coal, 17.3 per cent by petroleum and natural gas, and 11.1 per cent by tar sands and oil shale. The fraction consumed already is 4.1 per cent for coal, 10 per cent each for petroleum and natural gas, and zero for tar sands and oil shales.

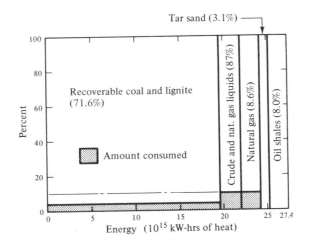

Fig 2.52. Total World Energy of Fossil Fuels

§2.15.3 United States Reserves. The corresponding data for the United States are given in Figure 2.53. The total ultimate reserves of energy from the fossil fuels in the United States is about 8.7×10^5 kilowatt-hours, or about one-third of the world total. Of this, 78 per cent is represented by coal, 16 per cent by oil shale, and 3 per cent each by petroleum and natural gas. The amount consumed already is about 3 per cent for coal, 38 per cent for petroleum and 22 per cent for natural gas.

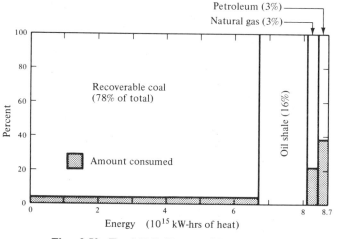

Fig 2.53. Total U.S. Energy of Fossil Fuels

§2.15.4 Energy from the Fossil Fuels in Perspective. To summarize the data that we have assembled on the energy supply from the fossil fuels, the world's total supply of energy from these sources, including that already consumed, amounts to about 27×10^{15} kilowatt-hours, of which about one-third occurs in the United States exclusive of Alaska. Of this energy supply, both for the world and for the United States, about three-quarters is represented by coal and one-quarter by petroleum, natural gas, oil shales, and tar sands.

The energy content of the fossil fuels consumed by the end of 1961 amounted to only about 4.7 per cent of the ultimate reserves for the world, and 5.0 per cent for the United States. However, the smallness of these figures tends to be deceptive and to lead to a false sense of security, because, as we have shown heretofore, with only a modest additional increase in the present rates of consumption, the peak in coal production for both the world and the United States will occur in about 200 years.

Since the reserves of petroleum and natural gas are much smaller than those of coal, and the ratio of their rates of consumption to their total reserves is much higher, it follows that these fuels will be much more short-lived than coal. In fact, the culmination in the world production of petroleum is expected to occur by about the end of the present century. In the United States the culmination in the production of crude oil is expected to occur before 1970, and that of natural gas before 1980.

This does not imply that the United States is soon to be destitute of liquid and gaseous fuels, because, as we have seen, there are still large reserves of oil shale and still larger reserves of coal from which such fuels can be produced, if necessary.

However, in keeping with the historical perspective with which we began this review, it is well to consider the exploitation of the fossil fuels in a span of history extending for some thousands of years before and after the present. On such a time scale the exploitation of the fossil fuels from the beginning to ultimate exhaustion, as is shown in Figure 2.54, will comprise but a brief episode.

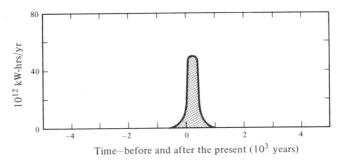

Fig   2.54. Total World Production of Fossil Fuels in Time Perspective

The total length of time during which a fuel may be exploited to some trivial amount is not a significant figure; the significant time span is that during which the cumulative production increases from, say, 10 per cent to 90 per cent of the ultimate reserves. For coal this figure promises to be only about 350 years. For the world's petroleum reserves, since only 10 per cent have been consumed up to now and the culmination is expected in about 40 years, it is estimated that an additional 40 per cent of the initial reserves will be produced between 1960 and the year 2000 and another 40 per cent between 2000 and 2040. Thus, about 80 per cent will be produced during the 80-year period between 1960 and 2040 A.D. The corresponding period during which 80 per cent of the petroleum and natural-gas reserves of the United States will be consumed will be somewhat shorter. The United States cumulative production of crude oil reached 17×10^9 bbls, or about 10 per cent, of the ultimate reserves by 1935. It is expected to reach 50 per cent by 1970 and 90 per cent by about 2005. The middle 80 per cent will accordingly be produced during the approximately 70-year period from 1935 to 2005. As compared with the production rate during this central period, that during the first and last 10 per cent of the ultimate reserves is relatively unimportant.

§2.16 Continuous Sources of Power.   We now direct our attention to continuous sources of power, or sources which, if exhaustible, represent so great a reserve of energy that, for time periods of a few thousand years, they may be treated as if they were inexhaustible.

For other sources of energy appropriate for power production either in parallel with, or as successors to, the fossil fuels, we again direct our attention to the terrestrial flux of energy depicted in Figure 2.1. The thermal power influx from the solar radiation of $178,000 \times 10^{12}$ watts dwarfs the inputs of 32×10^{12} and 3×10^{12} watts from the other two principal sources, geothermal energy and tidal energy, respectively. It also is approximately 31,000 times the world's 1969 rate of production of industrial energy of 5.7×10^{12} watts.

§ 2.17 Solar Energy. The first of these is solar energy. As we have already pointed out in §2.2.1, solar energy is intercepted by the earth at a mean rate of about 17.2×10^{16} watts, which is about a million times greater than the installed electrical-generating capacity of the United States in 1959.

At present only two channels of the flux of solar energy are available as large-scale sources of energy for human utilization. The first is the biological channel, beginning with photosynthesis; the second is the heat-engine channel, which produces the atmospheric and oceanic circulations and the hydrologic cycle, leading to wind power and water power.

§ 2.17.1 Biologic Channel. At the conference on energy held in New York on July 19–20, 1961, the energy flux of the biologic channel was reviewed briefly by G. Evelyn Hutchinson of Yale University.

Professor Hutchinson pointed out that the rates of the photosynthetic process in terms of the fixation of carbon per year are presently estimated to be as follows:

	Grams/yr of fixed carbon
Forests	$12 \ \times 10^{15}$
Agricultural lands	5.1×10^{15}
Grass lands	4.6×10^{15}
Total for land areas	21.7×10^{15}

The total amount of fixed carbon involved is about 1 to 3×10^{17} grams. The energetic efficiency of the process is only about 0.2 per cent.

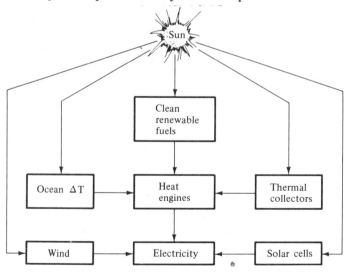

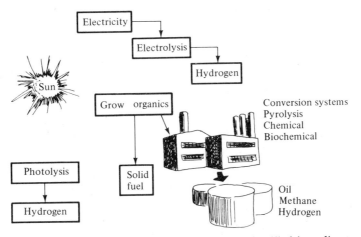

Fig 2.55. Electricity from the Sun. Solar energy can be classified into direct solar radiation and solar-derived energy. Furthermore, the energy may be converted to electricity, to heat or refrigeration, or to fuel. Figure 2.1 illustrates these processes. Solar radiation may be converted to electricity by solar cells or may be concentrated by collectors and converted to heat which can be utilized in conventional heat engines (such as steam turbines). Similarly the solar-generated energy as it appears in wind may be converted, e.g., to electricity by a propeller coupled to a generator. Also, solar radiation absorbed by the ocean and stored by raising the surface temperature of the ocean 30–40°F above the temperature at 2000 ft. depth can be utilized to produce electricity via a low-temperature-difference heat engine.

Fuel can be produced by direct solar radiation or from solar-derived energy. Solar radiation, with an appropriate catalyst, may decompose water to hydrogen and oxygen in a series of reactions. The hydrogen which turns back to water would be used as fuel. Solar radiation causes the growth of plants and other organisms such as algae, which can be used after suitable processing (drying, chipping, or grinding) for fuel. Such organic material can also be converted to methane, hydrogen, or oil by destructive distillation (pyrolysis), fermentation, or by high-pressure chemical processing.

Fuels such as hydrogen may be used in a fuel cell to produce electricity, or be burned along with plants to produce heat and electricity by standard means.

Solar cells might be deployed in space to produce electricity which would then be converted to microwaves, beamed to Earth, and reconverted to electricity. The solar radiation would be available almost 24 hours each day throughout the year. If feasible, eighty large stations, each 13 square miles in size, might satisfy the United States mainland electric power needs in 1985, or more than 3 times our present requirements.

Thus, while the biological efficiency in the capture of solar energy is low, the aggregate quantity is very large. The annual fixation of carbon on land by this process being about 7 times the fixed carbon in the fuels consumed per year.

The oceanic fixation of carbon per year is not accurately known, but could be as high as 35×10^{15} grams/year.

There is evidence that the greatly increasing use of the fossil fuels, whose material contents after combustion are principally H_2O and CO_2, is seriously contaminating the earth's atmosphere with CO_2. Analyses indicate that the CO_2 content of the atmosphere since 1900 has increased 10 per cent. Since CO_2 absorbs long-wavelength radiation, it is possible that this is already producing a secular climatic change in the direction of higher average temperatures. This could have profound effects both on the weather and on the ecological balances.

In view of the dangers of atmospheric contamination by both the waste gases of the fossil fuels and the radioactive contaminates from nuclear power plants, Professor Hutchinson urges serious consideration of the maximum utilization of solar energy.

§2.17.2 Wind Power.

The historical background on the development of power from both water and wind has been reviewed in §2.3. Wind power is essentially limited to comparatively small units and is suitable for such special uses as pumping well water and charging batteries for local household electrical uses, but it does not offer much promise of competing with other prime movers in producing large-scale electric power. Even for the traditional uses such as the propulsion of sailing ships and the Dutch windmills for pumping water from the Dutch polders and for grinding grain, the use of power from the fossil fuels and water power has almost completely displaced wind power.

§2.17.3 Water Power.

The only channel of solar energy which lends itself to large-scale power production is water power, which is made possible only by the fact that natural streams are a means of concentrating very large amounts of power in small areas. Yet it was not possible to utilize power in such quantities at a single locality before the development of the means for generating power electrically and transmitting it over large areas for utilization. Thus, while water power is one of the oldest and most important sources of industrial power, individual water-power units rarely exceeded a few tens of kilowatts in size prior to the introduction of electrical generation and distribution. Now sites are being developed in which individual installations have power capacities measurable in hundreds of megawatts.

Unlike the fossil fuels, water power is a rate of production rather than a quantity of energy. The long-term history of the development of water power accordingly should be represented by a logistic type of growth. The installed capacity must start at a very low level, increase with time, at first slowly and then more rapidly, and finally level off to a maximum when all available water power is being utilized.

When all available power is thus being used, power can be generated at this maximum rate more or less indefinitely, provided the climate does not change significantly, and also provided that a steady-state method of desilting the reservoirs can be devised. At present rates of deposition of silt, most of the large reservoirs will require only the order of a few centuries to become filled with sediment. Unless this sediment eventually is removed from the reservoir at the same rate as it is added, the power capabilities will be greatly diminished.

The significant quantities pertaining to water power in any given area are the maximum potential water power available and the amount of this that has been utilized up to any given time. A summary of the developed and potential water power of the world has been compiled by Young[53] of the United States Geological Survey. Using this as basic information, Francis L. Adams[48] of the Federal Power Commission presented a comprehensive review of water power before the Committee's conference on energy in New York on July 19, 1961. According to Adams the Federal Power Commission assumes a power capacity equal to 60 per cent of the U.S. Geological Survey's estimate of power at mean rate of flow at 100 per cent efficiency. Using this factor, Adams estimated the ultimate potential water-power capacity of the United States to be 148,000 megawatts, of which the amount already installed by the end of 1960 was 33,000 megawatts, or 23 per cent of the ultimate. A logistic curve of water-power development for the United States is shown in Figure 2.56.

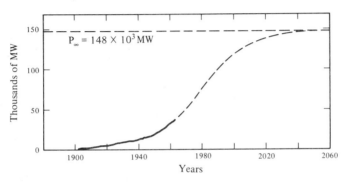

Fig 2.56. U.S. Installed and Ultimate Hydroelectric Power Capacity

Adams did not give data on the potential water power of the world in terms of megawatts of capacity, but rather in terms of the energy which could be produced per year expressed as kilowatt-hours per year. Using his ratio between installed power capacity and annual energy produced for the United States, it is possible to estimate the potential power capacity for the various areas of the world and the extent to which this has already been developed.

It will be noted, whereas the United States has an ultimate potential water-power capacity of 148×10^3 megawatts, of which 23 per cent is already developed, the world has a potential capacity of 2.86×10^6 megawatts, of which only 152×10^3 megawatts, or 5.3 per cent, has been developed.

Also, it is interesting to note that Africa, with a potential water-power capacity of 780,000 megawatts, has the largest water-power resources of any continent; and South America is second.

If the water power of the world were fully developed, the electrical energy produced per year would be about 12.0×10^{12} kilowatt-hours,[48] which would be about 6 times the electrical-power production of the world in 1959. The coal required to produce this amount of power would be about 25×10^9 short tons per year, or about 10 times the world's coal production in 1959.

According to these data, the world's total potential water-power capacity is about 3×10^{12} watts, which is of the same order of magnitude as the world's present rate of industrial-energy consumption. However, only 8.5 percent is developed at present, and this principally in the highly industrialized areas of North America, Western Europe, and Japan. The two continents with the largest potential water-power capacities are Africa and South America with capacities of 780×10^3 and 577×10^3 megawatts, respectively. These capacities are also the least developed of all of those of the world's major areas. Moreover, to develop and utilize such sources of power requires a simultaneous industrialization of the respective regions, and this, in turn, requires quantities of industrial metals roughly proportional to the power developed. Whether the earth's resources of metallic ores will be adequate for such an expansion remains to be seen.

Viewed superficially, water power has the appearance of being an essentially inexhaustible source of energy, or at least one with a time scale comparable to that required to remove mountain ranges by erosion. Actually, this may not be so. Most water-power projects involve the damming of streams and the construction of large reservoirs. The time required for most of these reservoirs to become filled with sediment will only be from one to a few centuries. Hence, unless a satisfactory solution to this problem should eventually be found, most of the world's water-power capacity may prove to be comparatively short-lived.

§ 2.17.4 Direct Conversion of Solar Energy.

§ 2.17.4 Direct Conversion of Solar Energy.   The fact that a large fraction of the total solar power occurs as direct solar radiation in desert and semidesert areas in tropical to middle latitudes makes an intriguing problem of somehow capturing this energy for human uses.

In briefings before the federal Task Force on Energy[95] Dr. Adams and Marjorie Pettit Meinel of the University of Arizona states that thermal conversion of solar energy into electrical power at efficiencies approaching 30% is possible using current technology borrowed from USAF, NASA, and AEC programs and that the economics of solar power appear to be comparable to the generation of power in 1973 using natural gas as a fuel. Their solar power concept offers a new energy option with the same target date as the breeder reactor; namely, the first half of the decade of the 80's. Although solar power has the potential of sustaining the entire power needs of the United States within reasonable uses of land area for this purpose. They recommended that solar power be considered today as a regional solution until the long-distance power transmission technology has been developed. There are many ways to convert solar energy into power. When you look at the efficiencies of conversion in 1973 silicon cells head the list. They are currently very expensive and need a breakthrough to reduce their cost, in order to make direct conversion economically attractive as a bulk energy alternative.

Accelerated biological conversion is technically possible, but here the problem is the low conversion efficiency of algae or crops, on the order of 1% or 2%. The economics of growing, gathering, and converting a crop to an energy fuel also look dim at this time.

Although the actual efficiencies attained by prior efforts at thermal conversion have been low, this low efficiency was due to operation of the system at low temperatures. These temperatures were so low that even a perfect Carnot thermodynamic cycle engine could not extract work efficiently from the collected solar heat.

It appears that thermal conversion offers the potential for the greatest improvement in the art of solar energy conversion. It also offers the largest absolute conversion efficiency of any method. The question then becomes: What technology might be available to increase the operating temperature of a solar collector? Thermal conversion using a thermodynamic cycle is promising in terms of potential efficiency, land uses, and possibility of *economic* power.

Before considering a simplified system block diagram, two critical facets of solar energy must be considered. First, the sun shines only part of the 24 hours and not on cloudy days, yet civilization needs electrical power on a continuous and reliable basis. Second, solar energy suffers from the fact that in winter the shortness of the day and the frequency of cloudiness conspire to create an energy deficiency. If you attempt to meet this problem by sizing the solar collecting area big enough for the winter season, then you have an immense excess of power in spring and summer. There are good solutions for both problems.

A practical solution to the intermittency problem seems in hand. Utilities are reliability oriented. In anything new—as in the early days of nuclear power—they like a reserve supply and traditional boiler. If such a standby boiler and fuel supply were part of a solar power facility, then the obvious solution to the winter deficit would be to use the summer excess to make a storable chemical fuel to be burned as a winter supplement. One then would need a smaller system to meet the annual power demands.

At this moment it is not entirely clear whether thermal storage is the only answer to averaging the intermittent aspects of sunlight. It is clearly preferable to attempting to store electrical energy in batteries. It appears preferable to pumped storage, as is currently used by electrical utilities to store energy for peak-shaving. The question is whether chemical fuel storage is better. If one needs chemical storage for the winter season, then why not use it every night also? The tradeoffs are complex and involve such questions as fuel cells vs thermal burning of the stored chemical fuel. The ultimate guideline is not how to accomplish energy storage and recovery: it is economics. Let us take a moment to look at a broader aspect of economics and how it affects solar power.

Economic factors need special attention when one is dealing with solar energy. Economics is the basic fact of life in the domestic energy field. To consider energy as an option, one must insist that the cost be commensurate with today's power costs. Civilization needs inexpensive energy.

There is a point in regard to solar power economics that should be stressed early in the game: Solar power farms are *capital intensive* projects. The need for legislation that will permit fast writeoffs of such projects should be met. If one is forced to amortize a capital-intensive project at today's high interest rates over the traditional 30 years established by law for utilities, then the cost of solar power will be approximately double. When one studies the relationship between average power cost and amortization time, one finds that the shorter the time the lower is the average cost to the consumer. Fifteen years looks much more reasonable than 30 years (see Chapter 11).

The Meinels have identified a number of key technology areas that can be combined to make the system shown in Figure 2.56.1, holding strictly to established arts and using engineering parameters that are well within the state of the art in each area.

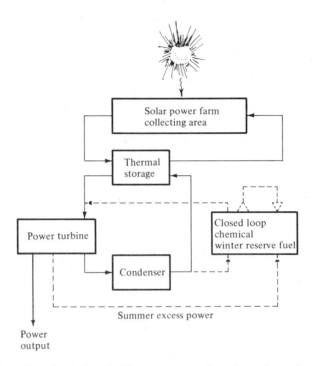

Fig 2.56.1. The Meinels' Proposal. The system consists of a solar collection area called a "solar power farm" for reasons that will be obvious later. The central subsystem is the thermal storage portion. This portion holds enough thermal energy for overnight operation, or for one or two cloudy days. The thermal storage portion is vital to the practical operation of a solar power scheme since it "decouples" the power turbine from the variable heat input from sunlight during the day. For example, if the power turbine were directly coupled to the solar collectors, the turbine would have to be three times as large, say 3000 MWe to produce power equal to that of a 1000 MWe turbine combined with thermal storage.

The key technologies applicable to solar power farms are:

Solar collector	Optical interference thin films
	Bulk-effect optical multilayer films.
	Continuous processors (HVD).
	Chemical vapor deposition (CVD).
Heat transfer fluids	Liquid metals.
	Salt eutectic fluids (HTS).
	High-pressure gas.
	"Heat pipes."
	Organic fluids (biphenyls).
Thermal storage	Salt eutectics.
	Liquid sodium.

	Liquid hydrogen.
Chemical storage	Synthetic hydrocarbons.
	Metals.

| System construction | Integrated manufacturing methods. |

Most technologies are from the AEC; the balance are from NASA, USAF, and ARPA.

There is, one particular topic that is key to the efficient extraction of power from solar energy.

How can one get a surface to the high temperature desired. The temperature goal is approximately 1000°F (540°C), the temperature of a modern high-pressure steam power turbine. The technology one requires is *optical film coatings* that absorb sunlight but prevent heat loss in the infrared. A surface that is "black" to sunlight but

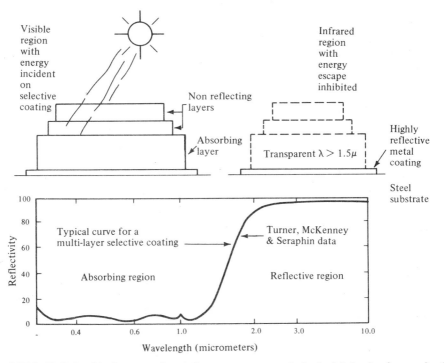

Fig 2.56.2. Relationship between the performance curve and physical behavior for a selective coating. Total coating is about 0.00001 inch thick. The type of selective coating shown above is particularly simple to understand. The absorbing layer is ordinary silicon. Silicon is opaque to sunlight but quite transparent in the infrared. If one places a highly reflecting metal coating between the steel substrate and the silicon, sunlight never penetrates to it because the sunlight is absorbed in the silicon. In the infrared both the silicon and the nonreflecting layers added to the silicon to improve collection efficiency are invisible. As a result, the surface attempts to emit as though it were a highly reflective metal, and since bright metallic coatings have very low infrared emissivities the surface traps the heat. The actual curve of reflectivity as a function of wavelength is shown in the lower portion of the diagram.

looks like a perfect mirror in the infrared. The art is not new—it is 35 years old—but only 10 years ago the cost of such selective coatings would have been prohibitive. These coatings are only about 1/100,000 of an inch thick and are deposited within a vacuum tank. Until Libby-Owens-Ford pioneered the continuous high-vacuum eva-porator for coating architectural plate glass, multilayer thin films cost in excess of $1000 per square meter. In 1973, the cost was only a few dollars per square meter, and in large quantities of a single type, the cost may eventually fall even lower.

The major question concerning selective coatings is not whether they can be made, or made economically, but their operating lifetime.

Selective coatings by themselves do not reach the desired operating temperature. One needs to concentrate sunlight on the coatings. The Meinel's have shown in previous publications that one must exceed a certain dimensionless parameter to obtain the proper set of operating conditions. This parameter is the product of the optical concentration of sunlight, X, and the selectivity a/e, which must be on the order of 100 for efficient extraction of energy from sunlight:

$$Xa/e \approx 100.$$

The measure of efficiency of a selective coating is given by the absorptivity in the visible divided by the emissivity in the infrared, a quantity denoted by a/e. Many selective coating types have been made by researchers over the years.

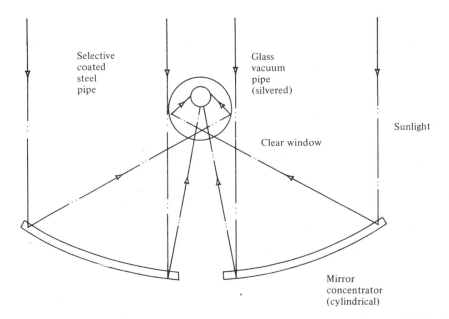

Fig 2.56.3. Cross section diagram of a typical collector PANEL using a cylindrical rear-surfaced mirror as the concentrator. A cylindrical mirror concentrates sunlight so that the focused beam enters an evacuated glass pipe through a transparent region. The remaining portion of the interior of the glass pipe is silvered to produce a high reflectivity. The sunlight entering this enclosure impinges on a steel pipe either directly or after a single reflection at the walls. The steel pipe is coated on its exterior with a selective coating of the type that we have

described. The enclosure is evacuated so that no heat is lost by convection. The result is that the steel pipe gets very hot, and if no way of extracting the absorbed energy is available, the pipe will rise in temperature to the order of 1600°F (900°C).

In operation, one extracts heat from this system by flowing a heat transfer medium inside the steel pipe. Either liquids or gases can be used. Likewise, one could imagine a heat pipe using something like potassium or mercury vapor as a means of extracting heat. Much more engineering study must be done before one can decide on the best means. There may be no clear-cut choice.

A key question is: How much of the energy can be extracted and converted into electrical power?

The Meinel's predict that approximately 50% of the solar energy incident on the selective coating can be converted into power. The actual conversion into power never is equal to the theoretical Carnot efficiency, but modern steam turbines of the type that would be used in the system are approximately 75% that of a perfect turbine.

There are expressions of concern over the amount of land required by solar power farms. The total area of solar farms needed to produce a major fraction of the total U.S. needs in the year 2000, say 1,000,000 MWe, is about 5000 square miles of collectors distributed on 10,000 square miles of land area. If all the collectors were gathered into one area, this area would be equivalent to a square about 70 miles on each side. If this area were now distributed over the 6 to 8 states having abundant sunshine, the total impact in a state or any region within a state would be small.

When one talks about land use priorities, it helps to establish some degree of perspective. For example over 500,000 square miles of land are used to produce agricultural crops, so the land needs for solar power constitute only 1% as much land as is used for crops. Crops produce about 1% of our energy needs, but the solar farms will produce a major fraction of the remaining energy needs.

Arizona Public Service Company, operators of the Four Corners Power Plant, noted that their coal-fired plant will require strip mining of a considerably larger area of land than would be needed to operate it with solar energy—and the Navajos' sheep could safely graze under and between the rows of collectors.

Some people worry that solar power farms will alter the climate of the desert. According to the Meinel's, they will not change the climate. As a matter of fact, the energy balance is almost perfect.

There is a subtle point hidden in the achievement of local thermal balance at the solar power farms: We must return the unused portion of the turbine thermodynamic cycle heat to the local environment. We call it "waste heat" when the power plant is nuclear or fossil since it adds something to the local environment. In the case of solar power, it is more appropriate to call it "return heat" since we need it to keep the local climate in balance. Yet one must be careful how this is done in order to avoid damage to the immediate spot.

The "return heat" energy could be a resource.

The normal terrain absorbs about 65% of the incident sunlight, reflecting 35% back into space. When we add the solar power farm, the collectors absorb about 95% of the sunlight. This would normally cause a heating effect, but in fact we are exporting about 30% of the absorbed energy to the distant cities in the form of

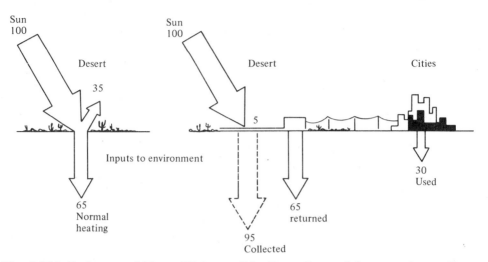

Fig 2.56.4. Environmental Thermal Balance—Solar Power Farms. Solar power farms will not change the thermal balance of the desert in which it is placed. Waste heat is only added to the environment in the cities where the power is consumed.

electrical power. The remaining 65% is returned to the environment by the power plant as waste heat of the thermodynamic cycle. Hence the thermal balance of the desert is unchanged.

There are certain maintenance and hazard questions relating to the solar power farms, but the Meinel's see no difficulty in meeting each with satisfactory answers.

Dust on collector	Periodic washing by automatic carriage that traverses length of a LOOP. Frequency of washing is estimated at once each 6 months in most locations.
Sand storms	Maintain natural desert cover undisturbed. Plant windbreaks of suitable trees. Turn panels parallel to ground during storms.
Hail and wind damage	Make components strong enough to meet the expected loadings.
Earthquake	Make all system interconnections of the flexible type. Avoid spanning fault zones with a farm.
Vandalism	Fence and provide a peripheral buffer zone.
War and sabotage	Dispersal of farms over a large geographical area. Make a single target a single farm.
Accidental spills	No problems in the gas-cooled portion. Physical removal of spill material in the salt eutectic portion. Passivate chemically, with physical removal of spill material in the liquid metal portion.

§2.17.5 Satellite Solar Power Station. At a briefing before the Subcommittee on Space Science and Applications and the Subcommittee on Energy of the Committee on Science and Astronautics, of the United States House of Representatives on May 24, 1973, Dr. Peter E. Glaser presented a summary of the *satellite solar power station (SSPS)* concept. The following material has been adapted from his testimony.

The sun radiates vast quantities of energy which reach the Earth in very dilute form. Thus, any attempts to convert solar energy to power on a significant scale will require devices which occupy a large land area as well as locations that receive a copious supply of sunlight. These requirements restrict Earth-based solar energy conversion devices which could produce power to a few favorable geographical locations, and even for these locations energy storage must be provided to compensate for the day-night cycle and cloudy weather.

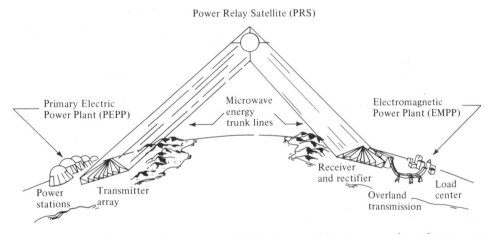

One way to harness solar energy effectively would be to move the solar energy conversion devices off the surface of the Earth and place them in orbit away from the Earth's active environment and influence and resulting erosive forces. The most favorable orbit would be one around the sun, but as a first approximation toward this very long-term goal, an orbit around the Earth where solar energy is available nearly 24 hours of every day could be used.

The SSPS represents an approach to power generation which does not use a terrestrial energy source. Thus, the environmental degradation associated with mining, transportation, or refining of natural energy sources is absent. Natural resources will have to be used to produce the components for the SSPS and the propellants for transportation to orbit. Nearly all the materials to be used for the components are abundant. For each SSPS, the rare materials required, such as platinum or gallium, would be less than 2% of the supply available to the United States per year.

§2.17.5.1. Environmental/Ecological Impact. The SSPS appears to have limited environmental and ecological impacts at the receiving antenna.

- Waste heat released by natural convection at the receiving antenna does not constitute a significant thermal effect on the atmosphere. If the antenna is located in desert regions where water is limited, there may be some slight modification of the plant community at and near the antenna site.

- With RF shielding incorporated below the rectifying elements, the receiving antenna operation can be compatible with other land uses, because there is only a small reduction of solar radiation received at the ground below the antenna. However, installation and maintenance of the antenna has to be planned carefully because extensive activities may be damaging to some ecologically fragile systems.
- Injection of water into the stratosphere and upper atmosphere by space vehicle exhausts during the assembly phase would be small in contrast to the natural abundance, of such water vapor, and does not appear to constitute a significant environmental effect.
- Noise from the launch operations would be of concern in the immediate vicinity of the launch facility and would have to be reduced by suitable design techniques or location of the launch facilities, but such noise is transient and of limited durations.
- There is substantial flexibility in choosing a suitable location for the receiving antenna. The area has to be contiguous but need not be flat terrain. The location can be in a region where the land is not suitable for other uses (e.g., desert areas, previously stripmined land, or near major electrical power users (e.g., aluminum smelters).

§2.17.6. Microwave Power Transmission. The viability of the satellite solar power station (SSPS) depends upon the ability to efficiently transfer the electrical energy captured in space to the terrestrial environment where it will be consumed. The only way known to do this is by means of an electromagnetic beam link. For this link to remain unbroken, it is necessary that the satellite solar power station be in synchronous orbit around the Earth. The distance of such an orbit—22,300 miles above the Earth's surface—precludes the efficient transfer of power by long-wavelength electromagnetic radiation. The necessity of efficiently penetrating the Earth's atmosphere under all meteorological conditions, including intense precipitation, precludes the use of wavelengths shorter than about 10 centimeters. The electromagnetic beam then must make use of wavelengths in the microwave region of the electromagnetic radiation spectrum.

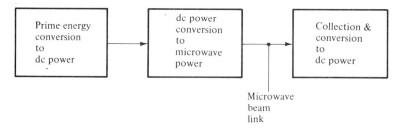

The use of a microwave beam to efficiently transfer power from one point to another is a concept of long standing.

However, the reduction of the concept to practice did not get under way until 1960 when microwave tubes reached sufficiently high power and efficiency levels to make them of interest for power transmission purposes. The first demonstration of efficient transmission of meaningful amounts of power took place at the Spencer Laboratory of

Raytheon Company on May 23, 1963, when the Company transferred power by microwaves over a distance of 25 feet and used this power to operate a 100-watt dc motor. Subsequently, Raytheon Company publicly demonstrated the flight of a small microwave-powered helicopter which was developed under U.S. Air Force contracts from 1964-1968. Microwave power was beamed from a ground station.

By 1968, the development of the microwave power transmission system concept had reached the point where it became of interest to the Space Science Laboratory of the George C. Marshall Space Flight Center in connection with the transfer of power between a manned space station and co-orbiting daughter satellites.

The transmission of microwave power has now reached the point where it appears that efficient transfer of large amounts of microwave power by microwave beam is technologically feasible and that it may be a practical solution to a number of problems in space, growing out of a need to transfer power where it is difficult or impossible to run wires. The latest application to be considered is the transfer of electrical power from a satellite solar power station to the Earth.

§2.17.6.1. Microwave power generation, transmission, and rectification. The power generated by the SSPS in synchronous orbit must be transmitted to a receiving antenna on the surface of the Earth and then rectified. The power must be in a form suitable for efficient transmission in large amounts across long distances with minimum losses and without affecting the ionosphere and atmosphere. The power flux densities received on Earth must also be at levels which will not produce undesirable environmental or biological effects. Finally, the power must be in a form that can be converted, transmitted, and rectified with very high efficiency by known devices.

Large amounts of power can be transmitted by microwaves. The efficiency of microwave power transmission will be high when the transmitting antenna in the SSPS and the receiving antenna on Earth are large. The dimensions of the transmitting antenna and the receiving antenna on Earth are governed by the distance between them and the choice of wavelength.

The size of the transmitting antenna is also influenced by the inefficiency of the microwave generators due to the area required for passive radiators to reject waste heat to space and the structural considerations as determined by the arrangement of the individual microwave generators. The size and weight of the transmitting antenna will be reduced as the average microwave power flux density on the ground is reduced by increasing the size of the receiving antenna and as higher-frequency microwave transmission is used. The size of the receiving antenna will be influenced by the choice of the acceptable microwave power flux density, the illumination pattern across the antenna face, and the minimum microwave power flux density required for efficient microwave rectification.

The ground site selection criteria will be greatly influenced by results of projected Earth resources studies, as well as social and political considerations. System aspects of site selection lend themselves to relatively simple and known analysis techniques.

Superconducting Transmission System

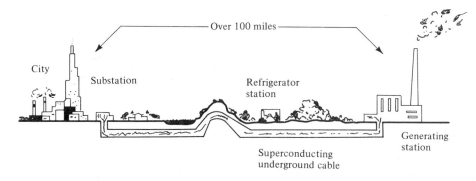

Highly efficient ground power distribution systems, such as the underground, cryogenic, high-voltage dc system will permit location of power consumption centers away from the power supply station. Low power flux density and efficient

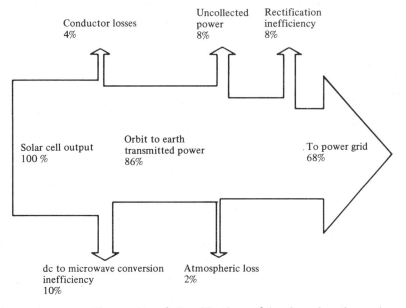

receiving antennae will permit a full utilization of land under the antenna. The biological effects of the non-ionizing radiation, subject to continuing long-term investigation, can become sufficiently well understood so that a straightforward SSPS biological effects program can be implemented to define criteria and limitations for several ground selection and multiple utilization options.

§2.17.6.2. The effect of the microwave beam upon man's environment. It has been pointed out that the major reasons for considering the satellite solar power station system as an alternative source of future power is that it does not consume any of the Earth's limited fossil fuel resources and that its impact upon the Earth's environment is minimal compared to alternative power sources. For example, its thermal effluent is minimal; with a rectenna (rectifying antenna) efficiency of 90%, the ratio of useful electrical power to waste heat is 9 to 1, as contrasted to a ratio of 4 to 6 for fossil and nuclear fueled electric power plants.

The satellite solar power station contributes no other pollution in the commonly used sense of the word. However, in the sense that we are immersed in a sea of man-made electromagnetic radiation resulting from radio, television, commercial communications, and radar, and to the degree that this may also be called pollution, then the microwave beam of the SSPS would add to this pollution. However, the beams would be tightly circumscribed by the areas devoted to the rectennas, and would have no significant impact outside of those areas. The rectenna would be surrounded by a guard-ring area and a fence to prevent animal life from entering the area.

Within the rectenna area and depending upon the power density of the incoming radiation, operating personnel might need protective shielding if they were in the open rectenna area. Domestic animals might be permitted in the area depending upon the array construction and the animal characteristics. The impact upon birds, flying aircraft or spacecraft would have to be examined, but it seems most likely that the power densities being considered would be several orders of magnitude removed from where they would have any direct impact upon aircraft.

It should be pointed out that the satellite solar power station system can be designed for a very wide range of microwave power densities incident upon the ground.

From the viewpoint of the general biological effects of microwave energy upon man, the only effect that has so far been established after many years of investigation is the heating effect. This is a relatively benign effect and man has the relatively high continuous-exposure tolerance of 10 milliwatts per square centimeter (approximately 10 watts per square foot) to this effect. This is the level of microwave radiation at the rectenna edges. For short periods of time he can tolerate much higher exposures. Our continuous-exposure standard in this country is currently set at that level primarily on the basis of the heating effect.

There is general agreement among biologists, however, that the investigation of the biological effects of microwaves should be continued, particularly with respect to any long-range or delayed effects. With respect to the possibility of delayed effects, it is noted that there are now tens of thousands of people who have been in the microwave industry over many years.

The effects of SSPS radio frequency interference on other users will be substantial. Unless the frequency is selected so as to avoid the more sensitive services close to the fundamental frequency, it is very likely that current users would be required to relinquish their frequency allocation to the SSPS.

Most sensitive radio astronomy services require fixed narrow bands for operation spread throughout the spectrum. The 3.3 GHz proposal for the SSPS fundamental operating frequency was selected primarily to maximize the capability of filtering SSPS noise at the existing radio astronomy frequencies and because filters can be designed and developed to achieve this objective.

Potential interference with current shipborne radar is possible.

Amateur sharing, state police radar, and radio location from high-power defense radar in the 3.23- to 3.37-GHz band will suffer interference. Thus, specific allocation for the SSPS would have to be negotiated with such users in mind.

§2.18 Tidal Power. In §2.2.3 it was pointed out that the total tidal power dissipated by the earth is about 1.4×10^{12} Watts, of which about 1.1×10^{12} Watts is accounted for by tidal friction in bays and estuaries around the world. It is the latter fraction which is susceptible to capture and conversion to electric power by suitable water-power devices.

The world's first major tidal-power installation is that of the La Rance estuary in France, which began operation in 1966. This had an initial capacity of 240 Mw which is planned to be increased to a total of 320 Mw.

Tidal power is similar to water power except that it is derived from the alternate filling and emptying, with the semi-diurnal period of 12 h 24.4 m of the synodical lunar day, of a bay or estuary that can be enclosed by a dam. When a tidal basin is enclosed, the maximum power obtainable would be by a flow cycle that permitted the basin to fill and to empty during brief periods at high and low tide. For such a cycle, the energy potentially obtainable per cycle would be given by

$$E_{max} = \rho g R^2 S, \tag{1}$$

where ρ is the water density, g the acceleration of gravity, R the tidal range, and S the area of the enclosed basin.

The average power for a complete cycle would be, therefore,

$$\bar{P} = E_{max}/T = \rho g R^2 S/T, \tag{2}$$

where $T = 4.46 \times 10^4$ sec is the tidal period, which is also the half period of the synodical lunar day. Of this, the theoretical maximum amount realizable in engineering calculations has commonly been taken to be from about 8 to 20 percent. However, in the French La Rance project the power realized is reported to approach 25 percent of this maximum amount.

In Table 2.1 a summary is given of the average potential tidal power, and the annual amount of energy that could be obtained from the world's most favorable tidal-power sites. These include individual sites with potential power capacities in the range of from 2 to 20,000 megawatts each.

Table 2.1. Tidal-Power Sites and Maximum Potential Power[a]

Locality or region	Average potential power $(10^3 kW)$	Potential annual energy production (10^6 kWh)
North America Bay of Fundy (Nine sites)	29 027	255 020
South America Argentina San Jose	5 870	51 500
Europe England Severn	1 680	14 700
France (Nine sites)	11 149	97 811
USSR (Four sites)	16 049	140 452
Totals	63 775	559 483

[a] Sources: Trenholm[84]; Bernshtein[85].

The total potential power capacity of all of these sites is 64,000 megawatts. This represents 2 percent of the total tidal energy dissipation of 3×10^{12} watts. It is also only 2 percent of the world's potential water-power capacity.

§2.19 Geothermal Energy.   As was pointed out in §2.2, the temperature in the earth increases with depth, in consequence of which heat is conducted from the earth's interior to its surface. An additional amount of heat is convected to the earth's surface by the gases and lavas of volcanos, and by hot springs in regions which have been heated above normal by volcanic activity.

The mean rate of increase of temperature with depth in areas remote from volcanic disturbances is about 1°C per 30 meters, or about 33°C per kilometer of depth. Hence, within drillable depths of 5 to 8 kilometers, temperatures as high as 150°–200°C above surface temperatures may be expected.

Superficially, it would appear that with such temperatures at drillable depths, earth heat sufficient for significant power generation could be obtained anywhere. Actually this is not the case. Rocks are very poor conductors of heat; thus the heat that could be obtained in this manner is negligible. The only situations in which earth heat can be used on a large scale are those at which hot volcanic rocks are comparatively near the surface and either volcanic, or circulating, ground waters act as heat collectors from large volumes of rocks. Since these hot rocks are finite in quantity and have finite contents of heat, it follows that the amount of energy

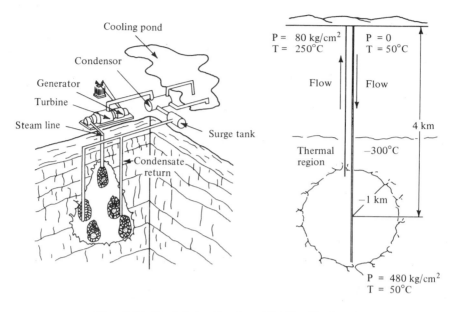

Fig 2.57. Heat Extraction from Hot Dry Rock

extractable from such a source must also be limited. Unlike tidal power and water power which depend upon continuing sources of energy, geothermal power depends essentially upon the "mining" and eventual depletion of temporarily stored quantities of volcanic heat. Such sources of thermal energy are accordingly associated with the volcanic regions of the world.

In 1969, the total installed capacity of geothermal electric power installations in the world amounted to 828 Mw. Most of the geothermal-electric plants at that time were comparatively small—in the range of 1 to 20 megawatts. The development of geothermal power at Larderello, Italy was begun in 1904, and has reached a capacity of 370 Mw. Power production at The Geysers in California was begun in 1960 with a small 12.5 Mw unit. Capacity was subsequently increased to 82 Mw by 1969. The plant at Wairakai, New Zealand, began operation in 1958 and has a capacity of 290 Mw, which is reported to be about the maximum limit for this locality.

As to the world's ultimate geothermal capacity, after extensive studies of the approximate magnitudes of the world's major geothermal areas, Donald E. White[86] of the United States Geological Survey has estimated that the stored thermal energy in such areas to depths of 10 km amounts to about 4×10^{20} thermal joules. With a 0.25 conversion factor this would yield 1×10^{20} joules of electrical energy, or about 3×10^6 Mwe-yrs. Then, if this amount of energy were to be withdrawn during a period of 50 years, the average electric power produced would be 60,000 Mw, which is about 60 times the present installed geothermal-electric capacity.

It thus appears that the maximum magnitude of geothermal-electric power capacity will probably be in the range of tens of thousands of electrical megawatts, but that this source of energy will probably be largely depleted in less than a century. While 60,000 Mw is a significant quantity of power, a better idea of just how significant can best be obtained by comparison with other sources. It is approximately equal to the world's potential tidal power, but only about 2 percent of the world's potential water power. Also we may compare the initial quantity of 4×10^{20} joules of geothermal energy with that of the fossil fuels. This represents less than 2 percent of the initial energy of the fossil fuels, and only 20 percent of that of petroleum liquids and natural gas combined.

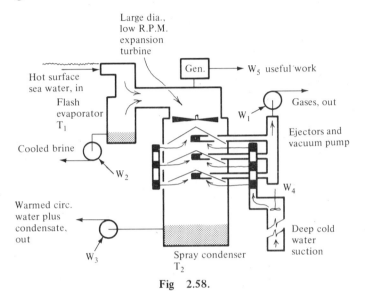

Fig   2.58.

§ 2.20 Nuclear Energy.   We come now to the most recent source of energy to become available for human use—the atomic nucleus. Nuclear energy results from each of two contrasting processes, the fissioning of a few of the isotopes of heavy elements in the atomic scale, producing lighter elements; and the fusing of light elements near the lower end of the scale of atomic numbers to produce heavier elements. In each instance the mass of the reaction products is slightly less than that of the reactants and the lost mass is converted into energy in accordance with the Einstein equation relating mass to energy,

$$E = \Delta m c^2, \tag{2.19}$$

where E is the energy released, Δm the reduction in mass, and c the velocity of light.

Since the velocity of light is 3.00×10^8 m/sec, then, if Δm is 1 gram,

$$E = 10^{-3}\, kg \times 9.00 \times 10^{16}\, m^2/sec^2$$

$$= 9.00 \times 10^{13}\, joules.$$

§ 2.20.1 Energy from the Fissioning of Heavy Isotopes.   The only isotope naturally capable of fissioning is uranium-235, which comprises 0.7 per cent of whole uranium. The remainder of natural uranium is the isotope U-238.

It was found by J. Chadwick in England in 1932[61] that in certain nuclear experiments a strange particle having approximately the mass of the hydrogen atom, or the proton, but zero electric charge, was emitted. This was known later as the *neutron*. Further experiments during the 1930's showed that normally nonradioactive elements, when bombarded with neutrons, can be made artifically radioactive. Finally, in January 1939, O. Hahn and F. Strassmann in Germany[61] reported obtaining barium from the neutron bombardment of uranium. Since barium is an element remote from uranium in the atomic scale, it could not have been produced by any simple radioactive transformation. This led to the surmise that the barium plus a complementary atomic particle must have been produced by the fissioning of uranium. This surmise was verified within the next few weeks in several different laboratories in the United States.

Subsequent studies showed that the fissionable uranium isotope was the comparatively rare U-235, and that the products from numerous fissionings comprise a wide scatter of isotopes, many highly radioactive, in the mid-range of the table of atomic numbers. The energy released per fission was found to have an average value of 200 million electron volts, or 8.90×10^{-18} kilowatt-hours.

From Avogadro's Number, there are

$$\frac{6.02 \times 10^{23}}{235} = 2.56 \times 10^{21}$$

U-235 atoms per gram. From this it follows that an energy released upon the fissioning of 1 gram of U-235 must be

$$2.56 \times 10^{21} \times 8.90 \times 10^{-18} = 2.28 \times 10^4 \text{ kw-hr}$$

$$= 8.21 \times 10^{10} \text{ joules}$$

This is equal approximately to the heat of combustion of 3 tons of coal or 13 barrels of crude oil.

The reduction in mass of 1 gram of U-235 upon being fissioned is then obtained from the Einstein equation

$$\Delta m = \frac{E}{c^2} = \frac{8.21 \times 10^{10}}{9.00 \times 10^{16}} = 0.913 \times 10^{-6} \text{ kg}$$

$$= 0.913 \times 10^{-3} \text{ gm,}$$

which is very nearly 1 part per 1,000.

Hence, the fissioning of 1 gram of U-235 produces 0.999 grams of fission products and loses approximately 1 milligram of mass which is converted into 2.28×10^4 kilowatt-hours of heat.

In addition to radioactive isotopes, the fission products of U-235 also includes neutrons. No sooner had the fissioning of uranium been demonstrated than intensive investigations were begun in the United States in an attempt to obtain a sustained fission chain reaction. This would be a reaction in which, if a single fissioning occurred from a stray neutron, then the neutrons produced would cause still other fissionings to occur and so be able to sustain the reaction.

Such a reaction was first achieved by E. Fermi and associates[61] in Chicago on December 2, 1942, using a "pile" with a graphite matrix in which lumps of common uranium or its oxide were placed in a three-dimensional lattice. When the pile had been built up with about 6 tons of uranium, it reached the critical stage and a sustained chain reaction was achieved.

At just beyond the critical level the reaction could be controlled by the insertion or removal of neutron-absorbing cadmium strips, making it possible to start, stop, increase, or retard the reaction at will.

The object of the wartime experiment was to produce nuclear bombs. Our present interest is limited to the fact that, by means of variations of the original Chicago experiment, it is possible to produce and control sustained fission reactions, and that the heat released can be used to operate conventional steam-power plants.

A schematic flow diagram of the fissioning of U-235 in a chain reaction is shown in Figure 2.59. The material products produced by the fissioning of a single atom are two other atoms plus neutrons, whose combined weights are a little less than that of the U-235 atom. The fission product of a large number of separate fissions comprises a scatter of atoms in the mid-range of the table of atomic numbers. Many of these fission products are radioactive, some with half-lives of approximately 30 years.

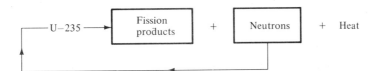

Fig 2.59. Schematic Representation of Nuclear-Power Reaction Involving the Fissioning of U-235

§2.20.2 The Breeder Concept. The difficulty posed by the use of U-235 for a power generation is its comparatively scarcity. However, it has been found that the two much more abundant isotopes, U-238 and Th-232 (which is essentially the whole of natural thorium), can be converted into fissionable isotopes by being placed in a nuclear pile powered initially by U-235. By this process, omitting intermediate details,

$$U\text{-}238 \rightarrow Pu\text{-}239,$$

and

$$Th\text{-}232 \rightarrow U\text{-}233;$$

and both plutonium-239 and uranium-233 are fissionable.

The nonfissionable isotopes, U-238 and Th-232, from which the fissionable isotopes, Pu-239 and U-233, are made, are known as *fertile* materials. The process of converting fertile isotopes to fissile isotopes is known as *breeding*. The process of breeding is illustrated for U-235 and U-238 in Figure 2.60. The same diagram would apply were Th-232 and U-233 substituted for U-238 and Pu-239.

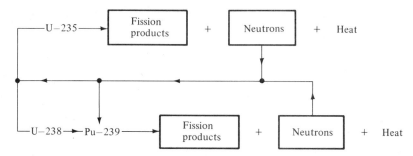

Fig   2.60. Schematic Representation of Breeder Reaction for U-238

By the breeding process, in principle all of uranium and all of thorium are potentially usable as nuclear fuels, instead of only the much scarcer isotope U-235. Since U-238 is 140 times as abundant as U-235 and thorium is geochemically about 3 times as abundant as U-238, it is evident that the available fuel is increased by a factor of about 400 if breeder reactors are developed. This, however, according to Alvin M. Weinberg, Director of Oak Ridge National Laboratory, is only a minimum of the gain potentially obtainable. The development of complete or nearly complete breeding changes the cost of the operation in such a manner as to make it economical to utilize rocks with low uranium or thorium contents. The fuel added in this manner is millions of times greater than that available when only U-235 can be used. Hence the energy gain contingent upon the development of breeder reactors is a very large factor.

The development of large-scale power by means of the fissioning of uranium and thorium and their derived isotopes reduces to three fundamental problems:

1.   the development of breeder reactors,
2.   an adequate supply of uranium and thorium, and
3.   proper disposal of the extremely dangerous fission products.

§ 2.21 Nuclear Power.   During the incredibly short period of 28 years since the first controlled fission chain reaction was achieved on December 2, 1942, the development of nuclear reactors, based principally on the fissioning of uranium-235, has progressed to the point where, by the end of 1969, 15 central-station nuclear power plants were in operation in the United States. These have individual capacities ranging from 22 to 575 electrical megawatts and a combined capacity of 3,482 Mwe. At the same time, according to the U.S. Atomic Energy Commission,[87] 82 additional plants with individual capacities mostly within the range of 500 to 1000 Mwe, and an aggregate capacity of 70,000 Mwe, were either under construction or contract. By 1980, the AEC estimates that the nuclear-power capacity of the United States will reach 150,000 Mwe, which will be approximately 25 percent of the total electrical-power capacity of the country.

An increase in 10 years from a power capacity of 3,482 to 150,000 Mwe represents an exponential growth rate of 37.6 percent per year with a doubling period of only 1.84 years.

In a joint report of January 1969,[88] the European Nuclear Energy Agency and International Atomic Energy Agency show that the central-station nuclear power capacity of the entire non-Communist world has also been increasing with a doubling period of about 2 years, and had reached by the end of 1969 a total capacity of 20,000 megawatts. It was assumed that after 1972 the growth rate will slow down, and that by 1980 the total capacity will be within the range of 220,000 to 340,000 Mwe.

§2.21.1 Uranium Requirements and Resources.

Estimations of the uranium requirements to meet this anticipated growth in nuclear-power capacity have been made by the U.S. Atomic Energy Commission and jointly by the European Nuclear Energy Agency and the International Atomic Energy Agency. According to Rafford L. Faulkner, Director of the Division of Raw Materials, AEC,[89] the U.S. cumulative requirements of U_3O_8 will reach 212,000 short tons by 1980, and 450,000 by 1985. For the non-Communist world outside the United States, he estimates that the cumulative requirements for the period 1969-1980 will be 212,000 short tons and that this will increase to 490,000 by 1985. The ENEA-IAEA in their joint report of 1969, estimate that the cumulative requirements of the non-Communist world by 1980 will be between 563,000 and 739,000 short tons of U_3O_8.

These are the requirements against which the resources of uranium must be weighed.

§2.21.2 The Supply of Uranium and Thorium.

Uranium and thorium are of widespread occurrence in the rocks of the earth's crust, but in very small amounts. The granites which are the principal parent rocks of the continents contain thorium and uranium in the approximate average amounts of 12 parts per million of thorium

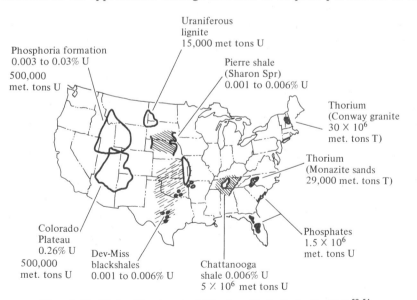

Fig 2.61. Major Uranium and Thorium Deposits in the U.S.[59,54]

and 4 parts per million of uranium. Since the sediments are principally derived from granitic rocks, and since the ocean waters contain essentially no thorium, it is expectable that the ratio of thorium to uranium in sediments should also be about 3 to 1. So far, however, nothing like this amount of thorium has been found. Uranium in sediments is fairly widespread, so we are led to suspect that some large concentrations of thorium in sediments, as yet undiscovered, will eventually be found.

Figure 2.61 is a map of the United States showing the locations and amounts of the principal known uranium and thorium deposits of the United States.

The significant fact is that the United States is estimated to have potential reserves from 700,000 to 2,300,000 metric tons of uranium in ores of comparable quality to those mined during the period 1948 to 1960. These ores have uranium contents ranging from 0.17 to 0.30 per cent, or contents ranging from 1,700 to 3,000 grams per metric ton.

With respect to reserves of much smaller concentrations in the range of 50 to 100 grams per ton, the quantities in this range in various black shales and phosphate rocks are very much larger.

According to a recent AEC news release, the U.S. reserves of U_3O_8 minable at less than \$8/lb amounted at the end of 1969 to 204,000 short tons. During 1969, as the result of 29.9×10^6 feet of drilling, 56,000 tons of U_3O_8 were discovered. This was at a discovery rate of 3.7 lbs/ft, which is less than half the average rate of 8 lbs/ft that had prevailed up until a few years previously.

For the reserves of U_3O_8 of the non-Communist countries, producible at \$10/lb, the ENEA-IAEA estimated a total of 700,000 short tons as of 1969. In a review published in 1970,[90] Robert D. Nininger of the AEC gave the estimates of the reserves of 17 different countries. His total estimate was 980,000 short tons. Of this, 843,000, or 86 percent, were accounted for by four countries, the United States, Canada, France and territories, and the Union of South Africa. The remaining 14 percent were distributed among 13 other countries.

From such estimates it appears that the uranium resources now known are adequate for the requirements of the non-Communist world for about the next 15 years. Present productive capacity, however, is not sufficient to meet the annual requirements beyond the year 1975. Although new deposits of uranium will undoubtedly continue to be discovered, the fact remains that uranium is a comparatively rare element, and its deposits producible in the low-cost range must occur in limited quantities. Hence, the uranium requirements of a nuclear-power capacity that doubles every two years can rapidly deplete all likely deposits of this nature.

For this reason, were the nuclear power reactors of the future to be of the same types as those at present, which consume only 1 or 2 percent of natural uranium, it is doubtful whether the episode of nuclear power could last for much longer than a century. It is imperative, therefore, if this consequence is to be avoided, that the present type of burner or low-conversion reactors based on the rare isotope uranium-235, be supplanted by breeder reactors capable of consuming completely the fertile isotopes uranium-238 and thorium 233—that is, the whole of natural uranium and thorium. Should this be accomplished, then not only would the producible energy from the present supplies of high-grade uranium be enhanced by a factor of 50 to 100, but that of the much larger supplies of lower-grade deposits of both uranium and thorium would become available. The present paper does not admit of a review of these low-grade deposits, though such a review has been made by McKelvey and

Duncan.[91] The following two examples, however, will convey some idea of the order of magnitudes of the energy involved.

In the United States, the Chattanooga shale of Devonian age crops out along the western border of the Appalachian Mountains in eastern Tennessee and neighboring states, and underlies at comparatively shallow depths much of the areas of a half-dozen midwestern states. In its outcrop area in eastern Tennessee, according to Vernon E. Swanson of the U.S. Geological Survey,[92] this shale has a uranium-rich stratum, the Gassaway member, which is about 5 meters thick and contains about 60 grams of uranium per metric ton. Even in Oklahoma, about 1,000 kilometers to the west, the stratigraphic equivalent of this shale is also uranium rich as determined by gamma-ray logging in oil wells.

In the Tennessee area, the uranium content of this shale would amount to 150 grams per cubic meter or to 750 grams per square meter of horizontal surface for the 5-meter thick stratum. Assuming the use of breeder reactors, the energy content per gram of natural uranium would be 8.2×10^{10} joules, or an amount equivalent to the energy content of 2.7 metric tons of bituminous coal, or to 13.7 barrels of crude oil. Therefore, the energy content of this shale per square meter of horizontal surface would be equivalent to that of about 2,000 metric tons of coal, or 10,000 barrels of crude oil. The energy content of an area 13 kilometers square would be equivalent to that of the world's resources of crude oil, and that of an area 60 kilometers square would be equivalent to that of the world's coal resources.

For a thorium resource, the Conway granite in New Hampshire, which crops out over an area of 750 km^2 and probably extends to depths of several kilometers, contains 150 grams of thorium per cubic meter.[54] This is equivalent to 400 tons of coal or 2,000 barrels of oil per cubic meter. The energy content of a surface layer of this rock 100 meters thick would be equivalent to that of 100 times the world's crude-oil resources, or to 4 times that of the world's coal.

From these data it is evident that the amount of uranium and thorium in concentrations of 50 grams or more per metric ton of rock at minable depths in the United States must be of the order of tens if not hundreds of millions of metric tons.

The significance of this will be apparent when the energy content of these nuclear fuels is compared with that of the world reserves of the fossil fuels. Assuming complete breeding, 1 gram of uranium or thorium upon fissioning will release 2.28×10^4 kilowatt-hours of heat.

The ultimate energy reserve of all the fossil fuels is about 28×10^{15} kilowatt-hours.

The amount of uranium or thorium required to produce this much heat would accordingly be

$$\text{Mass of U or Th} = \frac{28 \times 10^{15}}{2.28 \times 10^4}$$

$$= 12.3 \times 10^{11} \text{ grams}$$

$$= 1.23 \times 10^6 \text{ metric tons.}$$

Hence, the uranium and thorium reserves in the United States occurring in rocks having a content of 50 or more grams per metric ton must be of the order of hundreds to thousands of times greater than the world's initial supply of fossil fuels. Also, a rock having a uranium or thorium content of 50 grams per ton is energetically equivalent to about 150 tons of coal or 650 barrels of crude oil per ton of rock.

The world's supply of uranium and thorium in rocks containing 50 grams or more per metric ton is probably tens of times greater than that in the two examples just cited. Consequently, with the exclusive use of breeder reactors, the world's supply of energy appropriate for power production would probably be of the order of tens to hundreds of times greater than that of the fossil fuels.

§ 2.22 Waste Disposal of Fission Products.   The principal remaining problem is the development of means for economical and safe disposal of the fission products. Mention has already been made of the fact that when 1 gram of U-235 is fissioned 0.999 grams of fission products are formed, consisting of a wide spectrum of isotopes in the mid-range of the table of atomic weights.

According to F. L. Culler, Jr.,[57] Director, Chemical Technology Division, Oak Ridge National Laboratory, the fission products produced by 1,000 grams of U-235 with 30 per cent burnup, consist, after 100 days of cooling, of 230.00 grams of inactive isotopes, 15.93 grams of short-lived radioactive isotopes, and 16.61 grams of long-lived radioactive isotopes. The short-lived isotopes comprise fifteen different species with half-life periods ranging from seconds to 290 days. The long-lived isotopes consist of four species of which the two longest are cesium-137 and strontium-90 with half-lives of 33 and 25 years, respectively. These occur in amounts of 7.05 and 4.63 grams, respectively, and represent about two-thirds, by mass, of the long-lived isotopes.

All of these radioactive fission products are dangerous until they have decayed to the very low levels of tolerance prescribed for biological safety. A rule of thumb that has been used as an order of magnitude among the members of the Atomic Energy Commission's health physics division is that none of these materials can be considered to be safe for biologic exposure until a period of at least 20 half-lives has elapsed. For the short-lived fission products, this would be a period of the order of 20 years; for the long-lived isotopes the corresponding period would be at least 660 years, and possibly even 1,000 years.

On February 28, 1955, at the request of the Atomic Energy Commission, an Advisory Committee on Waste Disposal of the Division of Earth Sciences was established by the National Academy of Sciences—National Research Council. After a number of conferences with A.E.C. personnel and visits to Oak Ridge National Laboratory, the Committee issued a report dated April, 1957, in which, on page 3, the following basic principle was stated:

> Unlike the disposal of any other type of waste, the hazard related to radioactive waste is so great that no element of doubt should be allowed to exist regarding safety. Stringent rules must be set up and a system of inspection and monitoring instituted. *Safe* disposal means that the waste shall not come in contact with any living thing. Considering half-lives of the isotopes in waste this means for 600 years if Cs^{137} and Sr^{90} are present or for about one-tenth as many years if these two isotopes are removed.

After the preliminary conferences[58] the Committee concluded that the rate of generation of radioactive wastes at present is very small as compared with magnitudes which will be produced when the generation of power by nuclear fission begins its eventual exponential rate of growth. However, policies and practices initiated now should be of such a nature as still to be valid when the rate of production of wastes should be many times larger than it is at present. The total quantity of wastes was found not to be large, since if all the electric power produced in the United States at the present time were generated by nuclear-fission power plants, the fuel consumed and fission products produced per year would be only of the order of 100 metric tons.

With this in view the Committee reviewed the likely means of waste disposal, of which two were regarded with special favor: (1) in the salt mines or domes, preferably in solid form, and (2) in the form of heavy liquids in permeable sedimentary rocks in the bottoms of synclinal basins. It was pointed out, however, that none of the existing A.E.C. installations, and few of the proposed power plants, had been located at suitable waste-disposal sites, and it was suggested that eventually consideration should be given to locating power plants and waste-processing plants at suitable waste-disposal sites with either a regrouping of power distribution with respect to these sites, or else developing means for long-distance transmission of power to centers of consumption.

After five years, the Committee summarized its observations and recommendations in a letter addressed to John A. McCone, then Chairman of the Atomic Energy Commission:

Dear Mr. McCone:

On February 28, 1955, arrangements were formalized between the Atomic Energy Commission and the National Academy of Sciences-National Research Council to provide advisory services on geological and geophysical problems related to the disposal of radioactive wastes on continental areas. Your Academy-Research Council Committee on Waste Disposal has been active for some 5 years, has held an important conference attended by about 75 scientists and engineers, has closely followed the results of research on disposal problems, and has held numerous meetings, both at AEC installations and elsewhere.

Early in its deliberations, the Committee reached the conclusion which was later stated on page 3 of the report of April 1957 that no system of waste disposal can be considered *safe* in which the wastes are not completely isolated from all living things for the period during which they are dangerous. This period for high-level wastes containing the long-lived isotopes of Cs^{137} and Sr^{90} is at least 600 years. After an extensive review of possible disposal methods which would satisfy the stringent conditions of safety set forth above, your Committee, in light of the technology then existing, favored the following:

1. Disposal within chambers excavated or dissolved in rock salt.
2. Deep disposal in sands or other porous and permeable rocks near the lowest parts of synclinal basins.

While it is possible that other safe disposal methods may be developed, your Committee still regards these as the most promising methods, and feels that no worthwhile advantage will be gained by further delay in stating its appraisal of the present situation, namely:

No existing AEC installation which generates either high-level or intermediate-level wastes appears to have a satisfactory geological location for the safe local disposal of such waste products; neither does any of the present waste-disposal practices that have come to the attention of the Committee satisfy its criterion for safe disposal of such wastes.

The Committee's recommendations are as follows:

1. The Committee regards it as urgent that action be taken for the establishment of waste-disposal facilities at suitable geological sites where the accumulated wastes of the existing installations can be processed and safely disposed of.
2. Your Committee further recommends that approved plans for the safe disposal of radioactive wastes be made a prerequisite for the approval of the site of any future installation by the AEC or under its jurisdiction.
3. In particular, your Committee recommends that the Commission consider concentrating its chemical processing activities at a minimum number of sites located at satisfactory places for the disposal of radioactive wastes.

    Sincerely yours,

    H. H. Hess
    Chairman

Committee Members

John N. Adkins	William B. Heroy	Charles V. Theis
William E. Benson	M. King Hubbert	William Thurston,
John C. Frye	Richard J. Russell	Secretary

§ 2.23 Energy from Fusion.   Energy is obtained from fusion when hydrogen, or its heavy isotope deuterium, is combined into helium. The process is illustrated in Figure 2.62, showing three deuterium atoms combined to form one atom of Helium-4

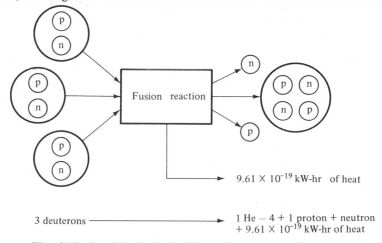

Fig   2.62. Possible Method of Producing Power by Fusion

plus one proton and one neutron with a release of energy of 9.61×10^{-19} kilowatt-hours of energy in the form of heat. Since the ratio of deuterium atoms to hydrogen atoms in water is about $1/6500$, and the deuterium can be separated at an energy cost which is a fraction of 1 per cent of the energy potentially obtainable from fusion, we may estimate about how much energy could be obtained from various amounts of sea water.

> The energy obtainable from 1 gram of water is about 3.30 kilowatt-hours of heat, or a little less than the heat of combustion of a pound of coal. The energy from 1 cubic meter of water is equivalent to that of 1,870 barrels of crude oil, and that from 1 cubic kilometer to $1,870 \times 10^9$ barrels of crude oil, or to 1-1/2 times the crude-oil reserves of the world.

These circumstances, including the abundance of water on the earth, and the fact that the end-product is common helium which is nonradioactive, make the achievement of controlled fusion potentially one of the most important goals in the history of mankind.

However, it may be well to stress that in the reaction shown in Figure 2.62 one neutron is produced with each atom of Helium-4. Since neutrons not only produce fissioning in fissile isotopes, but render many other elements artifically radioactive, a fusion power plant may be difficult to operate on that account. Certainly very heavy shielding will be required, and to accomplish this it might prove desirable to locate such plants at a considerable depth underground.

This problem was reviewed at a conference on energy by James L. Tuck of the Los Alamos Laboratory and his report was one of tempered optimiism. One point on which Mr. Tuck made a very strong plea was the need to prevent the waste of helium. Helium is absolutely essential in the cryogenic work to produce stong magnetic fields by means of superconductivity, and such fields appear to be indispensable as a container for fusion reactions.

At present, one of the more promising fusion reactions is that of deuterium and lithium-6. Omitting intermediate details, the end result of this reaction is the following:

$$_3^6\mathrm{Li} + {}_1^2\mathrm{D} \rightarrow 2\,{}_2^4\mathrm{He} + 22.4\,\mathrm{Mev}. \tag{2}$$

The use of this reaction will be limited either by deuterium or lithium-6, whichever may be the scarcer. The natural abundance of deuterium in sea water is approximately one atom of deuterium for each 6,500 atoms of hydrogen. From this it may be computed that one cubic meter of sea water contains 10^{25} atoms of deuterium, and the entire oceans, 1.5×10^{43}. A large fraction of this could be extracted by methods now available at an energy cost of but a small fraction of that releasable by fusion.

Lithium deposits, on the other hand, occur on land. Lithium is produced from the geologically rare igneous rocks known as pegmatites, and from the salts of saline lakes. The known lithium resources of North America and Africa were recently reviewed with James J. Norton, the specialist on this subject of the United States Geological Survey. According to Norton, these contain a total of about 1.6×10^6 metric tons of $\mathrm{Li_2O}$. Thomas L. Kesler, a geologist for the Foote Mineral Company, the largest producer of lithium in the United States, has published estimates[93,94] of about 2×10^6 metric tons for the United States, 390,000 for Canada, and 180,000 for Africa. Allowing for future developments, Norton considers 10×10^6 metric tons of elemental lithium to be a good order-of-magnitude figure for the presently known resources of lithium for these areas.

Lithium-6 represents only 7.42 atom percent of natural lithium, and one metric ton of lithium-6 contains 1.0×10^{29} atoms.

The resources of Li_2O in the United States, Canada, and Africa, based on the data of Norton and Kesler include a total of 19.6×10^6 metric tons of Li_2O of which the lithium-6 content is 675×10^3 metric tons which contain 6.75×10^{34} atoms of lithium-6. It appears, therefore, that even if we limit the extractable deuterium in sea water to 10 percent of the total content, the quantity of lithium-6 available is only about 10^{-8} of that of deuterium. Therefore, the energy of 22.4 Mev of the deuterium-lithium-6 fusion reaction may be entirely ascribed to the lithium-6. On this basis, the fusion of the 6.75×10^{34} atoms of lithium-6 of the estimated lithium reserves of the United States, Canada and Africa will produce 2.4×10^{23} joules of thermal energy. This is about the same as the total energy of the fossil fuels.

§ 2.24 A Review.   From the data we have just considered it should be clear that our modern industrial civilization may be distinguished from all prior civilizations, and from all contemporary civilizations in the so-called underdeveloped areas of the world, in its dependence upon enormous quantities of energy obtained from sources other than the contemporary biologic channel, and upon correspondingly large quantities of other mineral products, particularly the ores of the industrial metals. We have also seen that, although this development has had its beginning in the remote prehistoric past, most of it has taken place within the last two centuries, and principally since the year 1900.

§ 2.24.1 Rates of Growth.   We have seen how the progressive manipulation of the world's energy flux by the human species and, more recently, the tapping of the large stores of energy contained in the fossil fuels have continuously upset the plant and animal ecological equilibria, and almost always in the direction of increase of the human population. Consequently, during the last century or two—the period of history with which we are most familiar—the pattern of change which we have observed, and in which we have participated, has been of almost continual growth— growth of the world population at an increasing rate which has now reached 2 per cent per year, growth of the United States population from the first census in 1790 until 1860 at 3 per cent per year, growth in the world rate of industrial energy consumption for nearly a century at 4 per cent per year, and of United States consumption at 7 per cent per year.

Yet when reviewed in historical perspective, we have seen that these recent developments have had no precedents in human history, and that the rates of growth we have been witnessing, instead of being the "normal" order of things, are in fact the most abnormal in human history—the usual, or normal, state of affairs being one in which the magnitudes of various human activities have been subject to an almost imperceptible rate of change.

That such rates of growth are essentially ephemeral, and cannot be continued into the future indefinitely, can be seen by noting that the earth on which we live is finite in magnitude; whereas no physical quantity, whether the human population, the rate of energy consumption, or the rate of production of a material resource such as a metal, can continue to increase at a fixed exponential rate without soon exceeding all physical bounds.

For example, during most of the nineteenth century the rate of production of pig iron in the United States increased at 6.4 per cent per year. At such a rate of growth the production rate doubled in 11 years and increased 10-fold in 36 years. By 1900 the production rate had reached 15.4 million metric tons of pig iron per year. With eight more 10-fold increases the rate of pig iron production would be increased by 100 millionfold, or to 15.4×10^{14} metric tons of pig iron per year. At a steady rate of increase of 6.4 per cent per year this would take place during eight 36-year periods, or in 288 years.

The figure of 15.4×10^{14} metric tons is approximately the estimated total iron content (at 4.7 per cent average iron content by weight) of the rocks of the United States to a depth of 2,000 meters, or 1.2 miles. It is manifestly a physical impossibility to continue the nineteenth-century rate of growth until production rates anywhere near this magnitude have been reached. The growth rates not only must decline, but in all instances where exhaustible resources are concerned they must eventually become zero and then negative, as is shown in Figure 2.17.

For a renewable resource, such as water power, instead of the quantity of energy involved having some definite amount, it is the power which is finite. The growth curve with which we are then concerned is the amount of this power that is brought under control and converted to human uses as a function of time. Such a curve would be that of installed water-power capacity. This must start at zero; and then, after a period of growth, it must eventually level off asymptotic to some maximum amount, which may approach but cannot exceed the water power naturally available in a given area. This is the type of growth represented by the logistic growth curve of Figure 2.56.

Then we have the growth curves of biologic populations, of which that of the human population is only a particular example. Since the normal ecologic state is one in which biologic populations are nearly constant, or else oscillate with nearly constant amplitudes, as is the case with annual plants and insects, it follows that any rapid departure from this state must be due to some major disturbance.

It is well known, and has been shown experimentally in detail by Raymond Pearl,[70] that when a population sample of any biologic species is isolated from its ecological system and placed in a favorable artificial environment, this population will increase spontaneously at an exponential, or geometrical, rate. However, because of the finite size of the space in which this experiment must be performed, the geometrical rate of increase can continue only for a limited number of doublings before the rate of increase begins to slacken, and decreases ultimately to zero. The population itself increases in the manner of the S-shaped logistic growth curves shown in Figures 2.22 and 2.23 a type of growth which is described analytically by equation (2.7). In fact, the name "logistic curve" was first given to this type of curve, and its basic theory derived by the Belgian mathematician, P.-F. Verhulst,[73, 74, 75] in a series of celebrated memoirs on the law of population increase.

In case the food supply, rather than space, is a limiting factor, the population may reach a maximum and then decline and stabilize at some lower level. Or, of course, if the food supply fails it can decline to zero.

In a natural ecological environment,[69] conditions are much more complex. In a near-equilibrium state populations tend to remain nearly constant or to change very slowly with time. However, in response to some major disturbance *all* populations of the complex undergo rapid change (Figure 2.63). Some increase by a positive logistic growth to some higher number than before; others decrease and level off at some lower number; some may even become extinct.

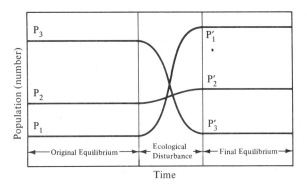

Fig 2.63. Population Changes Due to Ecological Disturbance

The significance of this to our present inquiry is that the whole biologic complex of the earth is at present in the midst of one of the greatest ecological upheavals known in geological history. The various biological populations are about mid-range in their transitions from their earlier near-equilibrium states to new equilibria at markedly different levels. In this transition some populations, notably that of man, are increasing; others, including most of the familiar wild animals and most native plants, are decreasing; some have already become extinct.

Because the earth is of finite magnitude, it is unavoidable that the present abnormal rate of increase of the human population must eventually slow down and ultimately become zero or even negative. The population itself may level off asymptotic to some maximum number, or it may overshoot and stabilize at a lower, more nearly optimum figure. Or, in the event of a general cultural degeneratin, it may be forced back to some level that could be sustained by the industry of a more primitive culture.

The alternatives faced by the human population at the time of the inevitable cessation of growth, as was pointed out by Frank Notestein during an informal panel discussion at Northwestern University on the occasion of its Centennial Celebration in 1951, are the following:

When the population growth ceases, the birth rate and the death rate (number of births per thousand per year, and number of deaths per thousand per year) must become equal. From the point of view of one who has to be a member of the population at that time, the question might be asked: What would be a desirable condition under which to live? Notestein suggested that a high standard of public health might be a major attribute of a desirable condition for existence.

However, a high standard of public health implies a long expectancy of life, which in the United States and Western Europe at present is about 70 years. With a life expectancy of 70 years, and an equilibrium birth rate and death rate, 1/70th of the population would have to be replaced per year, which, on a per-thousand basis, would be 1000/70, or a death rate and birth rate of about 14 per thousand per year.

In case this low birth rate should be unacceptable to the population, and its members insisted upon breeding at the biological rate of 40–50 per thousand per year, then the death rate would have to rise to the same figure. As a result the life expectancy would be reduced to 20–25 years, characteristic of a very low state of the public health. Hence, such a population could choose to have either a high standard of public health or a high birth rate, but it could not have both.

§ 2.24.2 Nonfuel Mineral Resources.

§ 2.24.2 **Nonfuel Mineral Resources.**   We have discussed in detail in the present report the nature of the supplies of the fossil fuels, and have shown that they can be expected to serve as principal sources of industrial energy only for a period of about 300–400 years. During this period petroleum and natural gas will be the earliest of the fossil fuels to approach depletion, with their span of greatest usefulness lasting less than a century. We have not made a corresponding review of mineral resources other than energy, since this is the subject of a companion report by Dean F. Frasché.[67] Nevertheless, since our modern industrial complex depends upon large supplies of both energy and nonfuel minerals, mention of the latter needs to be made in our appraisal of our present position and possible future evolution.

Like the fuels, the nonfuel mineral resources are distributed over the earth in a highly inequitable manner. The principal industrial minerals until now have been coal and iron ores, and the world's regions of industrialization have been limited to the areas of the northern hemisphere, where large quantities of coal and iron ores have occurred in proximity to one another. In countries like Brazil, which has large reserves of iron ore but almost no coal, significant industrialization has so far been impossible. Brazilian iron ores have been transported to the United States and other industrial centers where coal is available.

The mining of metallic ores customarily proceeds from deposits of highest grade, and, as these are exhausted, either mining must cease or else ores of progressively lower grades must be produced. In the United States the high-grade iron ores (50 per cent iron content or better) of the Lake Superior region have already been largely exhausted and mining of the lower-grade (30 per cent iron content) taconite ores is proceeding. The average grade of the copper ores mined in the United States has been declining for some decades, and today ores with a copper content as low as 17 pounds per ton, or 0.8 per cent, are being mined. A century ago most copper producers required ores with an average copper content of not less than 10 per cent; today the world average is 1.5 per cent or less.[72]

The mine production of lead in the United States reached a peak rate of 684,000 short tons per year in 1925 and 1926, and by 1960 this rate had declined to 244,000 tons, or to 36 per cent of the peak rate. Similarly, the United States production of zinc reached two peaks of about 775,000 short tons per year each, the first in 1926 and the second in 1942 during the war. By 1960 the production rate had dropped to 432,000 tons per year, or to 56 per cent of the peak rate.

The approximate world situation as of 1956 for six of the principal industrial metals—aluminum, iron, zinc, copper, lead and tin[71] is significant in that for only two of the metals, aluminum and iron, is the number of years supply of estimated exploitable reserves larger than 100 years. The years of supply of the other four metals range from 19 to 35 years. The ratios of the total content of each metal to the estimated exploitable reserves range, however, from 600,000 to 29 million.

These data emphasize two basic facts of the mineral industry:

1. The estimated world supply of metallic ores of grades now capable of utilization for most minerals is measurable at present rates of production in decades rather than in centuries.

2. The total amount of each metal occurring within minable depths is, on the average, the order of a million times larger than the amount of metal contained in currently exploitable grades of ore.

In principle, it is possible to mine and extract the metals from rocks having much lower metallic contents than present ores, but to do so would require much higher expenditures of energy per unit produced than is required at present, and would also require a much more sophisticated technology, particularly in the direction of large-scale industrial chemistry.

§ 2.24.3 Mineral Requirements to Industrialize Undeveloped Areas.

A problem closely related to that of the mineral and energy requirements of the presently industrializd areas of the world is the question of how much larger these requirements would be if the world were to be industrialized to the extent that has now been reached in the United States. An approximate answer can be given to this question by noting that the United States, with 6 per cent of the world's present population, consumes approximately 30 per cent of the world's total current production of minerals.

Let M_1 be the present rate of mineral production, and M_2 the rate that would be required to give the total world population the same per capita mineral consumption as that in the United States. Let P be the world population, and C the United States per capita consumption.

Then the per capita consumption for both the United States and the world would be

$$C = \frac{0.3 M_1}{0.06 P} = \frac{M_2}{P}.$$

Solving this for M_2 then gives

$$M_2 = 5 M_1.$$

In other words, if the whole world were industrialized to the same level as is the United States, the annual drain on the world's mineral resources would be about five times what it now is.

This neglects the fact, however, that before any area can reach the per capita energy and mineral consumption rate of the United States, it must first build up its industry to that level. Were the whole world to have done this, the minerals and energy required would have been about five times the present cumulative production of the world. At such a world rate of consumption the middle 80 per cent of the world's supply of crude oil and natural gas would be consumed during a period of about 15–20 years, and the corresponding period for coal would be reduced from about 350 years to less than a century. Moreover, the presently estimated world supply of the ores of most industrial metals, producible by present technology, would have been exhausted well before such a level of industrialization could have been reached.

Hence, so long as the world depends on the fossil fuels as its principal source of industrial energy, there appears to be little ground for the humanitarian hope of significantly improving the standard of living by industrialization of the underdeveloped areas of the world. For the same reason, there is not very much promise that the activities of the highly industrialized areas can be maintained at anything like present levels for more than a few centuries, and there are possibilities that shortages may develop before the end of the present century.

§ 2.25 Nuclear Energy. If a world-wide industrial collapse due to the exhaustion of the fossil fuels and the high-grade ores of metals within the next few centuries is to be forestalled, there appears to be no possible way of accomplishing this except by a newer and larger supply of energy suitable to the requirements of large-scale industrial operations. We have already observed that, while solar power is of this magnitude, it does not offer much promise of concentration such as to provide the power for large electric-power networks. Water power is of a lesser magnitude, but still large and capable of providing power in the hundreds-of-megawatts range in many parts of the world. It still, however, is not large enough, and besides it requires prior industrialization before it can be developed and used.

The only remaining source of energy that does have the proper magnitude and does lend itself to large industrial uses is nuclear. We have already seen that the supplies of uranium and thorium in the United States alone, occurring in concentration of 50 grams or more per metric ton of rock within a depth of 2,000 meters, have an energy content at least hundreds of times, and possibly thousands of times, greater than that of all the fossil fuels in the world. We have noted further that the energy content of 1 metric ton of such rock is equal to that of 150 tons of coal or 650 barrels of oil. Therefore, even if the extraction of this uranium or thorium should require energy equal to a few tons of coal or barrels of oil per ton of rock, the net amount of energy obtainable per ton of rock should still be many times greater than that from an equivalent mass of any fossil fuel.

The resources of fission energy, uranium and thorium, and of fusion energy, deuterium or heavy hydrogen, are quite as exhaustible as the fossil fuels, but the quantities are so large that it is doubtful if any significant diminution of the total reserves could be effected by industrial uses within the next thousand years. Hence, for all present purposes, nuclear energy may be regarded as being essentially inexhaustible in terms of human usage.

With such quantities of energy available, it then would become both possible and practical to work the lower and lower grades of metallic ores, and in so doing to begin to realize a part of the potential million-fold increase in reserves which might become available,[71] thus forestalling the otherwise imminent shortages of many of the industrial metals. With a source of energy of this magnitude, and the additional quantities of metals which would thus become available, the dream of improving the standards of living of all the races of man no longer appears so visionary.

§2.26 Time Perspective.   The present state of human affairs can perhaps more clearly be seen in terms of a time perspective, minus and plus, of some thousands of years with respect to the present, as depicted in Figure 2.64. On such a scale the phenomena of present interest—the growth in the rate of consumption of energy, the growth of the human population, and the rise in the standard of living as indicated by the increase in the per capita rate in energy consumption—are all seen to represented by curves which began near zero and rose almost imperceptibly until the last few centuries. Then, after an initial gradual increase, each curve, as the present day is approached, rises almost vertically to magnitudes many times greater than ever before.

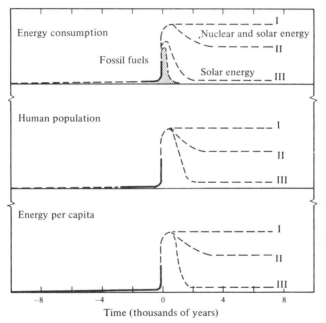

Fig   2.64. Human Affairs in Time Perspective

On this time scale the consumption of fossil fuels is seen to rise sharply from zero and almost as sharply to descend, with the total duration of the period of consumption representing but a brief interval of the total period of human history.

As to the future, if we disallow imminent annihilation by nuclear warfare, three distinct possibilities appear to exist. These are represented on the graphs as Courses I, II and III. One possibility, shown as Course I, is that we may be able to maintain our present scientific and technological culture, using the fossil fuels as an essential intermediate step in the transition to ultimate dependence upon the large-scale use of nuclear energy. Should this be successfully accomplished there appear to be no physical reasons why we should not be able to level our activities off asymptotically to some maximum level which could be maintained for many centuries.

There is also a possibility, indicated as Course II, that we may not succeed in overcoming the cultural lag between our inherited folkways and our present requirements in time to prevent a serious overshooting of the world population above a manageable magnitude. After a temporary state of chaos we might still be able to stabilize our population and the magnitude of our industrial activities at some lower and more nearly optimum level which could be maintained for a long period of time.

Finally, there is the possibility, indicated as Course III, that we could go into a state of confusion and chaos, including nuclear warfare, from which we might never be able to recover. In that case we could suffer a cultural decline and return to our former agrarian and handicraft level of culture. At what level the population would become stabilized in this event it is not possible to state with any assurance, but since modern medicines and techniques of public health are a by-product of our present culture, and not otherwise possible, it appears doubtful whether a population nearly as large as that of the present could be sustained.

Which of these three possibilities may be the one actually realized depends largely upon the foresight that can be exercised with respect to the guidance of human afaairs, and in large measure on whether the cultural lag can be sufficiently reduced between the inhibitory sacred-cow behavior patterns which we have inherited from our recent past and the action requirements which are necessitated by the socio-industrial complex with which we have to deal. If such impediments can be overcome it is entirely possible that, with only minor extensions of our present knowledge of the physics, chemistry, biology and geology of the world in which we live, we shall be able to make the transition to a stabilized industrial civilization with a decent standard of living and a high standard of health for all the world's human inhabitants. If we are unable to make this transition, and if we do permit ourselves to go into a cultural decline, then, as Brown, Bonner, and Weir[65] have pointed out, it is doubtful whether we shall ever be able to arise again.

§ 2.27 Outlook and Recommendations.

From this review of the world's energy resources appropriate for power production, two realizations of outstanding significance emerge. One is the brevity of the time, as compared with the totality of past human history, during which the large-scale production of power has arisen; the other is that energy resources sufficient to sustain power production of present magnitudes for at least a few millenia have now become available.

The limiting factors in power production are, therefore, no longer the scarcity of energy resources, but rather the principles of ecology. The production of power and its associated industrial activities are quite as much components of the world's ecological complex as are the populations of plant and animal species—witness, for example, the decline of the population of horses with the rise of motor vehicles. As we have observed, our industrial activities have been characterized by large exponential rates of growth only for most of the last two centuries, and the present rate of growth of nuclear-power capacity, with a doubling period of only two years, represents the most spectacular growth phenomenon in the entire history of technology. However, it is no less true with power capacity and automobiles than with biologic populations that the earth itself cannot sustain or tolerate any physical growth for more than a few tens of doublings. Therefore, any individual activity must either cease its growth and, as in the case of water power, stabilize at some maximum or intermediate level that can be sustained, or else it must pass a maximum and decline eventually to extinction.

If our future evolution is to follow one of the more desirable paths—one characterized by a high per capita utilization of energy, a general state of individual well-being, and a high standard of public health—then it is clear that a number of essential steps must be taken, some sequentially and others in parallel. Some of the more important of these are set forth in the following sections.

§ 2.28 The Growth of the World's Population Must be Brought Under Control.
While this is a problem of formidable magnitude, it is not intrinsically more difficult than the control of disease, in which the medical profession has already achieved marked success. The present flare-up of the world population is in fact the consequence of this success. Until comparatively recently, as we have noted, the world population was almost stationary. The birth rate and death rate were nearly equal but also near the biological maximum of 40–50 per thousand per year, with a life expectancy of 20–25 years.

During the last few centuries, and particularly during the last few decades, the death rate, world-wide, has been dropped spectacularly to a present average value of about 20 per thousand, while the birth rates of most of the world's population have been but little reduced. The difference between the birth rate and the death rate is a direct measure of the rate of population growth.

If the desirable objectives mentioned above are to be achieved, it is essential that the death rate and birth rate must be equalized at a low level compatible with a high standard of public health (about 15 per thousand). If this is not done, assuredly they will eventually become equalized at a level corresponding to a low standard of public health.

§ 2.29 New Sources of Energy Must be Developed.
As sources of energy for the world's future needs, the fossil fuels are exhaustible, solar power cannot practicably be concentrated, and water power, though large, is inadequate. This leaves us ultimately with only nuclear energy as the source which is both adaptable to large-scale power generation and of sufficient magnitude to meet the world's potential requirements.

§ 2.29.1 Fossil Fuels.
Since the fossil fuels are adequate to meet the limited needs of the presently industrialized areas of the world for the next few centuries, there is obviously no immediate emergency as to energy supplies for these areas. Neither, however, are there grounds for complacency, because most of the areas of the world are not industrialized and, so long as we depend upon the fossil fuels, are not likely to become so.

Among the fossil fuels themselves, because petroleum and natural gas are the least abundant and coal the most abundant, it is evident that in the comparatively near future a transition must be begun from crude oil and natural gas to the more abundant reserves of oil shale and tar sand, and ultimately to coal, for our supplies of liquid and gaseous fuels. This transition, utilizing the large research establishments of the petroleum industry, will probably be made in an orderly manner as rapidly as new sources of liquid and gaseous fuels are required.

In the coal industry proper, however, there is need for increased integrated research in all phases of production, processing, and transportation. Research of this kind would particularly benefit by a reorientation in which coal is regarded as an energy and organic-chemical raw material rather than as just a fuel, much as crude oil is regarded by the petroleum industry. (See Chapter 3.)

Although coal represents nearly 80 per cent of the energy reserves of the fossil fuels in the United States, it has been a depressed industry, largely because of a displacement by oil and gas, since World War I. One of the largest bottlenecks in present coal utilization arises from the prohibitive costs of railroad transportation. Promise of eliminating this bottleneck is now afforded by the recent successful developments in the transportation of coal in the form of a coal-water slurry by pipeline at a greatly reduced cost.

A needed restriction in the uses of coal should also be mentioned. Of all the coal reserves in the United States only a small fraction is suitable for the manufacture of metallurgical coke, which is particularly essential for the smelting of iron ore. Much of the coking coal has already been indiscriminately mined and burned as fuel. A control is needed whereby only noncoking coals are burned as fuels, reserving the more valuable coking coals for the metallurgical industry.

§ 2.29.2 Nuclear Energy. The eventual dependence upon nuclear energy as the principal source of industrial power in areas which now have abundant fossil fuels, and the immediate needs in other areas, makes it essential that research and development in this field should be vigorously pursued. With regard to fission energy, there are two very important problems. The first is the development of power reactors based upon complete or nearly complete breeding. This will permit the utilization not only of common uranium-238 but also of thorium-232. More importantly, it will make it economical to consider rocks with uranium and thorium contents as low as 50 grams per metric ton as practically utilizable ores, and so will enormously enhance the magnitude of the reserves of nuclear fuels.

The second problem associated with fission reactors is that of safe disposal of radioactive fission-product wastes. These wastes, some of which are toxicfor the order of a thousand years, must be completely isolated from the biological environment for their periods of danger.

The control of the fusion reaction—deuterium to helium—possibly represents the greatest energy goal now known. The problem is one of very great difficulty and may never be solved. Nevertheless, what is most needed at this stage of development is systematic, long-range, fundamental research of the type now being pursued, rather than some kind of a crash program.

§ 2.29.3 Synthesis of Chemical Fuels. Automotive vehicles for both highway and air transportation are dependent for their energy supply upon the energy stored chemically in the form principally of liquid fuels, and, so far as can now be seen, will continue to be so. Heretofore these fuels have been obtained almost solely from the fossil fuels in which the energy was originally stored by photosynthesis. On the other hand, it has long been known to be possible to manufacture simpler but equally useful fuels by means of the schematic chemical reaction

$$Energy + CO_2 + H_2O \rightarrow Fuel + O_2.$$

This has not been done because the energy required for the reaction would have to be obtained by burning already synthesized fossil fuels.

With the advent of nuclear energy this situation is drastically changed. Here, with an almost unlimited supply of energy potentially available, it would be a comparatively simple matter to synthesize any desirable quantity of liquid and gaseous fuels from common inorganic substances such as water and limestone. Were this eventually to be done, our remaining fossil fuels, comprising already synthesized complex organic molecules, could be more effectively used as the raw material for an increasingly versatile chemical industry (see Chapter 3).

§ 2.30 Eventual Dependence Upon Low-Grade Deposits for our Principal Supplies of Industrial Metals, and of Other Nonfuel Mineral Products, Must be Anticipated. Since this has been covered in a companion report on "Mineral Resources" by Dean F. Frasché,[67] it will not be further discussed here. It is mentioned only to emphasize the fact that the nonfuel mineral resources, together with the energy resources and the population problem, constitute a triumvirate of perhaps the foremost problems now confronting the human race.

§ 2.31 References and Notes

1. Commerce, Department of, 1961, *Statistical Abstract of the United States, 1961:* U.S. Govt. Printing Office, 948 p.

2. Jeffreys, Harold, 1952, *The Earth, Its Origin, History and Physical Constitution:* Cambridge Univ. Press. 376 p.

3. Landsberg, H., 1945, "Climatology,': p. 928–997 in *Handbook of Meteorology:* New York, McGraw-Hill Book Co., 1056 p.

4. Childe, V. Gordon, 1954, "Early Forms of Society," p. 38–57 in *A History of Technology,* v. I: New York, Oxford Univ. Press, 803 p.

5. Dickinson, H. W., 1958, "The Steam-Engine to 1830, p. 168–198 in *A History of Technology, v. IV*: New York, Oxford Univ. Press., 681 p.

6. Forbes, R. J., 1956, "Power," p. 589–622 in *A History of Technology, v. II*: New York, Oxford Univ. Press, 776 p.

7. Forbes, R. J., 1956, "Metallurgy," p. 41–80 in *A History of Technology,* v. II: New York, Oxford Univ. Press, 776 p.

8. Harrison, H. S., 1954, "Discovery, Invention, and Diffusion," p. 58–84 in *A History of Technology,* v. I: New York, Oxford Univ. Press, 803 p.

9. Jarvis, C. Mackenchnie, 1958, "The Generation of Electricity," p. 177–207 in *A History of Technology,* v. V: New York, Oxford Univ. Press, 841 p.

10. Nef, J. U., 1957, "Coal Mining and Utilization," p. 72–88 in *A History of Technology,* v. III: New York, Oxford Univ. Press, 721 p.

11. Notestein, Frank W., 1962, letter to John S. Coleman dated April 19.

12. Putnam, Palmer Cosslett, 1953, *Energy in the Future*: New York, D. Van Nostrand Co., Inc., 459 p.

13. Ritson, J. A. S., 1958, "Metal and Coal Mining," 1750–1875, p. 64–98 in *A History of Technology, v. IV*: New York, Oxford Univ. Press, 681 p.

14. United Nations, 1958, *The Future Growth of World Population*: New York, Department of Economic and Social Affairs, 75 p.

15. Hubbert, M. King, 1970, *Energy Resources for Power Production in Symposium on Environmental Aspects of Nuclear Power Stations*, International Atomic Energy Agency, Vienna.

16. Averitt, Paul, 1961, *Coal Reserves of the United States—A Progress Report January 1, 1960*: U.S. Geol. Survey Bull. 1136, 114 p.

17. Commerce, Dept. of, 1949 *Historical Statistics of the United States, 1789–1945*: U.S. Govt. Printing Office, 351 p.

18. _____ 1954, *Continuation to 1952 of Historical Statistics of the United States, 1789–1945*: U.S. Govt. Printing Office, 78 p.

19. American Gas Association, 1946–1962, Annual Reports of the Committee on Natural Gas Reserves of the American Gas Association.

20. American Petroleum Institute, 1937–1962, Annual Reports of the American Petroleum Institute's Committee on Petroleum Reserves.

21. American Petroleum Institute, 1959, *Petroleum Facts and Figures*: 457 p.

22. Blanpied, B. W., 1959, *Exploratory Drilling in 1958*: Am. Assoc. Petroleum Geologists Bull., v. 43, p. 1117–1138.

23. _____ 1953–1961, *Annual Editions of Statistical Abstract of the United States*: U.S. Govt. Printing Office.

24. Davis, Warren, 1958, *A Study of the Future Productive Capacity and Probable Reserves of the U.S.*: Oil and Gas Jour., v. 56, February 24, p. 105–116.

25. Hill, Kenneth E., Hammar, Harold D., and Winger, John G., 1957, *Future Growth of the World Petroleum Industry*: Paper for Presentation at the Spring Meeting of the API Division of Production, Rocky Mountain District, Casper, Wyoming, April 25 (Preprint), 42 p.

26. Hubbert, M. King, 1956, *Nuclear Energy and the Fossil Fuels*: Drilling and Production Practice, Am. Petroleum Inst., p. 7–25.

27. Interior, Dept. of the, 1956, *Impact of the Peaceful Uses of Atomic Energy on the Coal, Oil- and Natural-Gas Industries:* Background Material for the Report of the Panel on the Impact of the Peaceful Uses of Atomic Energy to the Joint Committee on Atomic Energy, v. 2, January, p. 68–89.

28. Levorsen, A. L., 1950, *Estimates of Undiscovered Petroleum Reserves*: Proceedings of the United Nations Scientific Conference on the Conservation and Utilization of Resources, 17 August—6 September, 1949, Lake Success, New York, p. 94–99.

29. McKelvey, V. E., 1961, Synopsis of Domestic and World Resources of Fossil Fuels, Radioactive Minerals, and Geothermal Energy in *Domestic and World Resources of Fossil Fuels, Radioactive Minerals, and Geothermal Energy:* Prelim-inary Reports Prepared by Members of the U.S. Geological Survey for the Natural Resources Subcommitte of the Federal Science Council, 20 p.

38. Moore, C. L., 1962, *Method for Evaluating U.S. Crude Oil Resources and Projecting Domestic Crude Oil Availability:* U.S. Dept. of the Interior, Office of Oil and Gas, 89 p.

31. National Petroleum Council, 1961, *Proved Discoveries and Productive Capacity of Crude Oil, Natural Gas, and Natural Gas Liquids in the United States:* Report of the National

Petroleum Council Committee on Proved Petroleum and Natural Gas Reserves and Availability, 40 p.

32. Netschert, Bruce C., 1958, *The Future Supply of Oil and Gas:* Baltimore, Resources for the Future, Inc., Johns Hopkins Press, 126 p.

33. Oil and Gas Journal, 1956, *Reserves Gain Less Than Billion Barrels:* 1955–56 Review—Forecast Issue, January 30, p. 178–179.

34. Oil and Gas Journal 1962, *Where U.S. Proved Reserves are "Stored":* 1961–1962 Review—Forecast Issue, January 29, p. 128–135.

35. Pogue, Joseph E., and Hill, Kenneth E., 1956, *Future Growth and Financial Requirements of the World Petroleum Industry:* New York, The Chase Manhattan Bank, Petroleum Dept., 39 p.

36. Pratt, Wallace E., 1942, *Oil in the Earth:* Lawrence, Univ. of Kansas Press, 135 p.

37. Pratt, Wallace E. 1944, *Distribution of Petroleum in the Earth's Crust:* Am. Assoc. Petroleum Geologists Bull., v. 28, p. 1506–1509.

38. Pratt, Wallace E. 1947, *Petroleum on Continental Shelves:* Am. Assoc. Petroleum Geologists Bull., v. 31, p. 657–672.

39. Pratt, Wallace E. 1956, *The Impact of the Peaceful Uses of Atomic Energy on the Petroleum Industry:* Background Material for the Report of the Panel on the Impact of the Peaceful Uses of Atomic Energy to the Joint Committee on Atomic Energy, v. 2, January, p. 89–102.

40. Terry, Lyon F., 1950, *The Future Supply of Natural Gas Will Exceed 500 Trillion Cu. Ft.:* Gas Age, October 26, p. 58–60, 98, 100, 102.

41. Weeks, L. G., 1948, *Highlights on 1947 Developments in Foreign Petroleum Fields:* Am. Assoc. Petroleum Geologists Bull., v. 32, p. 1093–1160.

42. Weeks, L. G., 1950, Discussion of *Estimates of Undiscovered Petroleum Reserves* by A. L. Levorsen: Proceedings of the United Nations Scientific Conference on the Conservation and Utilization of Resources, 17 August—6 September, 1949, Lake Success, New York, p. 107–110.

43. Weeks, L. G., 1958, *Fuel Reserves of the Future*: Am. Assoc. Petroleum Geologists Bull., v. 42. p. 431–438.

44. Weeks, L. G., 1959, *Where Will Energy Come from in 2059*: The Petroleum Engineer, v. XXXI, August, p. A-24–A-31.

45. Weeks, L. G., 1961, The Next Hundred Years Energy Demand and Sources of Supply: Alberta Soc. Petroleum Geologists Jour., v. 9, p. 141–157.

46. Zapp, A. D., 1961, *World Petroleum Resources in Domestic and World Resources of Fossil Fuels, Radioactive Minerals, and Geothermal Energy*: Preliminary Reports Prepared by Members of the U.S. Geological Survey for the Natural Resources Subcommittee of the Federal Science Council, 9 p.

47. Zapp, A. D., 1962, *Future Petroleum Producing Capacity of the United States*: U.S. Geol. Survey Bull. 1142-H, 36 p.

48. Adams, Francis L., 1961, *Statement on Water Power*: At Conference on Energy Resources, Committee on Natural Resources, National Academy of Sciences—National Research Council, Rockefeller Institute, New York, N.Y., July 19, 10 p.

49. English, Earl F., 1959, "Harnessing Geothermal Heat," p. 16–18 in *The Undiscovered Earth*: Proc. of the Conference, June 11–12, Birmingham, Alabama, Southern Research Inst., 56 p.

50. Penta, Francesco, and Bartolucci, Giorgio, 1962, *Sullo Stato delle "Ricerche" e dell' Utilizzazione Industriale* (Termoelettrical) del Vapore Acqueo Sotterraneo nei Vari Paesi del Mondo: Accademia Nazionale dei Lincei, Series VIII, v. XXXII, April, p. 1–16.

51. White, Donald E., 1961a, *Summary of Studies of Thermal Waters and Volcanic Emanations of the Pacific Region, 1920 to 1961*: U.S. Geol. Survey Unpublished Manuscript (transmitted to M. King Hubbert by letter of October 10, 1961), 11 p.

52. White, Donald E., 1961b, *Preliminary Evaluation of Geothermal Areas by Geochemistry, Geology, and Shallow Drilling*: United Nations Conference on New Sources of Energy, April 10, 12 p.

53. Young, Loyd L., 1955, *Developed and Potential Waterpower of the "ited States and Other Countries of the World, December 1954*: U.S. Geol. Survey Circ. 367, 14 p.

54. Adams, J. A. S., Kline, M. -C., Richardson, K. A., and Rogers, J. J. W., 1962, *The Conway Granite of New Hampshire as a Major Low-Grade Thorium Resource*: Houston, Rice University, Dept. of Geology, 18 p.

55. Atomic Energy Commission, 1958, "The Goals and the Problems," p. 228–243 in *Research on Power from Fusion and Other Major Activities in the A.E.C. Programs*: A.E.C. Semi-Annual Report, January–June.

56. Charpie, Robert A., and Weinberg, Alvin M., 1961, *The Outlook for Thorium as a Long-Term Nuclear Fuel*: Paper given following Dedication Ceremonies of the Canada-India Reactor, Bombay, India, January 17–18, 18 p.

57. Culler, F. L., Jr., 1955, *Notes of Fission Product Wastes from Proposed Power Reactors*: Report of Meeting on Subterranean Disposal of Reactor Wastes, Washington, D.C., November 15, 1954, 67 p.

58. Gorman, Arthur E., 1955, *The A.E.C.'s Views on Radioactive Waste Disposal*: Report of Meeting on Subterranean Disposal of Reactor Wastes, Washington, D.C., November 15, 1954, 18 p.

59. McKelvey, V. E., Butler, A. P., Olson, J. C., and Gottfried, David, 1961, "Uranium and Thorium Resources in the United States and World" in *Domestic and World Resources of Fossil Fuels, Radioactive Minerals, and Geothermal Energy*: Preliminary Reports cprepared by Members of the U.S. Geological Survey for the Natural Resources Subcommittee of the Federal Science Council, 20 p.

60. National Academy of Sciences—National Research Council, 1957, *The Disposal of Radioactive Wastes on Land*: Report of the Committee on Waste Disposal of the Division of Earth Sciences, Publication 519, 142 p.

61. Smyth, Henry DeWolf, 1945, *Atomic Energy for Military Purposes*: Princeton Univ. Press, 226 p.

62. Weinberg, Alvin M., 1959, *Energy as an Ultimate Raw Material*: Physics Today, v. 12, November, p. 18–25.

63. Weinberg, Alvin M., 1960, *Breeder Reactors*: Scientific American, v. 202, January, p. 82–94.

64. Weinberg, Alvin M., 1961, *The Problem of Burning the Rocks*: Paper given out at the Conference on Energy Resources, National Academy of Sciences—National Research Council, Rockefeller Institute, New York, July 19, 21 p.

65. Brown, Harrison, Bonner, James, and Weir, John, 1957, *The Next Hundred Years*: New York, The Viking Press, 154 p.

66. Cook, Robert C., Editor, 1959, *The Race between People and Resources in the ECAFE Region—Part I*: Population Bulletin, v. XV, p. 81–94.

67. Frasché, Dean F., 1962, *Mineral Resources*: National Academy of Sciences, Report to the Committee on Natural Resources, 33 p.

68. Hubbert, M. King, 1950, *Energy from Fossil Fuels*: The Smithsonian Report for 1950, p. 255–272. See also: Am. Assoc. for the Advancement of Science, Centennial, p. 171–177.

69. Lotka, Alfred J., 1925, *Elements of Physical Biology*: Baltimore, Williams ı Wilkins Co., 434 p.

70. Pearl, Raymond, 1925, *The Biology of Population Growth*: New York, Alfred A. Knopf, 213 p.

71. Pehrson, Elmer Walter, 1959, *Man and Raw Materials*: American Society for Testing Materials, Edgar Marburg Lecture, 1958, 18 p.

72. Pehrson, Elmer Walter, 1962, *Mineral Supply*: International Science and Technology, February, p. 23–27.
73. Verhulst, P. F., 1838, *Notice sur la Loi que la Population Suit dans son Accroissement*: Corr. Math. et Phys., T. S., p. 113–121.
74. Verhulst, P. F., 1845, *Recherches Mathématiques sur la Loi D'Accroissement de la Population*: Mémoires de L'Académie Royale, Belgique, T. XVIII, p. 3–38.
75. Verhulst, P. F., 1847, *Deuxiéme Mémoire sur la Loi D'Accroissement de la Population*: Mémoires de L'Académie Royale, Belgique, T. XX, p. 3–32.
76. Hubbert, M. King, *Energy Resources*, National Academy of Sciences-National Research Council, Washington (1962).
77. Hubbert, M. King, "Energy resources", Ch 8, *Resources and Man*, Freeman, San Francisco (1969).
78. Hubbert, M. King, *Nuclear energy and the fossil fuels*, Drilling and Production Practice (1956) 7.
79. Hubbert, M. King, *Degree of advancement of petroleum exploration in United States*, Am. Assoc. Petroleum Geologists Bull. **51** 11 (1967) 2207.
80. Pow, J. R., Fairbanks, G. H., and Zamora, W. J., "Descriptions and reserve estimates of the oil sands of Alberta," Ch 1, *Athabasca Oil Sands* (Clark, K. A., Ed.) Information Ser. **45** (1963) 1.
81. Duncan, D. C., and Swanson, V. E., *Organic-rich Shales of the United States and World Land Areas*, U.S. Geol. Surv. Cir. 523, Washington (1965).
82. Daniels, Farrington, *Direct Use of the Sun's Energy*, Yale, New Haven and London (1964).
83. Federal Power Commission, World Power Data, 1967, Washington (1969).
84. Trenholm, N. W., *Canada's wasting asset—tidal power*, Elect. Eng. News **70** 2 (1961) 52.
85. Bernshtein, L. B., *Tidal Energy for Electric Plants* [English trans. of 1961 Russian ed.], Israel Prog. for Sci. Translations, Jerusalem (1965).
86. White, Donald E., *Geothermal Energy*, U.S. Geol. Surv. Circ. 519, Washington (1965).
87. U.S. Atomic Energy Commission, Annual Report to Congress for 1969, Washington (1970).
88. European Nuclear Energy Agency and International Atomic Energy Agency, Uranium Production and Short Term Demand January 1969, Paris and Vienna (1969).
89. Faulkner, Rafford L., *Uranium Supply and Demand*, Address before Uranium Committee, Am Mining Cong., San Francisco, Oct. 19, 1969, Atomic Energy Commission, Washington (1969).
90. Nininger, Robert D., *World Uranium Picture*, presented before Colorado Min. Assoc., Denver, Feb. 13, 1970. Atomic Energy Commission, Washington (1970).
91. McKelvey, V. E., and Duncan, D. C., "United States and world resources of energy," *Proc. Third Symposium of Development of Petroleum Resources in Asia and the Far East*, Mineral Resource Ser. No. 26, **II** (1969) 9, United Nations, New York.
92. Swanson, V. E., *Oil Yield and Uranium Content of Black Shales*, U.S. Geol. Surv. Prof. Paper 356A, U.S. Geol. Surv., Washington (1960).
93. Kesler, Thomas L., "Lithium raw materials," Ch. 24, *Industrial Minerals and Rocks* (Gillson, Joseph L., Ed.), Am. Inst. Min. Met. Petrol. Eng., New York (1960).
94. Kesler, Thomas L., *Exploration of the Kings Mountain pegmatitees*, Min. Eng. **13** (1961) 3.
95. Briefings Before the Task Force on Energy of the Subcommittee on Science; Research and Development of the Committee on Science and Astronautics, U.S. House of Representatives, 92nd Congress, Second Session, U.S. GPO 79-837-0, Washington, 1972.

3

FOSSIL FUELS ARE NOT JUST FOR BURNING

Victor John Yannacone, jr.

§3.1. Introduction. In today's energy crisis, most of the nation's attention is focused on the use of crude oil or natural gas as fuel. Will there be enough gasoline? How about fuel oil for heating? Will electric utilities be able to generate power? This is understandable since 94 per cent of these non-renewable natural resources are consumed by merely burning them for power or heat.

We tend to forget that an ever increasing proportion of our oil and gas is used to make many of the man-made materials essential to modern life. About 5% of the oil, and 10% of the gas, goes into the manufacture of petrochemicals many of whose names are unknown to most consumers. Key petrochemicals include ethylene, propylene, benzene and naphthalene, ammonia, methanol and acetylene.

These "hidden products" from petroleum are the raw materials for many of the industries which represent a vital part of the economy. Some of these man-made products derived from crude oil and natural gas were created during World War II to meet emergency demands. Others have been developed since to improve products and to supplement a growing shortage in natural materials such as hardwood, copper, wool or zinc.

There were no petrochemicals before 1920 aside from carbon black, which has been made in the United States from natural gas since 1872. The growth of petrochemicals since 1920 (Fig. 3.1) has been spectacular due mainly to the following factors:

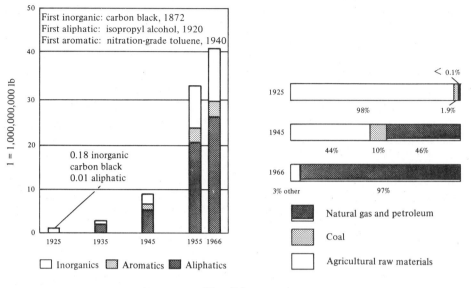

Fig   3.1.

1. The abundance and relatively, perhaps artificially, (See chapter 9) low cost of crude oil, natural gas, and natural gas liquids.

2. Advances made in the technology of petroleum refining, spurred especially by the demand for automobile gasoline and aviation fuel. These advances include more efficient fractional distillation and other separation processes, particularly for lower-boiling constituents, and the development of conversion processes to increase gasoline quality and quantity: including thermal and catalytic cracking, hydrocracking, catalytic reforming, hydrogenation, dehydrogenation, isomerization, and alkylation.

3. A demand for chemicals which in many cases could not be supplied in sufficient amounts, or with sufficient assurance of a steady supply and price structure, from other sources—coal, wood, or agricultural products.

4. The rapid growth in demand for synthetic fibers, plastics and resins, and protective coatings.

5. Research leading to products not before known commercially and to lower-cost processes for established products, often greatly expanding their application. A special characteristic of petrochemicals research has been the development of a series of related derivatives from the primary petrochemicals in order to establish the widest possible market.

§3.2. Petrochemical Manufacture. Petrochemical manufacture began during the 1920's. With the development of more efficient refining methods, particularly in the areas of distillation and thermal and catalytic cracking, researchers sought ways to process the by-products of these operations into useful chemicals. They went to work on the problem and found that ethylene, propylene, and a certain type of alcohol could be made by rearranging the carbon and hydrogen atoms of refinery gases.

Then came World War II, and petrochemical production received further stimulus. Plants sprang up almost overnight to produce the chemical ingredients for TNT, ammonium nitrate explosives, and synthetic rubber. With the end of the war, the boom continued. Today, about 3,000 different chemicals are processed from petroleum and natural gas and petrochemical manufacture is now a major industry.

Synthetic rubber was the first giant among man-made materials, a necessity created by World War II when natural rubber supplies to the United States were cut off. Today most of the world's rubber is synthetic. Very litttle still comes from rubber trees.

This evolution brought other benefits, above and beyond availablity. The first synthetic rubber, styrene-butadiene, increased the life of all rubber products. Introduction of polybutadiene meant more wear in tire treads. Marketing of polyisoprene offered extra heat resistance for heavy-equipment tires. Butyl rubber added new life to inner tubes.

Today virtually all passenger car tires are made from synthetic rubber. So is the chemical-resistant hose used in the radiators, power steering and brake systems of most vehicles.

Without synthetic rubber, the seals on washing machines would soon wear out. Millions of bathroom faucets would drip and waste water as their washers deteriorated. Vacuum cleaners and sewing machines would stop as drive belts broke.

It might be possible to return to natural rubber if absolutely necessary, but not easily. It takes about seven years for a rubber tree to start production. And those trees grow in Malaysia, Indonesia and Indo-China.

While the first plastic, invented in 1868, was made from cotton and camphor, plastics are now virtually all derived from oil or gas.

Today there are more than 40 chemical families of plastics. The largest volume consumed is polyethylene, an excellent electrical insulator which resists attack by chemicals. Next in commercial importance is vinyl plastic which is unaffected by water and resists damage from sunlight. Then there are polystyrene and ABS plastics which are tough and rigid with a glossy surface.

The absence of plastics would cripple communications. Construction would also suffer since all plywood is held together with plastic resins, as is the bulk of home insulation.

These are just a few examples of how plastics serve our needs.

Replacing plastics with other materials does not offer much in the way of a practical solution. One alternate material is synthetic rubber, but that, too, depends on petrochemicals. There is also a growing shortage of the natural materials that plastics have replaced in our society, and the manufacture of some alternate materials requires more energy than is needed to make most plastics.

The ability of the American farmer to produce more food and cotton per acre than any other farmer in the world is greatly dependent upon agricultural petrochemicals — fertilizers and pesticides.

In addition to the products already covered, petrochemicals are the raw materials for thousands of other products on which we have come to depend. The list is almost endless, including: hydraulic fluids, solvents, refrigerants, detergents, films, aspirin.

The dependence of many of these products on our limited supply of coal, oil and natural gas is hidden from the consumer. Consider the antifreeze which keeps our car engines from freezing. All antifreeze today is based on petrochemicals. What did we do in the past? We poured wood alcohol or grain alcohol into the radiator. Unfortunately, these chemicals boiled off rapidly and many motorists ended up with cracked motor blocks.

Then there's paint. Synthetic resins, made from petrochemicals, make up 75 per cent of the paints and coatings we use. Vinyl latex and acrylic latex paints are all products of modern chemistry. Our automobile paints and appliance enamels also depend upon resins.

Without their modern coatings, metal washing machine parts would quickly rust under the constant attack by soaps, bleaches, detergents and water and our automobile junk pile would grow even faster.

The wax for candles, milk containers, and waxed paper comes from petroleum. Cold cream, hand lotion, lipstick, perfume, hair tonic, mineral oil, and ointments, all are made from or contain ingredients derived from petroleum. Almost all dry-cleaning fluids come from oil. And so does the pigment for black ink and the carrying fluid for printer's ink. Many insecticides are made from petroleum or use oil products as a base. Protective treatments for wool in carpets, for wool or cotton in clothing, and for leather goods come from petroleum. Materials for industry, from pipes and paints to molds and insulations, and for the home, from dishes and garden hoses to squeeze bottles and toys, are manufactured from petrochemicals such as polyethylene, derived from oil refinery gases. Polypropylene, a relatively new and very versatile cousin of polyethylene, not only is the base for similar articles but also is a raw material for many additional film, fiber, sheeting, and other plastic products.

From oil come household and industrial detergents; coating to protect fabrics against mildew; and the raw materials for such man-made fibers as nylon and Dacron.

The tremendous output of petrochemicals requires only a small fraction of our petroleum supply. Less than one half of 1% of the oil produced in the United States could supply all the rubber the nation is now using; another fraction of 1% would take care of the industrial alcohol requirements. It is estimated that the number of different molecules that can be separated or made from crude oil and natural gas may run to half a million, each of them a potential building block for some new product. The possibilities of crude oil and natural gas raw materials, not just fuels, have only begun to be realized.

§3.3. Petroleum refining. From the highway, an oil refinery appears to be made up mostly of large, chimney-like columns, towering cylindrical structures, squat brick buildings, and smaller pieces of apparatus, all connected by a maze of pipes, and operating day and night, month in and month out—a classic example of continuous-flow operation.

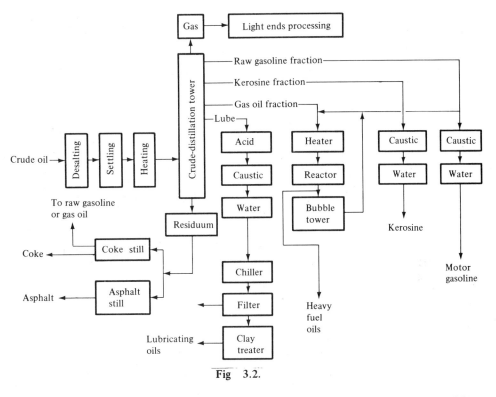

Fig 3.2.

There were 250 refineries operating in the United States on January 1, 1972. Refineries ranging in size from small plants capable of processing only 300 barrels of crude oil daily, to modern complex giants with daily crude capacities of more than 430,-000 barrels. At the beginning of 1972 the total crude oil capacity of all United States operating refineries amounted to more than 13 million barrels daily.

Fig 3.3.

Many factors influence the choice of a refinery location. Some refineries are built close to oil fields; others are constructed near large consuming areas. Frequently a controlling factor is easy access to water transportation.

§3.4. Chemistry of Hydrocarbons. To understand some of the processes of modern refining, it is necessary to know something about the nature of crude oil.

Crude oil, as it is delivered to the refinery, is a mixture of thousands of different hydrocarbons—compounds of hydrogen and carbon. The mixture varies widely from one oil field to another. Not all hydrocarbons are present in every crude.

The assortment of hydrocarbons in a crude oil and the proportions in which they are mixed determine its particular character and type. Crude oils are generally classified into three basic types. *Paraffin base* crude oils contain a high degree of paraffin wax and little or no asphalt. Besides wax, they also yield large amounts of high-grade lubricating oil. *Asphalt base* crudes contain large proportions of asphaltic matter, and *mixed base* crudes contain quantities of both paraffin wax and asphalt.

This is why crude oils do not always look alike. Some are almost colorless, or pitch black. Others can be amber, brown or green. They may flow like water or creep like molasses. Some crudes, containing relatively large amounts of sulfur and other mineral impurities, are called "sour" crudes. Others, having a fairly low sulfur content, are called "sweet" crudes.

Crude oil's basic unit is a molecule of one carbon atom linked with four hydrogen atoms. This is the molecule of methane, or marsh gas. Theoretically, millions of variations on this are possible, and millions of different hydrocarbon compounds can be formed.

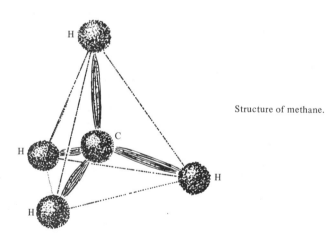

Structure of methane.

This is where the petroleum chemists step in. It is their job to rearrange and juggle the number of atoms to make new combinations which can open up a whole new world of products.

The refining process performs on a large scale what the chemist has already done in the laboratory. By a series of processes, crude oil is separated into various hydrocarbon groups. These are then combined, broken up or rearranged and, perhaps, other ingredients added.

Among the several refining operations known today, there are three major processes: *separation, conversion,* and *treating.*

§3.5. Separation. The most common separation processes are *solvent extraction, adsorption, crystallization* and *fractional distillation,* of which the last is the most important.

In *solvent extraction,* different components of a mixture are separated from one another by a liquid solvent that dissolves certain compounds but not others. *Adsorption* is quite similar to solvent extraction. The important difference is that a solid, rather than a liquid solvent is used.

The solid substance must be porous in order to adsorb, or hold, the undesired petroleum components on its surface. In *crystallization,* cooling the mixture causes some of its compounds to solidify or crystallize and separate out of the liquid.

§3.6. Fractional distillation: the fundamental process of refining. In distillation, the volatility characteristics of hydrocarbons are particularly important—that is, the ease with which the liquid vaporizes. Volatility depends on the boiling points of the various hydrocarbon compounds of which the crude is composed. Hydrocarbon boiling points range from over 250°F below zero up to several hundred degrees Fahrenheit above zero. Some crudes contain significant amounts of material boiling above 1400°F. Because different hydrocarbon compounds have different boiling points, they condense at different temperatures.

In the early days, the various hydrocarbons were separated by putting crude oil into a tank still and applying enough heat to vaporize part or all of it. The vapors were then condensed and collected as liquids.

As the demand for petroleum products increased, this simple "batch" process proved to be too slow and inflexible. In its place, refiners developed the pipe still. This was a continuous process in which crude oil was heated in pipes running through a furnace. Modifications of this process are still used today.

In modern refineries, crude oil is pumped through rows of steel tubes inside a furnace, where it is heated to as much as 725°F depending on the crude type. The resulting mixture of hot vapors and liquid passes into the bottom of a closed, vertical tower, sometimes as high as one hundred feet. This is a fractionating, or "bubble" tower.

As the vapors rise, they cool and condense at various levels in the tower. The liquid "residue" is drawn off at the bottom of the tower to be used as asphalt or heavy fuel. Higher up on the column, lubricating oil is drawn off at a lower temperature. Next come fuel oils, including gas oil, light heating oil, and light diesel fuel at still lower temperatures. Kerosine condenses still higher in the column and gasoline condenses at the top. Those gases which do not condense are carried from the top of the column.

The condensed liquids are caught by a number of horizontal trays, placed one above the other, inside the tower. Each tray is designed to hold a few inches of liquid, and the rising vapors bubble up through the liquids. At each condensation level, the separated fractions are drawn off by pipes running from the sides of the tower. The fractions obtained by this distillation process are known as "straight run" products.

§3.7. **Conversion.** At the beginning of the 20th century, the market for petroleum products began to change radically. Automobiles became more popular, and the demand for gasoline increased. But only a relatively small amount of gasoline can be distilled from the average crude oil. So refiners had to find some way of producing more gasoline from each barrel of crude processed. This problem was overcome by the development of conversion processes which enabled refiners to produce gasoline from groups of hydrocarbons that are not normally in the gasoline range.

These basic conversion processes are *thermal cracking, catalytic cracking,* and *polymerization.*

§3.7.1. **Thermal Cracking.** William M. Burton, a young chemist, introduced the thermal cracking process in 1913. With this process, the less volatile heating oil fractions are subjected to higher temperatures under increasing pressure. The heat puts a strain on the bonds holding the larger, complex molecules together, and causes them to break up into smaller ones, including those in the gasoline range.

With this discovery, refiners also found that cracking not only increased gasoline *quantity* per barrel of crude oil processed, but also produced a substantial improvement in its *quality.* The product obtained by thermal cracking was found to be far superior in antiknock characteristics than the gasoline obtained by straight distillation.

§3.7.2. **Catalytic Cracking.** Thermal cracking awakened refiners to what could be done by altering the petroleum molecule. This led to extensive probing into the physical and chemical properties of hydrocarbons.

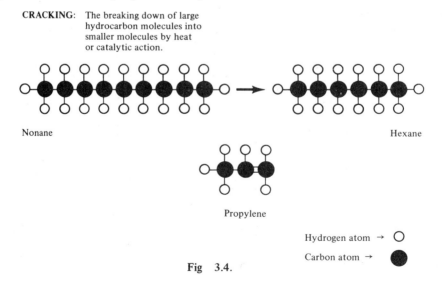

CRACKING: The breaking down of large hydrocarbon molecules into smaller molecules by heat or catalytic action.

Nonane

Hexane

Propylene

Hydrogen atom →

Carbon atom →

Fig   3.4.

Catalytic cracking was the next great advance. It was brought to the United States in 1937 by Eugene Houdry, a Frenchman.

Catalytic cracking does essentially the same thing as thermal cracking, but by a different method. A catalyst is a substance that causes or accelerates chemical changes without itself undergoing change. Unfinished heating oils are exposed to a fine granular catalyst and the result is the break-up of these heavier hydrocarbons into light fractions, including gasoline. The catalyst enables this break-up to be accomplished with only moderate pressure.

Catalytic cracking produced a gasoline with an even higher octane (antiknock) rating. This became extremely important in meeting the special needs of World War II. Today there are many versions of this process, using catalysts in the form of beads, powders, or pellets. Such catalysts range from aluminum and platinum to acids and processed clay.

§3.7.3. Polymerization. Polymerization is the reverse of cracking. Developed during the late 1930's to utilize refinery gases which were often wasted or burned as fuel, it is a method of combining smaller molecules to make larger ones.

Polymerization: The linking of similar molecules; joining together of light olefins.

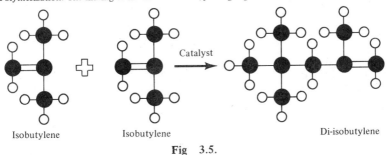

Isobutylene   Isobutylene   Di-isobutylene

Fig   3.5.

These gases are now subjected to controlled high pressures and temperatures in the presence of a catalyst, which forces them to unite, or polymerize, and form liquids called polymers. Polymers are essential components of high-octane motor and aviation fuels.

Other refining methods which alter the structure of hydrocarbons are also widely used today. *Alkylation, isomerization,* and *catalytic reforming* are three such methods which rearrange petroleum molecules to form high-octane products.

Alkylation: The union of an olefin with an aromatic or paraffinic hydrocarbon.

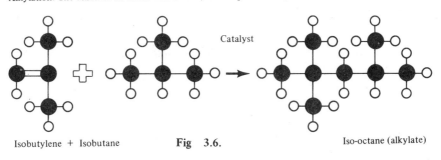

Isobutylene + Isobutane   **Fig   3.6.**   Iso-octane (alkylate)

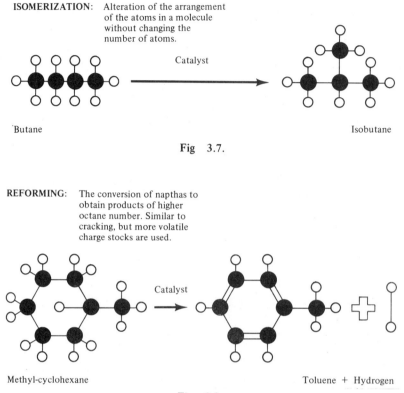

ISOMERIZATION: Alteration of the arrangement
of the atoms in a molecule
without changing the
number of atoms.

Catalyst

Butane

Isobutane

Fig   3.7.

REFORMING: The conversion of napthas to
obtain products of higher
octane number. Similar to
cracking, but more volatile
charge stocks are used.

Catalyst

Methyl-cyclohexane

Toluene + Hydrogen

Fig   3.8.

As a result of all of these advances, petroleum refiners in 1973 can now obtain an average of 19 gallons of gasoline from each 42-gallon barrel of crude oil refined. Far higher yields could readily be obtained if the market called for more gasoline relative to fuel oil and other petroleum products. In 1920, the gasoline yield from each barrel of crude was only 11 gallons. If no improvements had been made in gasoline refining techniques since 1920, refiners would have to process over 3.5 billion barrels of additional crude oil per year to meet the gasoline requirements of 1973.

§3.8. Treating. The crude oil discovered in Ohio during the mid-1880's was at first considered almost useless due to its high sulfur content. The kerosine it yielded under normal distillation smelled and smoked heavily. Somehow the sulfur and impurities had to be removed before the "sour" crude oil could find a market.

A chemist named Herman Frasch found the answer. He treated Ohio kerosine with a copper oxide and successfully removed the sulfur. Thus Frasch laid the foundation for "sweetening" the sulfur-bearing crudes found in many parts of the world today.

Frasch's experimentation led to a wide group of processes for eliminating impurities in petroleum products. Other sweetening compounds have been developed to change corrosive or odorous impurities into harmless, odorless substances. Acid treatments remove offensive materials from products, improving their resistance to deterioration.

There has been a trend, which is accelerating, to remove or reduce sulfur and sulfur compounds from motor gasoline, aviation fuels, diesel fuels and burner fuels. Sulfur compounds in gasoline limit the effectiveness of additives, especially the anti-knock additive. Sulfur compounds in aviation and diesel fuels increase wear and maintenance problems. Sulfur compounds in burner fuels affect operation but, more importantly, they must be removed in response to the need to reduce sulfur levels in the atmosphere.

§3.8.1. Petrochemical sulfur.

§3.8.1. Petrochemical sulfur. Sulfur is obtained by the oxidation of hydrogen sulfide, which occurs in some natural gas and most refinery gases, particularly in the off-gas from processess to reduce sulfur content of petroleum liquids (for example, hydrodesulfurization).

Sulfur in crude oil and natural gas is an impurity which is removed by various treating processes. Once "scrubbed" from petroleum, sulfur has many valuable uses, either in its natural state or as sulfuric acid. Among them are its use in petroleum refining and in the manufacture of paper, steel, synthetic fibers, pharmaceuticals, antifreeze, dyes, explosives, snythetic detergents, fertilizers and, of course, matches.

Sulfur production in all forms is about 9,300,000 long tons annually. Petrochemical sulfur from natural and refinery gas is nearly 1,300,000 long tons. This is about 14% of all sulfur production in the United States. Almost 86% of all sulfur is consumed as sulfuric acid in the manufacture of fertilizer, other chemicals, and steel. The remaining 14% is used in the pulp and paper industry, the manufacture of carbon bisulfide, or as ground or refined sulfur.

§3.8.2. Fuel oil desulfurization.

§3.8.2. Fuel oil desulfurization. With natural gas in short supply, untreated coal often inadequate to meet environmental requirements, coal desulfurization and coal gasification years away from significant commercial development, *oil hydrodesulfurization* a present reality, capable of providing desulfurized fuel oil appears to be a short-term solution to the energy-environmental dilemma, particulary in large urban centers.

As the nation approaches its deadline for environmental regulations and the demands for "clean" energy compound an already serious energy shortage, the important role that oil and hydrodesulfurization processes must play is becoming apparent. Thus, an examination of the oil situation, the capabilities of its desulfurization technology, and the quantity of its associated sulfur production is important to fully understand the effects they can make to an acceptably balanced energy-environmental position.

§3.8.3. Hydrodesulfurization.

§3.8.3. Hydrodesulfurization. The basic hydrodesulfurization scheme is shown in Figure 3.9. A high sulfur oil feedstock is pumped to an elevated operating pres-

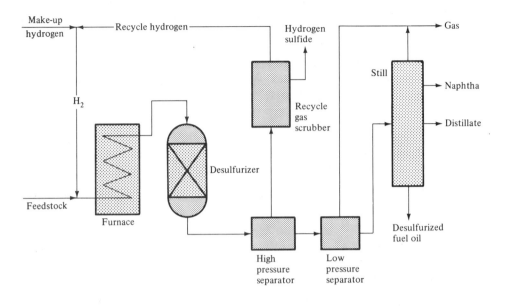

Fig   3.9.

sure. Hydrogen is added, the mixture is heated and sent to a desulfurizer containing a catalyst. In the desulfurizer hydrogen reacts with the sulfur in the oil to produce hydrogen sulfide. The desulfurized oil containing the hydrogen sulfide proceeds to a high pressure separator where the excess hydrogen and the hydrogen sulfide produced in the desulfurization are bled-off to a gas scrubber. In the scrubber hydrogen sulfide is separated from hydrogen. The hydrogen sulfide, is further processed to elemental sulfur in a Claus unit. The purified hydrogen is recycled to enter the process again with new feedstock. Makeup hydrogen equivalent to that utilized in the desulfurizer is added to the recycle stream.

From the high pressure separator, the desulfurized oil is sent through a low pressure separator where light hydrocarbon gases evolve. The remaining desulfurized oil may be utilized directly or sent to a still for separation into various fractions.

§3.8.4. Hydrodesulfurization technology; types of processes. Processes for the desulfurization of heavy fuel oil can be categorized as either direct or indirect. The feedstock for the direct process can be a whole crude, topped crude, an atmospheric residuum, or a vacuum residuum.

A typical material balance for a direct process desulfurizing a 4% atmospheric residuum is shown in Figure 3.10A. About 75% of the sulfur in the oil can be removed, and a fuel oil containing 1% sulfur can be produced.

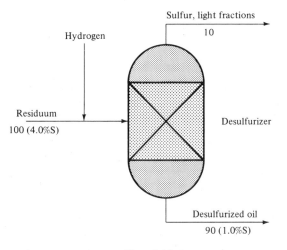

Fig   3.10.A

The indirect process for desulfurization is very similar to the direct process except that it is applied to a vacuum gas-oil fraction after which the desulfurized product is blended with the vacuum residuum. Figure 3.10B illustrates a typical material balance for an indirect process. An atmospheric residuum containing 4% sulfur is distilled in a vacuum to separate a vacuum gas-oil fraction containing 2.9% sulfur and a bottom vacuum residuum containing 5.8% sulfur. The vacuum gas-oil is desulfurized with hydrogen to produce an oil containing 0.2% sulfur. The desulfurized vacuum gas-oil is blended with the vacuum residuum to produce a fuel

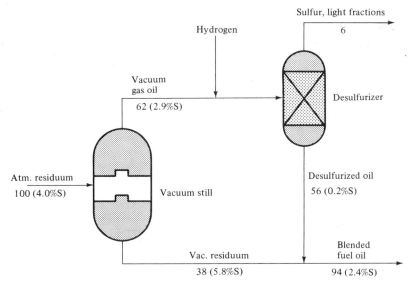

Fig   3.10.B

oil of 2.4% sulfur content. Since the sulfur level of the blended oil would be unacceptable in most locations, a number of other alternatives exist for handling the oils. The desulfurized vacuum gas-oil containing 0.2% sulfur could be marketed separately as a very low sulfur fuel. The high sulfur vacuum residuum could be sent to a coker for further processing into coke and oil, or the vaccum residuum could be blended with a low sulfur crude oil to produce an acceptable fuel oil. Finally, the vacuum residuum could be desulfurized separately and then blended with the desulfurized vacuum gas-oil.

Most hydrodesulfurization processes feature a fixed-bed type of catalytic reactor where liquid feed is brought into contact with a hydrogen rich recycled gas under conditions of elevated pressure and temperature. Generally, the operating pressure ranges from 800 to 3000 pounds per square inch (psi) in the direct processes which refine atmospheric residuum or vacuum tower bottoms. The indirect processes fed by vacuum gas-oil fractions, generally operate at 800 psi or lower. Operating temperatures normally range from 650° to 850°F. The temperature in the fixed bed reactor is controlled by injecting recycled hydrogen at various levels along the bed to act as a quench. Hydrogen consumption can range from 250 to 950 standard cubic feet per barrel of oil depending on the extent of sulfur removal required and the other impurities of the feedstock.

The catalyst is key to most of the processes and tailored for a specific oil. Some catalysts can tolerate deposition of large quantities of heavy metals and still maintain high desulfurization activity. Other catalysts need plugging-protection against suspended salts and particulates present in the feedstock. To protect the catalyst, these sediments are removed prior to entering the desulfurization reactor.

Cycle-length of the catalyst typically ranges from six to twelve months in fixed bed reactors. The period is determined by the initial catalytic activity, the extent of degradation caused by the metal content in the feed, and the severity of the operating conditions needed to achieve the required level of desulfurization.

The *H-Oil ebulating reactor bed* is said to overcome the problems encountered in hydrodesulfurization of residual oil in fixed-bed catalytic reactors. In the H-Oil process the feedstock is mixed with the hydrogen gas and both enter the bottom of the reactor. They pass through a distribution plate to evenly spread the oil and gas over the cross section of the reactor. The catalyst comprising the bed may be either in for form of 1/32 inch extrudates or a fine powder. The upward liquid velocity in either case is sufficient to keep the catalyst expanded and moving. The circulating motion eliminates the need for adding hydrogen quench at various levels along the reactor as is done with fixed bed units. Also any solids in the feed pass through the expanded bed. Thus, the reactor does not plug. Pressure drop experienced in commerically operating units has been constant for a period of a year. The catalyst is replaced continuously in small quantities to maintain a steady-state activity. Uniform catalyst activity can be maintained while processing a feedstock of varying metals content.

The trend in desulfurization technology indicates that improved sulfur levels are possible at design pressures of 400 to 800 psi and extended operating cycles of up to three years rather than the one year now common for catalyst regeneration.

Hydrogen partial pressure instead of total reactor pressure enhances the catalyst activity. Thus, the hydrogen-rich gas stream from the high pressure separator is treated in a scrubber to remove the hydrogen sulfide prior to recycling hydrogen back to the reactor. Hydrogen sulfide is regenerated from the scrubber liquor and can be processed into elemental sulfur in a Claus unit.

§3.9. Petrochemical Building Blocks. The family of petrochemicals consists of both organic and inorganic chemical compounds. Organic chemicals, which contain atoms of both hydrogen and carbon, trace their origin to living matter—the remains of plants and animals from which all hydrocarbons are derived. Inorganic chemicals are those whose origin is non-living.

Organic petrochemicals are classified into two basic types according to the molecular arrangement of their carbon atoms. The *aliphatics* are the straight-chain type—their carbon atoms are lined up in a row. The *aromatics* are the ring type—their carbon atoms are in a circular arrangement.

§3.10. The Aliphatic Petrochemicals. The most important aliphatic building blocks are ethylene, butylene, acetylene, and propylene. Hydrocarbons from these gases are chemically synthesized into a myriad of end products—from plastics and synthetic rubber to drugs and detergents.

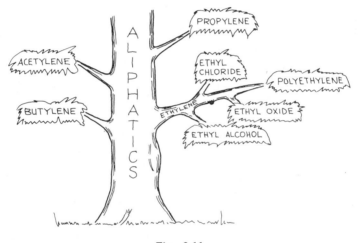

Fig   3.11.

From ethylene are derived ethyl alcohol, used in making acetaldehyde (an industrial chemical), solvents, pharmaceuticals, cleaning compounds, and polyethylene plastics, which have found applications in packaging, the manufacture of pipe, construction materials, and a host of household items. Ethylene oxide is also derived from ethylene gas and is used in making ethylene glycol—a prime ingredient in the production of automobile antifreeze, synthetic fibers, and films. Ethyl chloride is used to manufacture the gasoline anti-knock additive, tetraethyl lead.

Three of the four basic types of synthetic rubber have as their principal ingredient butadiene, a derivative of butylene. They are: GR-S or SBR, the all-purpose rubber which accounts for about 60 per cent of all synthetic rubber produced in this country, and which closely resembles natural rubber in general handling and wearing qualities; butyl, which is used in inner tubes because it holds air better than natural rubber; and butadiene-acrylonitrile, which resists oil and aromatic solvents, molds well, and is invaluable as a plasticizer. The fourth basic type of synthetic rubber is neoprene which is derived from acetylene. It resists cracking when exposed to sunlight, is resistant to oils, fats, water, and fire, and is easily processed.

Combined, these four basic types of synthetic rubber in 1973 accounted for 77 per cent of all new rubber used by industry, compared to natural rubber's share of somewhat less than 23 per cent.

Besides serving as the basis for neoprene rubber, acetylene has many other applications. This petrochemical building block, which is primarily derived from methane gas, is used in the manufacture of paints, adhesives, fibers and solvents.

Propylene, the fourth major aliphatic, is the basic ingredient for one of the newcomers to the plastics scene, polypropylene, one of the lightest and strongest of the petrochemical-based plastics, which also resists heat and chemical attack. Its principal use so far has been in the manufacture of plastic bottles. Other end uses of propylene include solvents, resins, drugs, drycleaning fluids, antifreezes, detergents, hydraulic fluids, and plasticizers.

Most ethylene is manufactured from thermal cracking of ethane and propane recovered from natural gas. Some is obtained from refinery thermal and catalytic cracking operations conducted primarily to increase the quantity and quality of gasoline produced, and some from thermal cracking of higher-boiling liquids, such as naphtha, gas oil, and natural gasoline.

Alternative routes for the production of certain chemicals including carbon black, ammonia, methyl chloride, ethylene oxide and ethyl alcohol, exist. For example, vinyl chloride and vinyl acetate are produced commercially from both acetylene and ethylene and acrylonitrile from acetylene and propylene. Dovetailing operations are illustrated by the synthesis of urea using carbon dioxide, coproduced with the hydrogen required to manufacture ammonia, and by the production and simultaneous utilization of hydrogen chloride in the manufacture of ethyl chloride and vinyl chloride.

Propylene is produced in large amounts from petroleum refining operations and as a coproduct in the manufacture of ethylene. The major use of propylene is still in the manufacture of high-octane gasoline components by alkylation and polymerization. However, since substantial quantities of propylene are required in the manufacture of other petrochemicals, it is expected that the share of propylene consumed in making gasoline will probably shrink in the future as chemical uses continue to grow rapidly. The principal petrochemicals derived from propylene are shown including polypropylene acrylonitrile, cumene, and propylene oxide. C_4 hydrocarbons (derivatives of butane and the butylenes) are available from natural gas liquids and petroleum refining operations. The saturated C_4 hydrocarbons, n-butane and isobutane, are primarily derived from natural gas and gasoline. Butanes are also formed during petroleum refining operations, and isobutane is intentionally produced by the isomerization of n-butane. The unsaturated C_4 hydrocarbons (butylenes) are formed as by-products during gasoline manufacture, and they are intentionally manufactured by dehydrogenation of the saturated C_4 hydrocarbons. Butadiene is the most important $C \times 4$ chemical derivative. The n-butenes are also used to make methyl ethyl ketone, an important solvent. Isobutylene is used in the manufacture of butyl rubber and polybutene, an important lubricating oil additive.

The oxo process (reaction of olefins with carbon monoxide and hydrogen) is an increasingly important process for the manufacture of alcohols. It provides aldehydes from olefins of one less carbon number, and the aldehydes may be converted by hydrogenation to their corresponding alcohols. The principal alcohols obtained are butanols and 2-ethylhexanol from propylene; branched-chain octyl alcohols from the heptene copolymer of propylene and butylene branched-chain alcohols from the heptene copolymer of propylene and butylene branched-chain alcohols from diisobutylene: and branched-chain decyl and tridecyl alcohols from propylene trinamer and tetramer, respectively.

§3.10.1. Methanol. Hydrogen has been suggested as a universal, nonpolluting fuel, since it can be produced from water and burns cleanly to water.

No consideration of petrochemicals can ignore methanol and the following sections have been adapted from a significant article in *Science* (**182**, 1299-1304, 28 December 1973) by Dr. T. B. Reed of the solid state division and Dr. R. M. Lerner of the communications division at the Lincoln Laboratory, Massachusetts Institute of Technology. For more detailed information on methanol, that article should be consulted together with the references cited by the authors.

Methanol, CH_3OH, can be thought of as two molecules of hydrogen gas made liquid by one molecule of carbon monoxide. It thus shares many of the virtues of pure hydrogen. It can be made from almost any other fuel—natural gas, petroleum, coal, oil shale, wood, farm and municipal wastes—so that a methanol economy would be flexible and could draw from many energy sources as conditions change. Methanol is easily stored in conventional fuel tanks trucks, and tankers; it can be transported through oil and chemical pipelines. Up to 15% of methanol can be added to commercial gasoline in cars now in use without engine modification. This

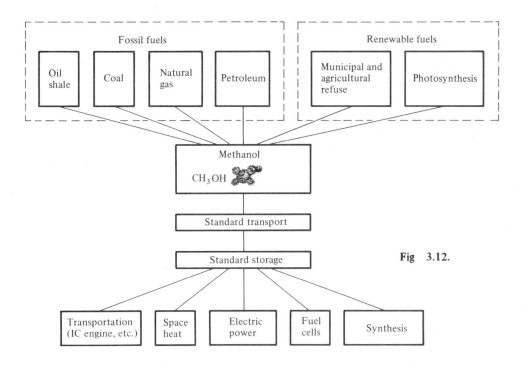

Fig 3.12.

methanol-gasoline mixture results in improved economy, lower exhaust temperature, lower emissions, and improved performance, compared to the use of gasoline alone. Methanol can also be burned cleanly for many other fuel needs, and it is especially suited for use in fuel cells.

Methanol, which is also called methyl alcohol, wood alcohol, or methylated spirits (7), is a colorless, odorless, water-soluble liquid.

It freezes at −97.8°C, boils at 64.6°C, and has a density of 0.80. It is miscible with water in all proportions, and spillages are rapidly dispersed. Methanol burns with a clean blue flame and is familar to most people as the alcohol used for heating food at the table or as the alcohol in Sterno. Mixtures with between 6.7% and 36% of air are flammable. The auto ignition temperature of methanol is 467°C, which is high compared with 222°C for gasoline This may account for the high octane number, 106, of methanol; a typical gasoline has an octane number of 90 to 100.

The energy content of a number of fuels is shown in Table 3.1 Hydrogen produces the most energy on a weight basis; hence its use in rockets where volume and cost are secondary considerations. Petroleum products such as gasoline have the highest energy of the fuels listed on a volume basis and are second highest on a weight basis; hence gasoline will long be preferred for airplanes where lowest weight is at a premium. However, of all the liquid fuels, methanol produces the second highest amount of energy on a volume basis.

Table 3.1 The energy content of some fuels, shown on the basis of weight and volume.

		Heat of combustion (low)*	
Fuel	Formula	Kjoule/g	Kjoule/cm^3
Liquids			
Hydrogen	H_2	124.7†	8.7
Methanol	Ch_3OH	20.1	15.9
Gasoline	C_8H_{38}	44.3	30.9
Solids			
Hydrides	VH_2‡	4.7	28.4
Coal	$C_{28}H_{.42}$	32.2	41.8
Wood	$C_{.32}H_{.46}O_{.22}$	17.5	14.2
Gases			
Hydrogen	H_2	124.7	0.0010
Methane	CH_4	61.1	0.0044

*Combustion to CO_2 and H_2O (gas).
†Conversion factors are: 0.948 kuoule = 1 Btu;
   2.10 kjoule/g = 10^{3}Btu/lb;
   0.27 kjoule/em^3 = 10^3 Btu/gallon;
   33.4 × 10^4 kjoule/em^3 = 10^6 Btu/ft^3.
‡ Vanadium hydride is given as an example.

Although methanol is not the cheapest fuel at this time, its properties make it competitive with the other fuels. Ethanol, which has many of the desirable fuel properties of methanol and could be used in many applications costs, in this country, about three times more than methanol, however in less industrialized countries, ethanol may be an attractive fuel because it can be produced from agricultural products by fermentation.

In the manufacture of methanol, the output of the plant can be increased by 50 percent if small amounts of other alcohols can be tolerated in the product. Such a mixture is called "methyl-fuel," and it contains more energy than pure methanol because of the presence of ethanol, propanol, and isobutanol. It can be produced in larger quantities at a lower price than pure methanol, and, in general, has superior properties as a fuel. Methanol and methyl-fuel can be considered synonymous in the fuel sense.

Although methanol is miscible with gasoline at room temperature, less than 10 percent is soluble in some gasolines at 0°C. However, the volatile constituents added to gasoline in cold weather to aid ignition increase this solubility. Also, the higher alcohols in methyl-fuel increase the solubility of methanol.

Methanol, although not highly toxic, can be lethal if ingested. Methanol vapors are also poisonous, but no more so than those of many other common substances. For example, the maximum allowable exposure to methanol vapor is 200 parts per million (ppm), while the value for ethyl alcohol is 1000 ppm; for benzene, 10 to 25 ppm; octane, 400 ppm (octane and benzene are typical constituents of gasoline); trichloroethylene, 100 ppm; and carbon tetrachloride, 10 ppm (36 Federal Register, 10504, 1971).

§3.10.2. Methanol; its manufacture. Methanol can be made from many sources.

Until about 1925 it was made (along with acetic acid and tars) by the destructive distillation of wood. Since that time, most methanol has been synthesized from CO and H_2.

$$CO + 2H_2 \rightarrow CH_2OH \text{ (gas)};$$
$$\triangle G + -90,800 + 229T \text{ (joule/mole)}$$

where $\triangle G$ is the free energy change and T is the temperature.

The CO and H_2 (synthesis gas) for manufacturing methanol can be obtained by partial oxidation of any carbonaceous fuel with oxygen or water. At present it is obtained almost exclusively from methane by partial oxidation with water, however, natural gas (methane) is already in short supply.

The natural gas which is flared (burned-off) at some middle eastern oil wells could be economically converted to methanol at the wellhead and shipped in conventional tankers to this country.

Methane gas is also produced biologically by the decomposition of natural wastes, such as pig and chicken manure and sewage. It has been claimed that such methane can be used for powering automobiles. It is also claimed that enough fuel could be made from this source to meet all present fuel needs in the United States, and that the use of such a process would reduce the problem of sewage and animal-waste disposal by at least 50%. In experiments with cars and trucks converted ot use methane, the U.S. General Services Administration has reported clean, reliable operation. However, the type of cylinder required to contain compressed gaseous methane severly limits the amount of fuel that can be carried; a six-cylinder sedan has a range of 80 km (50 miles), each cylinder measuring 6.7 m^3 and weighing 100 kg. Conversion of the organic wastes to methanol rather than methane would make this fuel source much more practical.

For the next few decades, coal is the most attractive candidate for methanol production. Coal has long been used for the production of synthesis gas, according to the endothermic reaction (with $\triangle$H being the heat change):

$$C + H_2O(gas) \rightarrow CO + H2\ 2$$
$$\triangle H = +\ 131.4\ kjoule$$

Although synthesis gas contains CO, which is poisonous, it is used for industrial power and for heating homes in many European cities without further conversion. Much work is in progress to develop methods to obtain methane and hydrogen from coal for use as pipeline gas. The same technology can be applied to the manufacture of methanol from synthesis gas. If lignite is used as the starting material instead of coal, the resulting ash may contain uranium equivalent to commercial uranium ore, as well as other valuable minerals such as molybdenum, vanadium, arsenic, manium, selenium, cobalt, and germanium. Efforts are being made to develop practical methods of gasifying coal in the gound, eliminating the need for strip mining and consequent land scrap destruction.

Some day we will run out of fossil fuels. By coupling the manufacture of methanol with the disposal of waste we could supplement our fuel supply and thereby prolong the existence of fossil fuels.

A recent patent describes an *oxygen refuse converter* that can dispose of our refuse and at the same time generate useful energy.

In a shaft furnace shown in Figure 3.13, unseparated trash or sewage sludge is fed into a hopper at the top. Low-cost oxygen (0.2 kg of O_2 per kilogram of refuse) is fed into this furnace near the bottom, creating a 1500°C zone that melts the metals and glasses found in refuse. These melts, drawn off as slag and metal, have 2% of the original refuse volume, while all other products are gaseous or water soluble. Carbon, burning in the high-temperature zone, produces CO, which rises through the furnace. The hot CO creates an intermediate-temperature zone where carbohydrates and plastics are broken down to a gas containing, typically, 47% CO, 28% H_2, 17% CO_2, and 5% CH_4 by volume. Finally, in the uppermost section the incoming refuse is dried as the gas mixture cools to about 100°C.

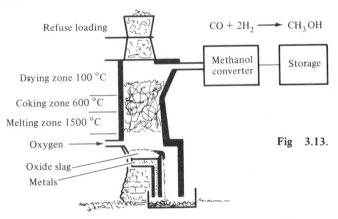

Fig 3.13.

This gas mixture stores 8.0 kjoule per gram of refuse (7×10^6 Btu per ton), or 76 percent of the original refuse energy. Because oxygen rather than air is used in burning, the output gas is high in heat content, low in volume, and relatively easy to scrub to remove fly ash and chlorine.

The United States produces about 1.8×10^{11} kg of solid refuse each year. The energy in the gas from this refuse is 1.4×10^{15} kjoule, or 2% of the 7.4×10^{16} kjoule (7.0×10^{16} Btu) consumed each year(32). If this gas were converted to methanol, it could supply about 8% of the fuel for our transportation needs. Although initially developed for refuse, converters of this type could also be used to convert farm waste and the waste from lumbering into more useful forms of energy such as methanol.

Forests, which are one means of capturing solar energy, represented the principal energy source for this country until about 1875.

Commercial forests now cover about 23% of the land area of the United States, or 2.1×10^{12} m^2. These forests intercept from the sun about 5.8×10^6 kjoule/m^2 per year, or a total of 1.2×10^{19} kjoule per year (33). If the conversion of solar energy with an efficiency approaching 1% could be achieved by improved forest management the annual energy harvest might be 1.2×10^{17} kjoule per year, more than our present energy needs of 7.4×10^{16} kjoule per year.

The advantage of utilizing forests for the production of methanol is that whole trees can be used, not merely those fractions that make good lumber or pulp. Calculations indicate that between 5 and 20% of our commercial forests operated as "energy plantations," could supply all of our electrical power.

§3.10.3. Historical uses of alcohol for fuel. During the last 50 years in the United States, methanol and other alcohols have not competed successfully with the abundant supplies of petroleum. Before this time, however, alcohols were used extensively as fuel. Alcohol became a popular fuel for lighting in about 1830, when it replaced malodorous fish and whale oils. In about 1880, kerosine replaced alcohol as a lighting fuel because of its sooty flame which gave more light; a clean flame produces no light without special additives. During the middle of the last century, France was partially on a methanol fuel economy. Wood was distilled in the provinces to give alcohol, which was burned in Paris for heating, lighting, and cooking. This was more economical than transporting wood to Paris and then disposing of the ashes.

During both World Wars I and II, when gasoline shortages occurred in Germany and France, vehicles of all sorts, including tanks and planes, used wood burners in the rear or in trailers. Wood chips were distilled to make alcohol vapors that included carbon monoxide and hydrogen; these vapors would (barely) drive the vehicle. In 1938, 9000 wood-burning cars were used in Europe. "Power alcohol" (ethanol) was also used by France and Germany to supplement gasoline supplies and stimulate alcohol production for anticipated use in munitions production.

In about 1920, manufacturers in the United States began to produce methanol for use as a solvent, for plastic manufacture, and for fuel injection in piston aircraft.

§3.10.4. Methanol and the internal combustion engine. It has been claimed that hydrogen is an ideal fuel for the internal combustion engine; (see §§2.29.1-3) and while hydrogen causes little pollution, it is difficult to store, costly, and difficult to burn efficiently in the engine without knocking and backfiring.

Methanol used as an additive or substitute for gasoline could immediately help to solve both energy and pollution problems.

A number of studies of methanol and ethanol have been conducted in the last 50 years to test their suitability as substitutes for gasoline in the internal combustion engine. Existing engines can be converted to use pure methanol by decreasing the ratio of air to fuel consumed from about 14 for gasoline to 6 for methanol, by recycling more heat from the exhaust to the carburetor, and by providing for cold starts. Compared with gasoline, the use of methanol in a standard test engine (without catalytic treatment of exhausts) yielded one-twentieth of the amount of unburned fuel, one-tenth of the amount of carbon dioxide, and about the same amount of oxides of nitrogen as gasoline. In these studies the reduced emissions were attributed to methanol being able to burn without misfire at an air-to-fuel ratio 25% higher than gasoline; exhaust temperatures with methanol were 100°C cooler; more spark retard was possible with methanol because of its higher flame speed. It was suggested that greater performance and economy could be expected in an engine designed specifically for methanol, and that such design should encompass higher compression ratios and a fuel injection system. In another study, on a one-cylinder research engine, it was found that 10 to 20% leaner mixtures could be tolerated with methanol than with gasoline. The amounts of unburned hydrocarbons, CO, and NO, produced were lower with methanol than with gasoline while the amounts of aldehydes produced were higher.

From these results it seemes clear that if gasoline becomes scarce or too expensive, we can design cars that will operate on pure methanol and cause less pollution. Specific fuel consumption would be higher on a weight or volume basis necessitating a larger fuel tank; but specific energy consumption (energy per kilometer) will certainly be lower because higher compression ratios and simpler pollution controls can be used.

The principal drawback to the immediate use of pure methanol as a gasoline number of unmodified private cars (year models 1966 to 1972) were tested and operated over a fixed course with varying concentrations of methanol. It was found that

    (i) fuel economy increased by 5 to 13%;
    (ii) CO emissions decreased by 14 to 72%;
    (iii) exhaust temperatures decreased by 1 to 9%; and
    (iv) acceleration increased up to 7%.

Methanol is said to have a "blending octane value" (BOV) of 130 (9, 10) defined by

$$BOV = [O_b - O_g(1 - x)]/x$$

where O_b and O_g are the octane numbers of the blend and the gasoline, and x is the volume fraction of methanol in the gasoline. From this, 10 percent of methanol added to gasoline with an octane rating of 90 would be expected to yield a fuel with an octane rating of 94, equivalent to the addition of 0.13 gram of tetraethyl lead per liter of gasoline. Ethanol has a BOV of 110 to 160, depending on the octane of the gasoline.

§3.10.5. Methanol; other uses. Although methanol has been suggested principally as a fuel for automobiles, it could also be used advantageously in most other fuel applications if it becomes sufficiently plentiful. It is a safe, clean fuel for home heating and can also be burned in power plants to generate electricity without polluting the atmosphere.

Methanol is one of the few known fuels suited to power generation by fuel cells. In principle, the fuel cell can convert chemical energy to electricity with much higher efficiencies than heat engines such as turbines. Although methanol is not as simple to use in a fuel cell as hydrogen, it can be stored and shipped more easily.

§3.11. The Aromatic Petrochemicals. Most of the organic chemicals thus far synthesized by chemists are derivatives of the aromatic class of organic petrochemicals, so named because of their rather pleasant odor. The unique ring structure of their carbon atoms makes it possible to transform aromatics into an almost endless number of chemicals.

The principal aromatics are commonly referred to as the *BTX group*—benzene, toluene, and xylene.

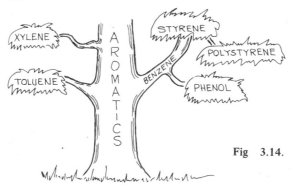

Fig 3.14.

Benzene is the primary member of the BTX group. Foremost among its uses is the making of styrene and phenol. From styrene come synthetic rubber, latex paints, paper coating, soles for shoes, and a variety of polystyrene plastic goods. Phenol, on the other hand, is used to make adhesives, insecticides, varnishes, resins, medicinal products, perfumes, and flavoring agents. Other direct uses of benzene include the manufacture of synthetic detergents, explosives, herbicides, plant hormones, dyes, photographic chemicals, and a host of other products.

Toluene's prime use during World War II was in the production of TNT and aviation gasoline. Today, it is largely used in the making of lacquer solvents, saccharin, aspirin, food preservatives, plastics, and the urethane foams used in protective padding, furniture cushions, insulation, and clothing.

The last member of the BTX group—xylene—finds its principal use in the manufacture of synthetic resins, vitamins, dyes, pharmaceuticals, paints, solvents, plasticizers, reinforced plastics, and synthetic fibers of the polyester type.

Up until 1940 the coal tar industry was the principal supplier of the aromatic compounds: benzene, toluene, xylenes, and their derivatives. Petroleum has become the chief source of cyclic organic compounds except for naphthalene, substantial amounts of which are still derived from coal tar.

Benzene, toluene, and xylenes are important high-octane components of gasoline. They are manufactured by catalytic reforming and recovered by distillation and extraction processes. Substantial quantities of toluene are hydrodealkylated to benzene. The more important of the cyclic derivatives are the high polymers (plastics and resins, fibers, and elastomers) with end products including nylon, polyester fibers and film, polystyrene, styrene-butadine rubber; epoxy resins, phenolic resins, and polyurethane from isocyanates.

Naphthenic acids are carboxylic acids of substituted cyclopentanes and cyclohexanes: they are present in crude oils. Owing to their oil solubility, naphthenates of appropriate metals are used as paint driers, fungicides, and lubricant additives. Naphthenic acid.

Cresylics (alkyl phenols, cresylic acids, and cresols) are made by refining by-product tar acids from coke ovens or gas works, and by recovery from refinery waste streams containing petroleum acids. The lower-boiling fractions of cracked petroleum contain xylenols and smaller amounts of cresols, ethylphenols, trimethylphenols, and methylethylphenols, which are produced at higher cracking temperatures. Cresols are also made synthetically by methylation of phenol to produce *meta-, para-,* and *ortho*-cresols and 2.6-xylenol. Their chief uses are in phenolic resins, wire enamel solvents, phosphate esters, ore-flotation reagents, oil and gasoline additives, antioxidants, and metal cleaners.

§3.12. The Inorganic Petrochemicals. 1 The principal inorganic members of the petrochemical family are ammonia, hydrogen, sulfur, and carbon black. Two— sulfur and hydrogen—are direct by-products of petroleum processing.

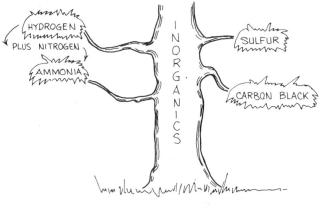

Fig 3.15.

§3.12.1. Hydrogen and Ammonia. Pure hydrogen, when drawn from a refinery's catalytic reforming unit, has many uses—in various refining processes, such as desulfurization and hydrogenation, as a liquid rocket propellant, in the manufacture of hydrochloric acid, hydrogen peroxide, and the hydrogenation of oils to make hard soaps, lard substitutes, varnishes, and lubricants. But it is when combined with

HYDROGENATION: The addition of hydrogen
to an olefin.

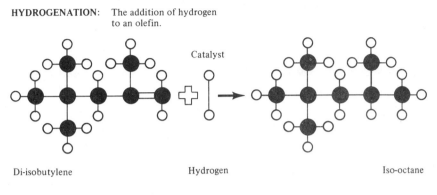

Di-isobutylene   Hydrogen   Iso-octane

Fig 3.15A.

nitrogen from the atmosphere, that hydrogen finds its greatest current use—the formation of ammonia, the third basic inorganic petrochemical. Combining the two elements to form ammonia requires intense heat, pressure, and a catalyst. Ammonia is initially produced from this reaction as a compressed gas which is then cooled to form a liquid. Most of the ammonia produced in this country goes into the manufacture of commercial fertilizers. The balance is used to make livestock feed supplements, insecticides, explosives, synthetic fibers, cleaning agents, pharmaceuticals, and for use in the processing of paper, pulp, rubber, and certain metals.

§3.12.2. Carbon black. Carbon black is produced by the burning of oil or natural gas in the presence of just enough oxygen to prevent all the carbon from being consumed. The end result is the blackest substance known to man. Its uses are wide and varied. They include the strengthening of rubber and cement; the manufacture of paint, ink, batteries, radio and television tubes, explosives, and steel alloys; and the polishing of lenses, prisms, and reflectors used in precision optical equipment.

§3.13. Petroleum Products. At the turn of the century, it was relatively simple to pinpoint the major markets of the petroleum industry. The principal uses were illuminants and lubricants—the major energy source was coal. Today, petroleum supplies the power for our transportation network, our factories and farms; the heat for our homes and large buildings; the lubricants to keep the wheels of industry turning; the material with which much of our highway, road, and street network is paved; not to mention the host of products derived from petrochemicals.

Approximately 3,000 products are currently produced wholly or in part from petroleum, in addition to the 3,000 or so petrochemicals.

Most crude petroleum is useful only as a raw material for the manufacture of a large number of products, such as fuels, lubricants, paving materials and base compounds for chemical manufacture. It is not unusual for a large oil company to list 200-1000 products for sale.

A normal crude oil will contain thousands of hydrocarbons, ranging in molecular weight from that of methane, CH_4, the chief constituent of natural gas with a molecular weight of 16 to those of asphaltic compounds with molecular weights of about 50,-000-100,000. In refining, such a crude oil is first distilled into fractions which cover its boiling point range. The lowest-boiling fractions are the normally gaseous hydrocarbons—methane, ethane, propane, and the butanes. These hydrocarbons are used in fuels, or as chemical raw materials, or as constituents of liquefied petroleum gas sold in cylinders.

The hydrocarbons from pentane to dodecane C_5 to C_{12} go into gasoline motor fuel, and this product, in all its varieties, takes a substantial part of the crude oil supply of the world. Gasolines for reciprocating engines cover a considerable range of chemical composition, with corresponding variations in properties, the more important being volatility and resistance to knocking. Minor uses of gasoline hydrocarbons are in cigarette lighter fluids, spot removers, dry-cleaner's naphthas, and other solvents.

Another substantial fraction of the crude oil is used for kerosine manufacture. The composition of the hydrocarbons in kerosine overlaps that of the heavy end of the gasoline fraction in hydrocarbon content, lying at about C_{10} to C_{11}: the hydrocarbons in conventional gas oil extend to about C_{20}: and heavy fuel oil may contain hydrocarbons up to C_{70} or C_{80}. Asphalts are compounds still more complex and essentially non-volatile substances together with complex compounds of sulfur, nitrogen, and oxygen.

The above major products make up 90% of the output of the oil industry. The remaining 10% is divided into a wide variety of materials, the most important being the lubricants. These extend from very light, low-viscosity oils employed for high-speed machinery, such as spinning and weaving equipment, through very heavy oils for reciprocating steam engines to semi-solid and even solid lubricating greases.

All petroleums except the condensates (very light crudes carried as vapor in deep gas reservoirs) contain lubricating materials, but the lower-grade crudes, for example, those of the Middle East, are contaminated with asphaltic materials and compounds of sulfur and nitrogen which cause difficulty in refining. Good lubricating stocks can be derived from such crudes with fair economy though the use of differential solvent extraction.

Before the development of solvent extraction methods, natural petroleums were favored for the production of lubricating oils, but they have undesirable characteristics which show up under severe operating conditions. Modern lubricants consist of a blend of refined petroleum products supplemented by additives—antioxidants, anticorrosion agents, dispersant-detergents, compounds which improve viscosity-temperature characteristics, and others as the needs of a particular application may demand. Where liquid lubricants cannot serve, oils thickened with soaps of sodium, calcium, aluminum, and lithium or with oleophyllic solids—modified silicas and bentonites—are employed.

Native asphalts are rarely satisfactory for use. They are often more expensive than those made from petroleum. The manufactured asphalts have largely replaced the native materials because their properties can be manipulated by selection of situable crude oils, by noncracking distillation, by oxidation at high temperature, by blending, and by use of additives.

Petroleum waxes, recovered in the making of lubricating oils from paraffinic crudes, fall into two classes: (1) refined waxes, macrocrystalline in type, essentially the normal paraffins C_{20} to C_{30}, used in candles, paper waxing, and household paraffin wax: and (2) microcrystalline waxes, the amorphous waxes of commerce, used widely in paper sizing, coating frozen food packages, insulating, and making petrolatums.

Small quantities of petroleum are used to make technical white oils, emulsifier sulfonates, insulating oils, insecticides, rubber extenders, hydrogen and town gas, synthetic detergents, and intermediates for the chemical industry. The raw materials vary a good deal but are taken largely from the heavier fractions of the crude oil.

§3.14. Petroleum Fuels. Among the products derived from petroleum are the fuels which now supply over three-quarters of all the energy consumed in the United States. Petroleum energy provides the power for supersonic jet aircraft, the fuel for small space heaters, and a multitude of uses in between.

§3.14.1. Gasoline. Finished motor and aviation gasolines are blends of straight-run gasoline (obtained by primary distillation), natural gasoline (one of the liquids processed out of natural gas), cracked gasoline, reformed gasoline, polymerized gasoline, and alkylate.

To these gasoline blends, refiners add a wide variety of chemicals, called additives, for the purpose of improving the quality of the gasoline as a fuel.

Antiknock compounds are one such gasoline additive. Their purpose is to reduce or eliminate the "knock" or "ping" that occurs when the fuel is not being properly burned in the engine. The measure of gasoline's resistance to engine knock is its octane number. This is determined by comparing a gasoline with fuels of known composition and knock characteristics under specified conditions. Two different types of test conducted in a single-cylinder laboratory engine yield antiknock additives, in which more costly hydrocarbon compounds are used to achieve antiknock quality.

Many refiners are now producing gasolines containing little or no lead-based antiknock additives, in which more costly hydrocarbon compounds are used to achieve antiknock quality.

Over the years, the octane numbers of gasoline have increased as automotive engineers have developed cars with higher compression ratios. In 1935, regular-grade gasoline had a Research Octane Number of 72. By 1972, it had risen to 94. Premium gasolines rose during the same period from Research Octane of 78 to almost 100.

Another gasoline additive is an antirust compound, which protects the fuel system against corrosion from the minute quantities of water which condense from the air in fuel tanks.

Anti-icing compounds prevent or protect the engine from ice which can form during engine warm-up in cold weather. Antioxidants and inhibitors prevent the formation of gum deposits in the engine.

Detergents added to gasoline remove and prevent deposits in fuel systems and carburetors. Ignition control compounds reduce deposits in the combustion chamber so that they will not preignite the gasoline, and also modify deposits on spark plugs to reduce spark plug misfiring.

Gasoline at this time is the petroleum industry's principal product (See Chapter 9).

§3.14.2. Aviation Gasoline. In the early days of aviation, airplanes were fueled with the same kind of gasoline burned in automobiles. It was not until World War I that serious research was begun on airplane engines and on aviation fuels and lubricants. In 1918, specifications were set up for a suitable aviation gasoline that was distinctly different from the fuel consumed in automobiles.

After the war, research continued on ways to develop increased power without enlarging the size of the airplane engine. Researchers realized that the key was a standardized fuel of high antiknock characteristics. Refiners went to work and developed an aviation gasoline of about 87 octane number. In 1934, an aviation fuel of 100 octane was developed. The engine's size remained the same, but its power output was considerably increased.

Commercial aviation began to grow significantly during the early thirties, but it was somewhat curtailed as the Second World War approached. The war generated an immediate military demand for large quantities of 100-octane aviation gasoline. The pre-war peak annual production of all aviation fuel had been 14.7 million barrels. By 1945, 100-octane aviation fuel production alone had reached a level of more than 124 million barrels.

After the war, commercial aviation came into its own and Americans took to the air. Mail and all kinds of freight were carried by air. Private aviation surged upward, as more and more business organizations bought and maintained their own airplanes. Small plane flying became popular and aircraft became important to farmers for seeding, dusting, and spraying their crops.

§3.14.3. Jet Fuels. When research first began on jet fuels, commercially available kerosine was used because of its relatively low volatility—an important jet fuel requirement. Today's commercial jet airliners still depend on a highly refined kerosine for most of their fuel supply.

Military jets require a somewhat more complex fuel to withstand the severe conditions of supersonic flight. To meet the military's jet fuel requirements, scientists have developed carefully compounded blends of kerosine and gasoline.

Jet aircraft consume enormous amounts of fuel. Some of the jumbo-sized jets in operation during 1972 consumed an average of 3,604 gallons of fuel during each hour of flight. Their fuel tanks can hold up to 47,210 gallons of fuel.

§3.14.4. Kerosine. In the early days of the petroleum industry, kerosine was the refiners' principal product, used primarily as an illuminant. While kerosine is used for cooking, space heaters, farm equipment, and as jet fuel, it has many other uses. Kerosine is an ingredient of insecticides, paints, polishes, and cleaning and de-greasing compounds. On farms, kerosine not only powers tractors and other equipment, but also provides the fuel for heaters used in the curing of tobacco.

§3.14.5. Diesel Fuels. In 1892, Dr. Rudolph Diesel patented what is today one of our most important sources of automotive power.

Diesel engines are fundamentally different from gasoline engines. A diesel's ignition is caused by the heat of compressed air in a cylinder, not by a sparkplug, as is the case in a gasoline engine.

Early diesel engines were massive, built to withstand tremendous heat and pressures. At first, they served primarily as stationary power sources in factories and ships and they burned almost any oil as fuel.

Large diesel engines are still important stationary power sources, but through the years diesels have gradually been adapted for other uses. Today, diesel engines provide economical power for heavy road equipment, such as trucks, buses, and tractors. In recent years, railroads have turned to diesel engines for locomotives and now represent a large market for diesel fuels.

Just as modern diesel engines have become specialized, the diesel fuels consumed to run them are highly refined for their specialized uses. Modern diesel fuels are manufactured in several grades, ranging from heavy oils to light kerosine-type oils.

§3.14.6. Fuel Oils. In the early 1900's, coal and wood were the primary sources of heat in homes and buildings across the land. Burner fuels were considered a by-product in the processing of gasoline.

In 1918, the first fully automatic oil burner appeared on the market, opening up a vast market for burner fuels and eventually spelling the end of furnace drudgery.

Today, these fuel oils are specially designed to meet the needs of residential and commercial heating, manufacturing processes, industrial steam and electrical generation, marine engines, and many other uses.

Fuel oils are generally classified as either *distillates* or *residuals.*

Distillates are the lighter oils, some of which are used in space heating, water heating, and cooking. But the major market for distillates is in the automatic central heating of homes and smaller apartment houses and buildings.

One of the big advances in home heating has been the development of the *Degree Day:* the system worked out by oil companies to determine just when each customer's oil tank needs filling, and enabling dealers to supply their customers on regularly established delivery schedules rather than on demand.

Besides heating homes, distillates are also used to make snow. Several ski areas are now using machines, powered by distillates, which turn compressed air and water into artificial snow.

Residuals are the heavier, high-viscosity fuel oils which usually need to be heated before they can be pumped and handled conveniently. Industry is residual oil's major market—it fires open-hearth furnaces, steam boilers and kilns. Large apartment and commercial buildings rank second as a residual oil consumer, and gas and electric utilities third. A heavy fuel oil product containing 0.5% or less of sulfur may be produced by direct use of the desulfurizing vacuum gas-oil or by blending in vacuum, atmospheric residuum or other refinery stocks. The vacuum residuum is the most difficult to desulfurize as it contains the highest amount of contaminants. Thus, maximum disposal is highly desirable. The amount of vacuum residuum that may be blended is clearly limited by its sulfur content. For example, only about 5 volume percent of high sulfur vacuum residuum may be blended with a desulfurized vacuum gas-oil of 0.3% to meet a heavy fuel oil specification of 0.5% sulfur.

Viscosity and pourpoint are other fuel quality standards that must be met and limit the amount of residuum that can be blended.

No. 6 oil requires perheating for burning and handling. No. 5 heavy may require preheating for handling as well as burning particularly in cold elimates. No. 5 light, however, may only require preheating before burning, but not for handling. Finally, No. 4 does not usually require preheating for burning or for handling.

The blending of desulfurized gas-oil components ro reduce sulfur content of residual fuel oils greatly reduces the viscosity of the product. Because viscosity may have changed with a reduction in sulfur content does not mean that the oils are actually higher grade since the pourpoint specification may not be satisified.

In addition, adjusting the pourpoint by the use of light African residuals can sometimes lead to a problem. These are paraffinic or waxy based oils of naturally low sulfur content. The high melting point waxes have a tendency to separate when mixed with other residua. Although the pourpoint may thus be properly adjusted, it no longer is a reliable guide as to the non-solidifying properties of the residual fuel oil at the time of use. Thus, the lines and filter screens ahead of burners could plug.

To make all these physical adjustments properly with limited quantities of desulfurized product at any one location is very difficult. Furthermore, other important qualities change as well with desulfurization. Consider an important one. API gravity and heat content change according to the amount of sulfur contained in the oil. The heat content varies inversely with the API gravity of the product. As the API gravity increases or the fuel becomes lighter, the sulfur content of a fuel decreases as well as its heat content. Therefore, the price of a desulfurized fuel is more costly in two respects. One, additional refining is needed to lower the sulfur content. Secondly, the accompanying phenomena of a lower heat content per gallon occurs. For example, the calcualted reduction in Btu content in desulfurizing from a 2.8% to a 1% sulfur fuel oil is about 5.1%. It is a reduction in heat content of 9.0% if desulfurized to an oil of 0.3%. Thus, an equivalent increase in volume of oil is required to achieve the same heat content. This factor subtly compounds the nation's energy shortage and the customer's fuel costs.

The selection of a desulfurization process depends largely on the nature of the available residuum feedstock and the desired sulfur level of the fuel-oil pool. Feed properties with important influence on the technology are: sulfur level, metals content, Conradson carbon or asphaltene content, nitrogen content, and oil viscosity. Most of these affect the economics and consumptive use of hydrogen. For example, oils of higher sulfur content require more hydrogen for desulfurization. The metals content and Conradson carbon both have a deteriorating effect on catalyst life. The evolution of large quantities of ammonia produced from the nitrogen content interferes with hydrogenation on active catalyst sites. The diffusion of reactants into the catalyst diminishes as viscosity increases.

§3.14.7. Petroleum Coke. Petroleum coke is almost pure carbon. As such, it has many useful properties — it burns with little or no ash, it conducts electricity, it is highly resistant to chemical action, it does not melt, and it has excellent abrasive qualities.

Coke's properties make it invaluable in the manufacture of electrodes for electric furnaces and electro-chemical processes. Carbon or graphite, often made from petroleum coke, is used in flashlight and radio batteries. Sandpaper and knifesharpening whetstones are made by fusing sand and coke.

But petroleum coke is primarily a fuel, valuable in refining aluminum, nickel, special steels, and chemicals. The fact that it is almost pure carbon reduces the chances of contaminating the metals or chemicals being refined.

§3.14.8. Liquified Petroleum Gas (LPG). Familiar to many as "bottled gas," LPG consists primarily of propanes and butanes, highly volatile gases that are extracted from refinery and natural gases. LPG has a unique double characteristic. Under moderate pressure, LP-Gases become liquids and can be easily transported by pipelines, railroad tank cars, or trucks. Released from its storage tank, LPG reverts to vapor form, burning with high heat value and a clean flame.

Part of the growing demand for LPG is as fuel for internal combustion engines, primarily those used in buses, tractors, forklift trucks, and other in-plant equipment. A principal use for the fuel is in the manufacture of petrochemicals, but it is also used extensively in the home—for stoves, refrigerators, water heaters, space heaters, furnaces, clothes dryers, and even air conditioners.

§3.15. Lubricating Oils. Finished lubricating oils range from the clear thin oil that is placed by a hypodermic needle on tiny compass bearings to the thick dark oil that is poured into the massive gears of giant machines. All equipment with moving parts requires lubrication.

As the needs of industry have become more complicated, lubricating oils have become more specialized. American companies manufacture hundreds of different oils to fill exacting requirements. Industrial lubricating oils must maintain their friction-fighting properties, without thickening or coagulating, for thousands of hours between shut-downs.

Lubricating oils for automobile engines are designed to maintain an even viscosity, neither thinning out in searing heat nor thickening in below-zero winter weather. They are designed to prevent formation of excessive carbon, corrosive acids, or sticky deposits. Chemical detergents enable these oils to hold in suspension harmful matter that would otherwise accumulate on engine parts.

Lubricants for the space age have been designed to meet the challenge of entirely new conditions, such as the absence of atmospheric pressure or gravity, cosmic radiation, and temperature extremes far beyond those experienced on earth. Space lubricants must also fill other requirements, including extreme cleanliness and durability.

§3.16. Greases. Greases are used to lubricate hard-to-reach places, such as bearing housings that cannot be made oil-tight, or areas where the splashing or dripping of fluid lubricants might contaminate products.

The essential ingredients of grease are lubricating oil base stocks, soaps which act as thickening agents, additives to improve performance and stability, and fillers to increase wearability.

Technologists have developed greases in hundreds of consistencies to meet innumerable performance characteristics. Some greases must resist the intense heat of steel rolling mills or the sub-zero temperatures encountered by high-altitude aircraft. Others must withstand acids or water. Still others must stand up to the grinding friction of railroad roller bearings.

§3.17. Waxes. The two types of wax derived from petroleum, *paraffin* and *microcrystalline,* are extracted from lubricating oil fractions by chilling, filtering, and solvent washing.

Paraffin wax is a colorless, more or less clear crystalline mass, without odor or taste, and is slightly greasy to the touch. Microcrystalline waxes do not crystallize like paraffin, and are thus composed of finer, less noticeable particles.

Wax is used for the most part in packaging. It waterproofs and vapor-proofs articles such as milk containers and wrappers for bread, cereals, and frozen foods. Wax also is used in casting intricate parts and components of machinery, jewelry, and dentures.

§3.18. Asphalt. For centuries, men obtained asphalt from natural deposits, or "lakes," where it survived as residue after the air and sun had evaporated the lighter petroleum fractions that existed with it. Solid or semi-solid at normal temperatures, asphalt liquefies when heated. It is a powerful binding agent, a sticky adhesive, and a highly waterproof, durable material.

Today, asphalt is an important petroleum product, extracted as a refining residue or by solvent precipitation from residual fractions. By careful selection of crude oils, controlled air oxidation, and blending, modern asphalt is given several added properties—inertness to most chemicals and fumes, weather and shock resistance, toughness, and flexibility over a wide temperature range.

There are a multitude of uses for asphalt. It is a major road paving material. It also surfaces sidewalks, airport runways, and parking lots. It goes into products such as floor coverings, roofing materials, protective coatings for pipelines, and under-body coatings for automobiles.

§3.19. Coal. The most abundant of the fossil fuels, both in the United States and in the world is coal.

Our actual coal reserves are much smaller than the total resource estimate, however, because only a fraction of the total resource can be considered minable under present economic conditions and with present technology. Most of the overall coal resource lies either at depths too great, in seams too thin, or at locations too remote from present-day markets, to permit economical recovery.

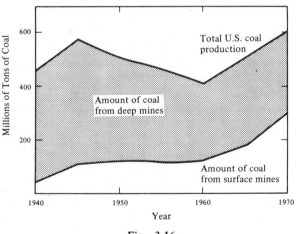

Fig   3.16.

Unfortunately, this apparent abundance of coal is illusory for at least two reasons. The annual consumption rate will probably rise sharply as shortages in the other fuels become more pronounced, as new uses for coal are developed, and as normal industrial growth occurs. And intensive exploitation of the coal resources of the United States may be limited by the stresses upon the environment which will result from coal production and consumption.

§3.19.1. Fossil fuel pollution problems. The combustion of fossil fuels for use as direct heat, mechanical motivation, and electricity generation results in unacceptable levels of air pollution even today. (see Figure 3.17) The future energy use projections indicate that this problem could reach disastrous proportions in another decade or so.

Fig   3.17.

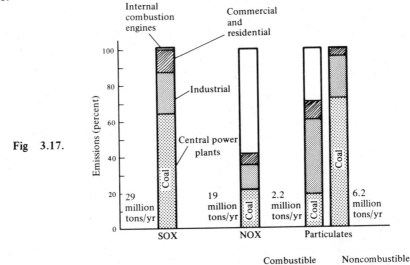

If the required technology is not developed and implemented in the near future, the sulfur dioxide emissions will double in about fifteen years. During this

Fig   3.18.

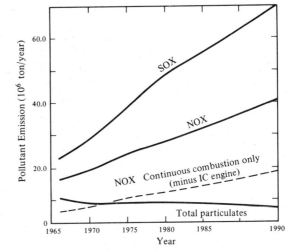

time period the nitrogen oxides will increase about 65%. Projections of potential emissions are shown in Figure 3.18.

As shown in Figure 3.19, most of the sulfur oxides emitted by central power plants originate from the combustion of coal, which contains, on the average, about 2.5% sulfur. Approximately half of the sulfur in the coal is in the form of pyrites, FeS_2 (inorganic sulfur); and the other half is chemically bonded to the coal matrix (organic sulfur). Most coal in the United States containing 1% sulfur or less is located in the Rocky Mountains area.

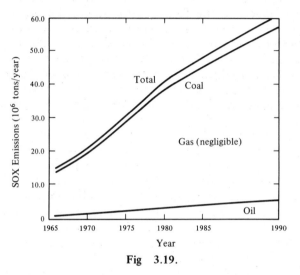

Fig   3.19.

The sulfur content of fuel oil varies with source.

Natural gas is the only fossil fuel that is substantially free of sulfur. However, this **fuel is** in relatively short supply.

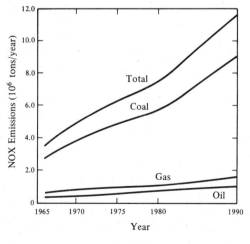

Fig   3.20.

Whereas sulfur dioxide emissions depend almost entirely on the amount of sulfur in the fuel, nitrogen oxides (projections shown in Figure 3.20) are formed by

§3.19.2. Refining coal. Coal is the largest known fossil fuel energy resource base in the nation. However, the use of raw coal has been limited because of its content of sulfur, non-combustible minerals and water. Environmental protection becoming more stringent and widespread, while the demand for energy and the increased demand for energy augur well for refining raw coal into environmentally acceptable energy products in the future. Fortunately, the nature and quality of refined coal can be tailored to future end use requirements. At the one extreme, coal can be just sufficiently refined to meet environmental standards when burned in large combustion units. At the other extreme, it can be converted to premium synthetic methane. Midway in the spectrum are hosts of possible products, including, improved supplemental fuels for blast furnaces, metallurgical coke produced by environmentally acceptable processes, superior quality electrode carbons, a wide range of chemicals, and feedstocks for fuels to power various types of vehicles.

Although no one energy type or technology is expected to completely resolve the energy-environmental issue, the merits of a coal refining industry become more and more apparent as the nation's environmental concerns intensify and its demands for "clean" energy compound an already serious energy shortage. A key in the establishment of a coal refining industry is the solvent refined coal process. The product, solvent refined coal, is likely to become an important fuel of the future.

The known U.S. coal resource base is larger than that of oil and gas combined. However, naturally occurring reserves of coal with less than 1% sulfur are only a fraction of the total. Although the resource base is large, productive capability limits the early availability of these low sulfur coals. New health, safety, and environmental requirements for both surface and underground mining deter the development of new coal supplies. Uncertainty regarding the economic feasibility of stack gas scrubbing technology have caused the coal industry to hesitate in opening new low sulfur mines at the expense of closing high sulfur mines yet unamortized. Assuming the low sulfur coal in the west will be developed rapidly, it is far from the population and industrial centers where the majority of coal burning power stations are located. Transportation costs to these centers would exceed the price of the coal in most instances. In addition, western coal typically contains high mineral matter and moisture, which deter efficient combustion in units designed for eastern coal. Removal of these inert materials by solvent refining at the mine would greatly reduce the shipping cost per unit of heat content and enhance the combustion characteristics of the product. Such factors augur well for a new coal refining industry wherein, coal from any number of mines might be refined into a uniform product which meets environmental requirements under a multiplicity of uses.

§3.19.3. Solvent refined coal. Coal has several undesirable characteristics. It is a solid; and, it contains sizable quantities of mineral matter, sulfur compounds, and water. Being a solid, coal becomes the most difficult fuel to handle. The special equipment required for coal handling is higher in capital, maintenance, and operating costs leading to a greater handling charge for coal. The mineral matter, sulfur the combination of nitrogen and oxygen from the air that is used to achieve combustion. Nitric oxide (NO) is formed at the high temperatures reached during combustion. Nitrogen compounds in the fossil fuels, particularly coal, also contribute to formation of nitrogen oxides. The nitrogen oxides produced may be reduced, but not eliminated, by modifying the combustion process to either reduce the peak combustion temperature or lowering the time that the nitrogen and oxygen are present at elevated temperature. There is no technology available for reducing nitrogen oxides to nitrogen and oxygen in the oxygen-rich flue gas. (In cayalytic converters for automobiles this reduction is readily accomplished because the combustion exhaust is oxygen-lean.)

Coal is the major source of particulate emissions from central power plants. There appears to be no inherent technical reason why the amount of carbon (combustible particulates) cannot be held to very low levels using known methods which enhance combustion. The quantity of fly ash emitted by modern boiler furnaces depends on the character of the ash, the furnace design and the conditions in the furnace during combustion. Emissions of fly ash are controlled by downstream methods such as electrostatic precipitators. An important factor in fly ash control, however, is that when the SO_3 level in the flue gas is reduced, conventional electrostatic precipitators become less efficient.

If coal-fired operations were uncontrolled, particulate emissions would range from 6 to 10 pounds per million Btu. At most installations, emissions range from 1 to 4 pounds per million Btu. The proposed federal standards would limit emissions to the atmosphere to no more than 0.2 pound of particulate per million Btu heat input. These limits were justified on the performance of existing electrostatic precipitators in 1973. Consequently, there is some basis to project a trend as presented in Figure 3.21 toward reduced particulate emissions beginning hopefully in 1980.

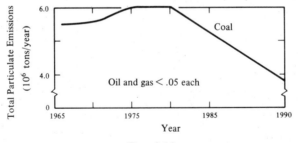

Fig   3.21.

compounds, and water in coal impart other undesirable conditions which limit its broader use. The transportation of the incombustible mineral matter and water along with the coal adds an undesirable additional cost to the coal. In energy conversion processes, the coal's moisture reduces the combustion temperature with an accompanying loss in efficiency. Furthermore, after the coal is consumed, the ash must be discarded. In processes where the ash remains finely divided and is expelled in the exhaust, the surrounding atmosphere becomes polluted. In addition, the sulfur compounds released in the energy conversion processes corrode the equipment in passing through and, when vented, defile the air. Thus, many a subtle price must be paid by consumer and society for the extraneous materials in the coal.

Typical physical properties and chemical analysis of solvent refined coal are as follows:

Physical Properties

Fusion Point . . . . 300–520°F

Heat of Fusion . . . . 50 Stu/lb.

Specific Heat . . . . 0.55 @ 400°F

Thermal Conductivity . . . . 0.06 @ 400°F

Specific Gravity . . . . 1.25 (1.1 @ 550°F)

Bulk Density . . . . 40 lbs./cu. ft.

Particle Sixe. . . . . 16–200 mesh

Fuel Value. . . . . 15,000 Btu/lb.

Chemical Analysis	**Weight %**
Carbon	. 88.41
Hydrogen	. 5.15
Nitrogen	1.84
Sulphur	< 0.78
Oxygen	3.72
Ash	0.10
	100.00

Solvent refined coal is free of water, low in sulfur, very low in ash, and sufficiently low in melting point that it can be handled as a fluid. Its heating value is 16,000 Btu per pound regardless of the original coal from which it was processed. It is a shiny black solid that at ambient conditions is hard and brittle, readily grindable into a fine powder.

Combustion tests indicate that solvent refined coal when fired in large boilers was equivalent to Bunder "C" oil.

The delayed coking of solvent refined coal has the potential of producing quality coke of electrode grade when further calcined. Solvent refined coal is also expected to make a good pitch binder suitable for forming electrodes for use in the aluminum industry.

Coke produced from solvent refined coal also appears to have the advantages sought in blast furnace operation, First, solvent refined coal has its mineral matter removed and sulfur content greatly reduced; thus, the fluxing stone requirements will be greatly reduced. Second, the solvent refined coal can be cast from its hot, liquid state into any shape or size. Ideal particles could be formed to permit maximum wind rate, throughput, and production in the furnace. As a supplemental fuel to the blast furnace, the low hydrogen to carbon ratio of solvent refined coal will permit greater savings in burden coke than when fuel oil or natural gas is used as a supplemental fuel.

Because of its physical and chemical characteristics, solvent refined coal has numerous advantages as a fuel. First, a uniform product can be produced regardless of the raw coal used for manufacture. The product has a higher heating value than the raw coal which leads to easing of fuel procurement, to savings in fuel transportation and storage facilities, and to reduction in furnace size. Because refined coal is essentially free of mineral matter, pulverizers will require less power and less maintenance. Ash handling and ash storage facilities should be less expensive and electrostatic precipitators may not even be required.

Solvent refined coal when hot can be fired as an oil. When utilized as a liquid, solids handling equipment can be replaced by liquid handling equipment at lower capital costs. Thus, when new power plants are designed to burn solvent refined coal as a liquid, capital and operating costs will be lower than those of an equivalent raw coal fired unit. Furthermore, the uniformity of solvent refined coal regardless of the raw coal used in its production will lead to a uniform boiler design and consequently economies in boiler production. At the present time, since boilers must be designed for a particular type of coal, when a different type of coal is substituted, severe operating problems usually result, as was the case in attempting to utilize low sulfur western coals in boilers designed for midwestern coal.

The use of low sulfur fuel is a positive guarantee that sulfur dioxide pollution would be controlled at all times. The power plant would not be dependent on reliable operation of a stack gas scrubbing unit to assure meeting emission standards.

§3.19.4. The solvent refined coal process.
Several methods of coal desulfurization have been examined and are in some stage of development. These include extraction of pyritic sulfur from raw coal, devolatilization or carbonization to a low sulfur char, conversion to hydrocarbon liquids plus a small amount of char, and gasification to either high or low Btu gas.

Pittsburg & Midway Coal Mining Company under the sponsorship of the Office of Coal Research, U.S. Department of the Interior demonstrated that solvent refining was technically feasible on a pilot scale in 1964.

Solvent refined coal is produced by first dissolving coal under pressure in a recycled solvent containing a small quantity of hydrogen. The coal solution is then filtered to remove virtually all of the mineral matter including the pyritic sulfur. The recycle solvent as well as a small quantity of light liquid product with a high content of benzene, toluene and xylene is distilled off. The remaining reconstituted coal is the solvent refined coal product with a uniform heating value of about 16,000 Btu per pound. The yield of solvent refined coal and other liquid products is about 90% of the original (MAF) coal. The mineral matter is reduced to approximately one tenth of one percent. Depending on the composition of the fed coal the sulfur level can be reduced to as low as 0.3%. The moisture content of any coal is completely eliminated.

Following the flow diagram shown in Figure 3.33, coal feed is crushed to a size smaller than 1/8 of an inch, and slurried with about three parts of a solvent fraction that is generated from the coal structure in the process. This recycled solvent fraction has a normal boiling range of about 550–850°F. The slurried coal together with hydrogen (2–2.6% by weight of MAF coal processed) is preheated and fed to a slurry dissolver. The dissolver operates at a temperature of about 825°F, and up to a total pressure of about 2000 psig. Under these conditions and in the presence of the solvent, about 95% of the coal (MAF) is dissolved.

The crude solution is passed to a receiver and reduced in pressure to about 150 psig, causing dissolved gases, such as hydrogen, hydrogen sulfide, methane and other light hydrocarbons, to be flash-distilled from the liquid. The hydrogen is separated in the gas treatment column and recycled. Hydrogen sulfide can be converted to elemen-

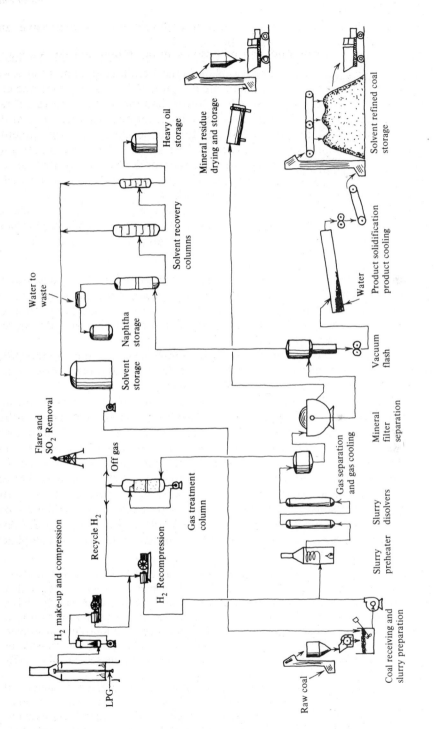

Fig 3.22.

tal sulfur. The remaining liquid slurry is filtered to remove the mineral matter. The filtrate is then vacuum flashed to recover recycle solvent to slurry new coal feed. Bottoms from the vacuum flash tower form the solvent refined coal product which solidifies at about 300°F. All of the pyritic sulfur and about 60% of the organic sulfur are removed. A total sulfur removal of 85% is not uncommon.

If the processing plant is adjacent to the power plant, the solvent refined coal may be maintained as a hot, viscous liquid and burned directly. If not, the solvent refined coal is solidified for subsequent storage and handling. Three distinct approaches to solidification have been studied. The process for product solidification illustrated in Figure 3.22 involves freezing the refined coal in a water bath. The product is then moved from the bath by conveyors.

The solids from the filters are discharged through an enclosed screw conveyor to a rotary dryer. The mineral matter as a filter residue contains a substantial quantity of wash solvent as an absorbed liquid. The wash solvent is removed as vapors in a rotary drum dryer by purging with inert gas. The dried solids are then transported to disposal.

When necessary a heavy ends distillation column separates high boiling material from the process solvent. The heavy oil product, if any, is expected to have an initial boiling point of about 700° to 900°F. This heavy oil product will be a low sulfur by-product of the process. It could be used as a fuel or blended back to lower the sulfur content in hgih sulfur fuels such as found in No. 6 residual oil.

A light liquid having a boiling range of abour 100° to 450°F and a yield of about 10 to 15% by weight of the coal charged is also expected as a by-product. The light liquid would contain about 15% to 20% by weight of phenol and cresylic acids which could be separated. The heavier fraction of the oil could be processed by catalytic hydrocracking followed by catalytic reforming, while the light liquids could be processed by hydrotreating and reforming. The reformate would have a high concentration of benzene, toulene and xylene.

§3.19.5. Coal desulfurization. With the expanding requirement for low-cost energy and projected increases in fossil fuel power generation, the need to eliminate sulfur oxide emissions from power generating stations is of major importance. The availability of low-sulfur coal would provide a straightforward solution to reducing the level of sulfur dioxide pollution with minimum impact on current industrial practices.

Removal of the sulfur from coal may be achieved by extraction of the sulfur using a method such as the TRW process, removal of the sulfur dioxide from stack gases, or by the conversion of coal to a gaseous form. The later process is the most expensive but permits the sulfur to be removed as hydrogen sulfide, which is relatively efficient.

Recently TRW has discovered and demonstrated a new low-cost technique for chemically extracting sulfur compounds from coal without significantly altering the physical form of the coal. This process is shown schematically in Figure 3.23. It uses iron compounds (ferric salts) to convert the inorganic sulfur pyrites into elemental sulfur and some sulfate. Thus, iron compounds are used to remove under relatively mild conditions, the sulfur which is bound chemically to iron in the coal. The organic compounds are then partially extracted from the coal with selective organic solvents.

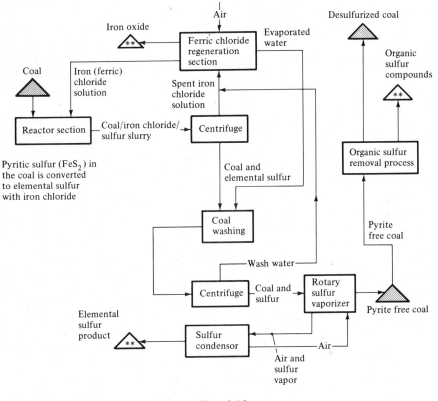

Fig 3.23.

Conceptually, the extraction of sulfur from coal without altering the physical form of the coal is simpler than removal of sulfur dioxide at the power plant site. The processing plants could be located in the general vicinity of the mine and clean coal shipped to various power plants if the extraction process is employed, The major waste product of the process, iron oxide, could be shipped back to the mine and used to prevent subsidence, or may be used as a source of iron. The current process for removing sulfur dioxide from power plants is based on scrubbing with limestone or limestone/dolomite. As shown in Figure 3.24, this requires the shipment of minerals to the power plant; the collection of wet calcium sulfate or other mineral sulfate and, finally, the disposal of these materials.

In natural gas processing, petroleum refining and, in the future, fossil fuel gasification, hydrogen sulfide is generated. Present processes which convert the hydrogen sulfide are only 90-95% efficient and exhaust large quantities of sulfur dioxide into the atmosphere. There is an obvious need for a reasonably-priced process to recover essentially all the sulfur from coal in a pollution-free manner.

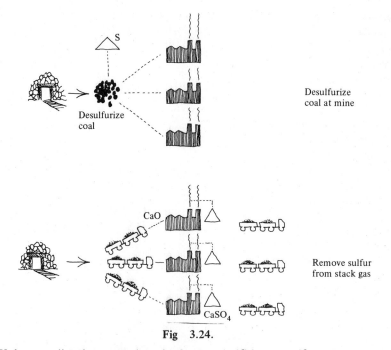

Fig   3.24.

TRW has outlined a complete hydrogen sulfide to sulfur recovery process which the corporation claims has the capability of reducing sulfur dioxide emissions

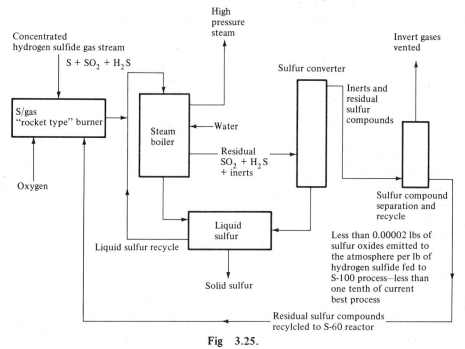

Fig   3.25.

to less than 50 ppm. This process is shown schematically in Figure 3.25. At this time the process is dependent on the combustion of hydrogen sulfide with oxygen, and the operating cost depends strongly on the cost of oxygen. In areas where pipeline oxygen can be made available (most refineries) the process is highly attractive for the implementation of pollution standards.

§3.19.6. Coal gasification. This well-known process produces a fuel gas from coal by liberating all volatile components and converting the carbon to carbon monoxide in a closed vessel. Oxygen (air) and steam must be fed in to provide essential chemical reactants and maintain a proper heat balance. Dissociation of the steam results in both hydrogen and methane appearing in the product gas. Any sulfur present in the coal appears as hydrogen sulfide in the product. Heavy ash particles drop through the grate which supports the fuel bed, while light particles exist with the product.

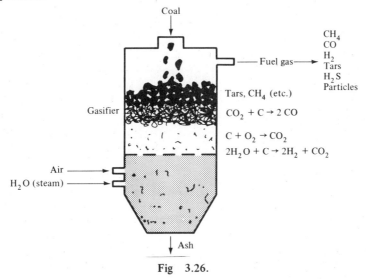

$$CO_2 + C \rightarrow 2\,CO$$
$$C + O_2 \rightarrow CO_2$$
$$2H_2O + C \rightarrow 2H_2 + CO_2$$

Fig 3.26.

The process is shown in Figure 3.26 as it takes place in a conventional fixed-bed gasifier. The counter current flow of coal and air through the device is ideal for proper heat and mass transfer.

Coal particles which are under a characteristic minimum size must be removed from the feed stream in order to prevent plugging of the fuel bed.

Add on SO_2 removal systems have increased the cost of generating electrical energy.

A practical fluidized-bed gasifier in which true counterflow between coal and gas streams can be achieved and in which the sulfur/sorbent reaction can be confined to an ideal temperature zone requires multiple fluidized beds properly connected. Such a design incorporating four beds, operating as different temperatures, can be incorporated into two pressure vessels as shown in Figure 3.27.

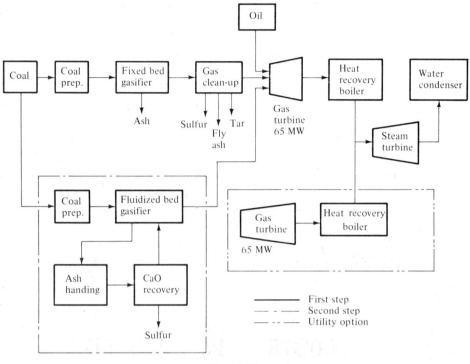

Fig   3.27.

§3.20. Petroleum, Petrochemicals, Planning and Priorities. In planning how to meet the nation's energy crisis, all elements must be considered. This is particularly true as we set priorities during periods of shortage. It does little good for the police car to have gasoline if there are no tires. The electricity generated by burning our limited, exhaustible and non-renewable supplies of coal, oil and natural gas is of little use if we cannot insulate the wires that carry the electricity into our homes and factories.

Although utilities can consider nuclear power as a source of heat to make steam and generate electricity, the products derived from petrochemicals are completely dependent at present upon oil and natural gas for their raw materials, In addition, only natural gas provides the process heat required at certain critical stages of the refining operations necessary to delicately rearrange nature's molecules.

Any rational and complete national energy policy should:

- insure that petrochemicals have an adequate supply of *raw materials — namely, crude oil and natural gas at this time.*
- maintain an adequate supply of critical energy products needed for petrochemical *processing.*
- insure that the petrochemical industry has a sufficient supply of energy products needed as *fuels.*

Part III

COSTS *VS* BENEFITS
BENEFITS *VS* RISKS

*

4

HOW SHALL WE GENERATE ELECTRICITY? CRITERIA FOR PUBLIC CHOICE.

Victor John Yannacone, jr.

§4.1. Costs and Benefits. Cost-benefit analysis as a technique for evaluating the economic desirability of a proposed industrial development project is not new. The concept of assessing the total costs and benefits of projects involving the generation, distribution and use of electricity, such as power plant siting, is not new, and such costs and benefits can be expressed in environmental, social, and economic terms. The task of evaluating costs and benefits involves hard trade-offs among conflicting interests, requiring that statements of costs and benefits must be made explicit. In order for assessment of costs and benefits to be useful, both the magnitude of the cost or benefit and the probability of the cost or benefit must be specified. The probability of cost or benefit is particularly important in evaluating potential costs/benefits and long term or indirect costs/benefits. Since analysis of costs and benefits is almost always somewhat subjective, the importance of making the methods of evaluation explicit cannot be over-emphasized.

It's the public that must make clear statements of priorities after costs and benefits have been specified. Statements of relative values of public and private goals, and constraints of the natural environmental system must guide all industrial location decisions.

§4.2. Environmental Impact Assessment. Environmental impact refers to the accumulated effects of a particular action on the natural, social and economic environment of the region within which such effects can be demonstrated.

In the past, environmental impact assessment generally has considered only effects on the natural environment measured as immediate effects upon existing organisms, or levels of undersirable materials added to the air and water.

In considering facilities for the generating and distribution of electrical energy, such as power plants and transmission lines, it is important to assess environmental impact by measuring, estimating and predicting the total direct and indirect effects on natural and human ecological systems both on and off the site over short and long time-spans. The task of measuring and estimating indirect effects is neither easy nor inexpensive, but the long-term benefits to society should justify the costs.

There is a pressing need to examine effects on the socio-economic environment concurrently with studies of effects on the natural environment since these systems strongly interact. The effects of industry location on social and economic systems must be considered not only in terms internal to those systems such as employment and income, but also in terms external to such systems particularly effects on the natural environment which accompany the location of a new industry.

The purpose of this Chapter is to provide the student and concerned citizen with sufficient background and information to evaluate legislation, administrative action, and regulatory proposals which may have adverse effects on the environment.

Three areas of particular concern to citizens considering the environmental impact of industrial development are:

1. the criteria used in evaluating the environmental impact of industrial development,

2. the information upon which the criteria are based, and

3. the procedures by which decisions are made and implemented.

The natural and socio-economic environment of any region ultimately rests in the custody of some governmental organization which responds to that collection of numerous and varied interests commonly lumped together as the "public." Decisions as to whether an industrial plant will be built and if so, where it will be built must be based on information concerning the effects of the decision on natural environmental systems and the socio-economic system of a region. The tradition of making industrial location decisions primarily on the basis of industrial economics and the immediate economic considerations of a political unit is changing under the National Environmental Policy Act (NEPA) and more specifically under similar state legislation. Benefit to the industry may often be obtained only at great expense to natural systems and the postulated benefits to any political unit must be carefully examined. Both direct and indirect effects of the plant siting must be analyzed over time.

Among the most urgent needs of those assessing the impact of industrial siting is the need for meaningful information. An important difficulty with present information on the socio-economic systems and natural environmental is its inaccessability to the public. Data scattered among a number of public agencies, or buried in technical reports are available only to a very limited number of decision-makers and generally only consider narrow areas of interest.

Concerned citizens should insist upon advance public disclosure of siting alternatives and energy use options, and sufficient lead time to review all the data listed in this section before full committment to any energy generation, distribution and utilization option or site selection is made.

The geographic boundaries within natural systems—watersheds, air sheds, biological communities, physiographic and geologic formations, and soil patterns—are seldom if ever coterminous with political boundaries. Watersheds, for example, may be contained within a single township or may be of international extent. Many important watersheds appear to be at least of multi-county size and some of those are multi-state in dimension. Citizens in any political subdivision which may be affected by a particular land use or energy policy decision should participate in the decision-making process.

§4.2.1. Direct effects. Probable direct effects on the socio-economic system of the region can be determined quickly from precise descriptions of the proposed plant (payroll, public facilities required, total value of plant, etc.). Probable direct effects on the natural environment can also be estimated from studies of the proposed site when the plant size, construction methods, production processes, resources required, and materials to be added to the natural system are known.

§4.2.2. Indirect effects. Effects on natural systems and socio-economic systems may occur over long time periods and wide geographic areas. Application of ongoing analyses of the systems involving the proposed site is the necessary means of determining the probable long term and spatially distributed effects. Studies assessing the significance of general system dynamics, materials cycling, synergism and economic multiplier effects must precede industrial location decisions. If well-defined models for predicting the behavior of systems are available, they may be the only substantial scientific framework to consider impacts of long term duration and wide geographic extent in both natural and socio-economic systems.

§4.2.3. Geographic extent of environmental impact. The geographic extent of the impact of industrial site selection on the natural and socio-economic systems of a region is a critical element of industrial site decision-making. The indirect social, economic and natural environmental effects of siting almost any major industrial facility extend beyond the municipality, township, or county in which the site is located. Clearly, decisions based on assessment of total environmental impact must be made by all governments whose area of geographic concern includes the area of significant environmental impact from the project.

Determination of the magnitude and extent of specific effects of industrial plant siting in a particular location still requires site-specific data collection and analysis. Collection of data on the natural environment necessary to determine the probable extent of environmental impact attributable to a particular industrial plant requires a minimum of two years in most of the continental United States.

Collection of social and economic data and analysis of the social and economic composition of a region is important in the determination of socio-economic impact, however such studies can usually be completed in a shorter time period.

§4.2.4. Systems analysis and modeling. Considerable emphasis in this Chapter is placed on the need to understand the system of interacting elements or component parts of the social, economic, and natural environments, because the most important indirect effects of generating, distributing and consuming electrical energy cannot be determined without a description of the behavior of the interacting elements in the system. For example, the detrimental effects of mercury released into the water system as industrial waste were not understood until systems analysis showed that the mercury was transformed biologically into an organic mercury compound which was then dangerously concentrated in the food chain. Similarly the total effect of the siting of a particular industry in a region cannot be determined until socio-economic analysis reveals the probable associated development which will accompany the new plant siting.

Since systems analysis includes a description in measurable terms of the behavior of a system over time it makes it possible to predict changes in the system over time and space.

One of the most important uses of systems analysis is in determining which data are most significant in determining the environmental impact of energy generation, distribution and utilization. Decisions regarding which data should be included in a comprehensive case study can be made more intelligently within the context of the systems analysis of a conceptual model.

§4.3. Energy and Materials. In order to consider the effects of environmental contamination certain basic ecological concepts must be considered, and while biological communities, for the most part, are inordinately complex, there do seem to be certain real and distinct levels on which biological communities are organized and on which they can be studied.

The study of living organisms can be organized on a series of different levels corresponding to the different ways in which they are studied. Organisms can be studied at the molecular level, or at the level at which molecules are organized into cells, or at the level where cells are organized into tissues and organs, or at the level of the entire individual organism. The study of ecological systems deals with the higher and much more loosely organized levels where individuals are organized into populations, populations into trophic levels, and trophic levels into ecosystems.

Two general processes impose a gross patterning and organization upon ecological systems: *energy flow* and *material cycling.*

Living systems depend on the flow of energy from the sun, and over any appreciably long period of time, subject to the Laws of Thermodynamics, the amount of energy entering a biological system upon the earth is roughly equal to the amount of energy lost as heat to outer space. The existence of both source and sink is essential since the earth must maintain this energy balance.

Some incident solar radiation is trapped by plants. The energy is then transferred from one part of the biological system to another, each transfer involving biochemical reactions during which thermal energy is liberated, so that eventually all of the energy first trapped by photosynthesizing plants leaves the system in the form of heat.

The earth is thus an open system with respect to energy, however, with respect to matter, the earth is essentially a closed system, the amount of matter being generally fixed, with merely the state changing. Energy flows throughout the system and in so doing establishes material cycles. The water cycle is an example.

For some chemical elements the natural cycle is fairly simple. For others, including many of the elements used by biological systems, the cycle is more complex. Man now uses almost all of the elements thereby complicating and disrupting natural material cycles.

The most complete cycles tend to be those in which there is a gaseous phase such as the hydrologic (water) cycle, the oxygen-carbon dioxide cycle, and the nitrogen cycle.

These cycles tend to be balanced and the amount of material in any one phase, such as oxygen in the atmosphere, tends to remain relatively constant. Small disturbances are corrected naturally and changes are usually only temporary. However, these cycles may not be able to return to the steady state following disturbances sufficiently large to cause severe disruption of the cycle.

Sedimentary cycles involve movement of materials from land to sea and back again. Elements moving through such cycles tend to take thousands or even millions of years to cycle, and the cycles tend to be less complete than those with a gaseous phase. In the sedimentary cycles materials are leached or eroded from rocks on land and carried by rivers to the oceans. Most of the material is deposited on the shallow inshore areas of the oceans and eventually returned to the land when these sediments are raised into mountains by the geological process of uplifting. There is some slight loss of certain elements such as calcium and phosphorus since small amounts of these elements are deposited in deep ocean sediments. Such losses in the geochemical cycle are probably permanent, at least as far as the time frame of mankind is concerned.

Man can use the enormous energy sources available to him to interfere in a massive way with the cycling of materials. In particular, by mining and drilling operations elements are removed from more or less concentrated forms and dispersed, a process which clearly cannot persist indefinitely for any element. This process of dispersion represents a substantial disruption of the cycling process. Copper, for example, once mined and dispersed as pipe and wire, will not recycle naturally to an extent again useful to human populations. Of course, we coud recycle it artificially within our industrial system.

The exploitation of naturally occurring elements by man leads to pollution. Pollution may be defined as the interference or perturbation of the natural cyclic processes so that materials accumulate where they shouldn't. With respect to some elements this results, in part, from the simple fact that because we extract materials from the earth at very high rates, the natural cyclic processes cannot cope with the increased rate at which these materials are returned to the system. Possibly the most widely discussed example of this breakdown in natural cyclic processes is eutrophication. In eutrophication, the rate of input is too high and degradation continues in the absence of oxygen by different organisms, than those which normally degrade waste using the oxygen dissolved in the water. The results are usually offensive to man. This change in the quantity of input results in a change in the quality of the biological system, illustrating the point that biological systems frequently respond to changes in the rates of input of materials or energy in a non-linear fashion, and often manifest threshold level effects. (See Chapter 5.)

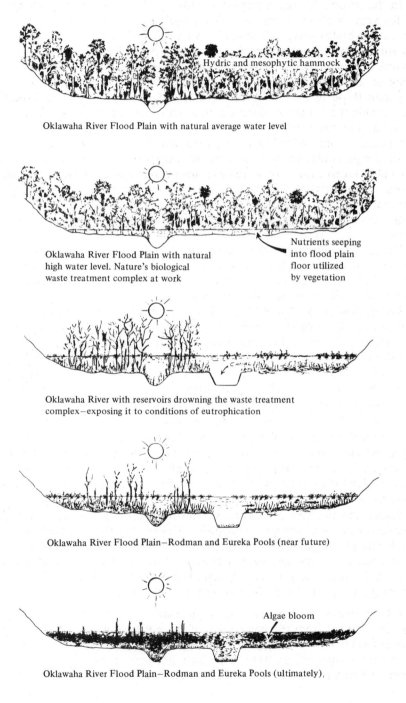

Oklawaha River Flood Plain with natural average water level

Oklawaha River Flood Plain with natural high water level. Nature's biological waste treatment complex at work

Nutrients seeping into flood plain floor utilized by vegetation

Oklawaha River with reservoirs drowning the waste treatment complex—exposing it to conditions of eutrophication

Oklawaha River Flood Plain—Rodman and Eureka Pools (near future)

Oklawaha River Flood Plain—Rodman and Eureka Pools (ultimately),

Energy flow and material cycling can be considered in greater detail by examining trophic levels.

In the oceans, energy is trapped mainly by phytoplankton, which provide the food for very small invertebrates called zooplankton, which in turn are eaten by larger invertebrates and small fish, which are attacked and eaten by larger fish and other animals, which in turn become the food or prey of still larger animals and man. Each link in this food chain may be treated as a trophic level—a group of organisms feeding on similar food sources. Typically the organisms in any trophic level are larger than those in lower trophic levels. Food chains, or more realistically, food webs, provide the pathways along which energy and nutrients move throughout an ecosystem, and the trophic levels represent the transfer points along the way at which nutrients are utilized and energy exchanged.

At any particular point in time, in any particular ecosystem, there is no doubt that trophic levels may exist in reality; however, many organisms, regularly, or at different stages of their life cycles are found in different trophic levels. Trophic levels are, in ecology, an abstraction, however useful in conceptualizing the structure of ecosystems and investigating the flow of energy and material throughout the system.

For any biological system to persist there has to be a basic trophic level which traps solar energy and fixes it in a form consumable by the organisms in higher trophic levels. This function is performed by green plants and this first level is generally referred to as the producer level.

Just as necessary for the completion of nutrient material cycles are organisms which can break down the complex organic molecules of plant tissues. Ultimately this function is performed by micro-organisms (bacteria and fungi).

Intermediate activities involving the greenplants of any ecosystem include those trophic levels in which the plants are consumed while alive or upon death but prior to decomposition by micro-organisms.

There are two general series of trophic levels commonly encountered in an ecosystem. One, the herbivore chain, has plants as the basic trophic level, herbivores (which eat green plants), predators (which eat other animals, particularly herbivores) and parasites. The other series, the detritus or decomposer chain, has dead materials as the basic trophic level. The distinction between the two food chains involved is not complete, since snails, for example, eat both living and dead organisms.

Plants, the primary producers, synthesize organic molecules from inorganic molecules in the environment utilizing solar energy in the process. This energy is fixed during photosynthesis, during which carbon dioxide combines with water in the presence of radiant energy and enzymes associated with chlorophyll, to form glucose and release oxygen. The largest proportion of the biomass represented by green plants consists of carbohydrates such as glucose.

All organisms need energy for maintenance, growth and reproduction. This energy is obtained from the conversion of the potential energy stored as chemical energy in food to kinetic energy, or the energy of work. It should be obvious that the total amount of solar energy fixed in the form of chemical energy represented by the biomass of green plants sets the upper limit to biological production and activity.

Solar energy is not trapped uniformly over the earth's surface nor is biomass produced uniformly. The most important determinants of productivity are the climate, including precipitation and the extent of incident solar radiation, soil structure and availability of plant nutrients, and the efficiency of a particular plant community in trapping the incident available solar energy. Production is generally defined as the amount of organic matter elaborated over a specified time period, whether or not it all survives to the end of that time period. Several points have a crucial bearing on the transfer of energy and materials throughout an ecosystem.

(1) During natural transformations of energy from one form to another form, some portion of the energy involved is liberated as heat. Organisms cannot generally use heat energy directly to do work so there is a loss of total energy available to a biological system at each transfer. For example, when an animal releases the potential energy contained in glucose some portion of the energy becomes available for growth and activity (work) and the rest is lost to the environment as heat. In doing work, energy is also transformed into heat.

(2) All organisms have maintenance, growth and reproduction requirements. Energy is made available for these purposes by breaking down large food molecules into smaller molecules which are then further degraded to produce energy plus waste products in the process called respiration. Part of the energy produced during respiration is immediately lost as heat, the rest of the energy made available is used. First to maintain the organism; then to synthesize new biomass as growth or in reproduction, and finally as mechanical energy available to perform the various functions of the organisms.

(3) The producer trophic level made up of green plants has a very different chemical composition than the higher trophic levels. In particular, a given quantity of plant biomass has a much lower energy (caloric) content than the same quantity of animal biomass. A great deal of plant biomass must be degraded to turn the energy into the concentrated form found in animals.

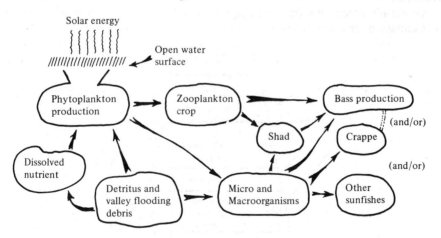

Generalized scheme of "well-ordered" open water lake or reservoir.

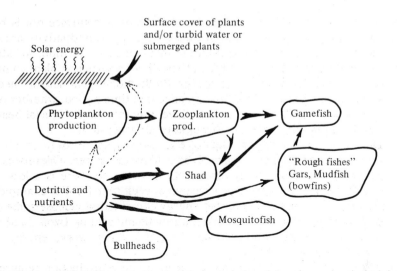

Generalized scheme of trophic level and species abundance in reservoir or lake after excessive plant invasion.

(4) Finally at each trophic level the production is either passed on as food to the next highest trophic level, or it is passed on to the decomposers. Thus the herbivore food chain is continually releasing larger amounts of energy to the decomposer food chain.

A number of general conclusions can be drawn from this basic information. The amount of energy available to any trophic level over a period of time must decrease as we move up trophic levels. The maximum transfer rates being fixed at low levels by thermodynamic and biochemical constraints. The production of biomass must therefore decrease at each higher trophic level. It also appears that the efficiency with which food is converted into biomass is lower for herbivores than for carniovres.

Generally accepted studies indicate the existence of a pyramid of production of new biomass over a period of time.

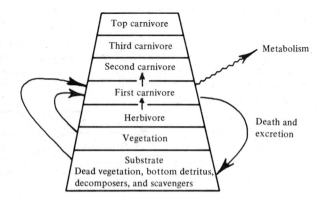

Schematic representation of the flows of DDT in the ecosystem.

By contrast with the herbivore chain, over long periods of time the decomposer food chain consumes all of the incoming production and completes the dissipation of biologically fixed energy in the flow of heat.

The detritus food chain is of crucial importance in nutrient cycling and the same thermodynamic and biochemical principles apply as in the rest of the ecosystem. A system of trophic levels can be outlined, although perhaps less clearly. The individual organisms and the species involved in the detritus food chain are small and inconspicuous by habit, however, tremendous complexity exists in the system.

There are two distinct aspects of the decomposition process and both are required for completion. In the first, large pieces of detritus (dead animals, trees, leaves, and the like) are fragmented and partly decomposed, generally by invertebrates; then these coarse organic particles are decomposed into less complex inorganic compounds by simple organisms such as bacteria and fungi. The particular contribution of each organism involved in the decomposition process varies with the environmental parameters of the area in which the process occurs.

§4.4. **Populations.** A population is a group of organisms belonging to the same species and living in the same area so that there is a possibility of interactions among its members.

Populations are characterized by birth rates, death rates, age distribution, genetic composition and other observable characteristics. Under certain circumstances ecosystems are considered as being organized at the population level and interactions within and among populations are of concern.

The general academic discipline concerned with the organization of ecosystems by populations is called *Population Biology,* and that dealing with population abundance is *Population Dynamics*.

A population cannot be considered to the exclusion of other aspects of the ecosystem; rather it forms a focus and other component parts of the ecosystem are studied with a view towards how they influence the population.

The numbers of organisms in a population are determined by three processes—the *birth rate (natality),* the *death rate (mortality)* and movements.

Although the elements of population movement: emigration, dispersal, migration, are important in considering population growth and limitations upon growth, principal consideration has generally been given to birth and death rates. Populations increase in density when births exceed deaths and decrease when deaths exceed births. For populations which are approximately stable in numbers the birth rate must equal the death rate. Growing populations become limited in overall numbers either by an increase in the death rate or a decrease in the birth rate or some combination of both.

Since populations grow by multiplications, they have the capacity to increase in numbers at a constant rate. The numbers can double at any constant interval even though the numbers get larger at each doubling. This potential for exponential growth fortunately is rarely realized in nature for a variety of reasons.

There are a variety of important characteristics of natural populations.

(1) Self regulation below the level of food destruction appears to be fairly common in some natural populations such as territorial birds and predatory animals.

(2) Changes in the quality of animals through time, by means of genetic change or physiologic change plays a role in determining abundance.

(3) In some systems, a species is more stable if it has a more complex or mixed age distribution.

(4) The history of a population is to some extent incorporated in its age distribution. The age distribution is a sort of population memory and the consequences of this history are carried into the future because of the existence of the age-distribution. This sort of demographic time-lag is well illustrated in certain human populations.

Populations do not increase indefinitely because all populations are constrained by limiting factors of some kind. Among the limiting factors and important considerations in population dynamics are the sources of mortality including enemies, competitors, food supply, available area, weather, internal population factors such as cannibalism, stress and waste build-up. The factors affecting natality include food supply, opportunities for reproduction, nesting sites, weather, and interference among members of the population, such as territorial behavior in certain animal populations, stress and the operation of a social hierarchy. Stable populations are generally defined as those which are stabilized by density dependent process. *Density dependent processes* are those regulatory factors which increase in effectiveness as a function of increasing population density. Ecologists tend to look to such factors as food, enemies, nesting sites and interactions among organisms in the population for density dependent effects.

The separation of populations into stable and unstable is rather arbitrary. It is probably the case that there is a continuous variation from populations which are very stable to those which are extremely unstable.

The importance of the concepts of density dependence, population regulation and stability goes beyond the realm of pure ecology into the area of environmental management. Although the problem of getting the optimum yield from an exploited natural population is complex, at its elementary level, for naturally stable stocks, exploitation should be density dependent over the long run to guarantee persistence of the population. Disregard of this obvious fact has led to the near extinction of certain species of whale. Similarly in the management of game-birds and deer, the aim is to kill the surplus each year by hunting and thus impose a density dependent mortality factor.

Many of the environmental problems of this generation are essentially population problems. Certain populations of small mammals noted for cyclic variations in numbers have been studied for decades and there are still several competing theories to explain the data. The reasons for the long times involved in population studies and their essential inconclusivensss reflect some of the unalterable characteristics of natural systems. Natural systems are complex and it is difficult to isolate the effects of interacting factors. Ecological systems, unlike physical systems are often *non-Markovian,* that is, the future state of the system cannot be predicted on the basis of the present state of the system but depends on the history of the system. Thus population events vary not only with locality, they are time-dependent as well.

§4.5. Communities and Ecosystems. Ecology can also be studied at the community level. Botanists in particular are interested in trying to describe communities in such a way that distinct assemblages of species can be categorized and classified together, while their limits can be defined and they can be separated from different communities. *Plant Ecology* was the earliest of the ecological disciplines and to many biologists, ecology is still essentially the study of plant communities.

It is often possible to recognize recurrent groupings of species which make up a large part of the community so that certain communities have come to have common descriptive designations generally representing the make up of the individual elements of the food web present in the area—the salt marsh, the deciduous woodland, the suburban community, the old field, the fresh water lake, the fresh water marsh or bog.

Another way of considering changes in communities is to study the changes as a function of time+*succession*. The succession of terrestrial communities under various circumstances has been fairly well described and occurs in an orderly straightforward way. Each stage or community tends to modify the environment making it more suitable for some other group of organisms, so that there is an interaction between the biological and physical elements of the system.

The production of plant biomass in early successional stages is high and the production/biomass ratio is high. As biomass accumulates and we move to a larger, older community with greater biomass, the relative rate of biomass production declines, the production/biomass ratio tends to decrease.

There is a trend during succession from an open to a more closed system with respect to nutrients. Climax communities tend to retain the nutrients they have and recycle them.

Ecosystems are linked by a flow of materials and energy and therefore are not truly independent of one another.

Fresh water bodies in particular are open systems. Thus streams gain nutrients in the run-off from surrounding terrestrial communities and there is a constant downstream drift of nutrients in the form of detritus and organisms.

A. A portion of the undisturbed Oklawaha
    Regional Ecological System

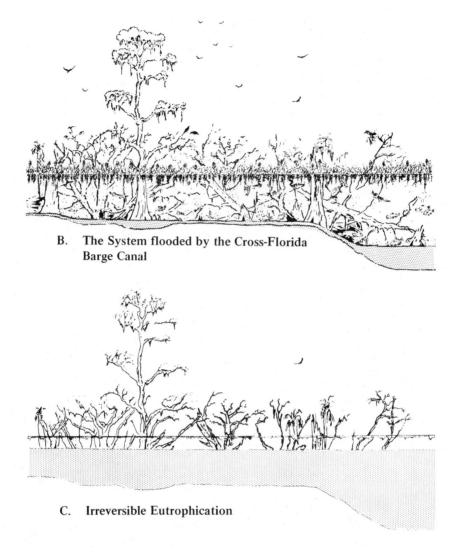

B. The System flooded by the Cross-Florida
    Barge Canal

C. Irreversible Eutrophication

Lakes receive inputs of potential energy and nutrients from upstream, and while the rate of flow is slowed within lakes, they still pass on nutrients downstream to the estuaries, the continental shelf, ocean plankton and the deep ocean floor.

In aquatic systems, however, not all of the material is passed downstream. In slow moving rivers and lakes, particles and dead organisms settle to the bottom where the organisms are decomposed, and the water body gradually fills up. A pond fills up and becomes a swamp during the process. As vegetation chokes the pond, the species of fish and other organisms change. The swamp is invaded by vegetation from the surrounding communities and terrestrial succession begins.

Ecosystems are dynamic, not only in that they are open systems, constantly receiving and losing materials, but because their character changes with time and they evolve into different kinds of systems.

Yannacone—The Energy Crisis—15

The concept that different components of ecosystems interact, and that perturbation of one component of a system is likely to be felt elsewhere in the system, can be extended to interactions among different ecosystems, since all ecosystems are essentially open and liked together.

Migratory species illustrate the classic case of liked ecosystems. Populations of salmon, for instance, are greatly influenced by conditions in the freshwater river systems where they spawn.

The links between coastal marshes and ocean ecosystems play a great part in determining the productivity of the ocean. These coastal wetlands are rich in nutrients and extremely productive. The young stages of many fish and shellfish feed there and then move out into more open waters. There is also a large movement of nutrients from marshes to coastal waters. Our seafood harvest depends upon the maintenance of such estuarine environments, yet such areas are extremely vulnerable to development and also represent a sink for the accumulation, concentration and ultimate distribution of pollutants.

§4.6. Systems analysis and ecology. Systems analysis has become one of the most valuable tools available for the study of ecological systems.

Much of the material in this section is taken from portions of *Systems Analysis in Ecology,* edited by Kenneth E. F. Watt, Academic Press, N.Y., 1966; in particular Chapter 1, "The Nature of Systems Analysis," by Dr. Watt, and Chapter 2, "Complexity of Ecological Systems and Problems in Their Study and Management," by Dr. David Pimentel; *Systems Analysis and Simulation in Ecology,* 2 vols., edited by Bernard S. Patten, Academic Press, N.Y. 1971; amplified by personal communications with Dr. Watt, Dr. Orie Loucks, and others involved in the early applications of systems ecology to environmental law. Footnote attribution of sources is omitted, with thanks to all, however, the author takes responsibility for any errors or oversimplifications which might appear.

Ecological systems are composed of many components which interact in a large variety of ways. Each biological component of the system is affected by the physical elements of the system, and all the variables change not only with respect to time, but from place to place since the environment is heterogeneous. Discrete components interact and each component of the system affects all the others in one way or another. The complexity of the system of interlocking cause-effect pathways confronts us with a superficially baffling problem, and systems analysis is precisely designed to handle such situations.

In general, the system is analyzed in terms of its components. The processes affecting each component are analyzed and described so that changes with respect to time and distance can be described and ultimately predicted. The interrelationship of the components of the system is also analyzed and a model of the system is developed.

Eventually the model is tested by attempting to simulate, generally with the assistance of a computer, the consequences of alterations in the state variables representing components of the system.

In the case of a real ecological system, no attempt at simulation can be truly complete. Indeed, the art of systems ecology is to determine the crucial elements and processes that govern the general behavior of the ecological system as a system. Systems analysis is particularly useful to citizens and legislators who have to make decisions on less than a total data base.

Viewing an ecological system as an interlocking complex of processes characterized by many reciprocal cause-effect pathways, it can be seen that one of the principal attributes of a system is that it can only be understood by considering it as a whole.

Systems analysis has its roots in military and industrial operations research, applied mathematics, probability, statistics, computer science, engineering, econometrics and biometrics.

There are some standard approaches to dealing with the great complexities inherent in the considerations of real systems such as the operating maxim that complex processes can be most easily dissected into a large number of relatively simple unit components, and that complex historical processes in which all variables change with time (evolve) can be dealt with most easily in terms of recurrence functions which express the state of a system at time $t + 1$ as a function of the state of the system at time t. Thus the system is considered not in terms of its entire history but rather in terms of the cause-effect relationships that operate through a typical time interval. This idea of the recurrence relationship is common throughout mathematics. Matrices of transition probabilities in Markov processes are merely stochastic versions of a recurrence relation. Difference equations, differential difference equations, dynamic programming, and the "loops" of computer programs are all based on recurrence relations in which the output from each stage in the computation is the input for the following stage. No breakdown in this approach occurs if the state of the system at time t is a function of the state of the system not only at time $t-1$, but also at time $t + 1, t + 2 \ldots t + n$. This merley increases the number of variables in the recurrence relationship and increases the dimensionality of the problem.

Another important basic principal of systems analysis is that optimization of processes is a central consideration. This premise brings to systems analysis the whole body of pure and applied mathematical theory related to the maximization and minimization of functions, the mathematics of extrema.

Combining the basic ideas of recurrence relations and optimization, leads to the concept that the purpose of systems analysis is to determine the optimal choice from among an array of alternative strategies at each of a sequence of times—the idea of the multistage decision process.

Multistage decision processes share two important basic similarities from a computational standpoint—high dimensionality and the need to be solved by some iterative process—requirements common to other types of problems that occur regularly in systems analysis: multiple linerar regression, iterative, nonlinear regression, gradient methods for finding maxima and minima, and simulation studies in general. All problems leading directly to electronic computers for assistance.

Feedback control is another concept of systems analysis that is also important in ecological systems, so that a realistic mathematical description of a process includes terms such that deflection toward the equilibrium state, or steady state, follows departure from it within certain limits. Interaction, like other features of systems models is easier to describe in terms of changes at an instant in time rather than changes over a period of time, so that models of interactions are typically conceived of in terms of differential equations rather than algebraic equations. It is easier to consider a process in terms of the rates of change at an instant in time rather than in terms of the history of the process over a considerable period.

Inequality contraints are encountered commonly in ecological systems analysis problems, as are thresholds and limits. Similarly, the common technique of computer programming in terms of a cyclically repeated routine or "loop" is suitable for consideration of ecological problems where historical processes unfold through the repetition of variants of the same basic cycle of events, and dispersal occurs through a parallel process, but in space as well as time.

Another important concept from systems analysis useful in some ecological systems studies in that of information. The amount of information is related to the degree of order or negentropy (negative entropy, see §1.19) in a system and this concept plays a role in studies of community organization. Modern digital computers are well suited for dealing with many of the computational problems of information theory.

The principal reason for using systems analysis in ecology is the complexity of ecological systems. Ecological complexity originates from a variety of causes: number of variables; number of different types of variables (ecological, genetic, physical chemical, endogenous, exogenous); different levels of organization of ecological systems (populations, communities, trophic levels, cycles) and the nonhomogenous and nonuniform distribution of system elements throughout time and space.

There is a definite sequence of steps required to properly investigate an ecological system for the purpose of describing its current condition or predicting its future course under different circumstances and the follow is an adaption of the method described by Dr. Kenneth E. F. Watt.

§4.6.1. Measurement. The first step in studying a complex system is to determine the variables and causal pathways that seem important in determining the function of the system. This list can be prepared from informaiton in the literature, *a priori* considerations, casual field observations, or formal pilot studies followed by a scientific program of sampling the values of the relevant dependent and independent variables. The technical problems encountered at this step involve the technique of measurement, sampling theory and application, logistics, and instrumentation.

§4.6.2. Monitoring. Finally, systems for continuous collection of some data over time are needed. In some cases basic elements for such data collection already exist and need only to be organized. Great quantities of social and economic data useful in pre-planning and determining the impacts of change are collected daily by units of local government. County registers of deeds, zoning commissions, county and municipal treasury offices and clerks, and many other agencies could be recording valuable information in coordinated form for automated data processing at a higher level of government. Such efforts could provide useful data at low cost for monitoring change and updating systems models. Other continuous monitoring efforts, such as studies of the effects of the release of specific industrial wastes into the environment, require specialized methods and technology, but such studies are essential in improving our ability to predict detrimental and costly effects in the future.

§4.6.3. Analysis. After all of the variables provisionally considered of importance have been measured, it is necessary to evaluate their real relative importance. The field program will ultimately be revised to discontinue sampling those variables which do not significantly contribute to the overall definition of the behaviour of the system, and only those variables which make a statistically significant contribution to the variance of the dependent variable need be included in the systems models. The ecologist concerned with analysis of a complex system is often led to the use of multiple regression techniques in conjunction with multiple analysis of variance. Additional complexities of interpretation arise if we use stepwise regression, or when certain variables enter the systems model non-linearly.

§4.6.4. Description. After it has been determined which variables need to be considered in order to fully describe the system, a model is structured. The first models are generally conceptual models which simply seek to fully describe the system and its behaviour qualitatively without making any attempt to predict its behaviour quantitatively.

A model is simply some method, usually a mathematical equation or set of equations, which can be used to describe the behavior of a system (a watershed, air mass, etc.). Complex mathematical models require the use of an electronic computer for solution of these equations. Model development includes comparison of predictions based upon the model with observed system behavior. If calculated behavior does not correspond closely enough to observed behavior, appropriate changes are incorporated in the model to make it more realistic.

A wide variety of models are available to describe the movement of water and substances contained in water, movement of materials in the atmosphere and the accumulation of substances in individual organisms or communities of organisms.

The importance of models is that they can often be used to predict the consequences of certain events or actions well in advance, thus allowing the public to consider positive and negative results before embarking on a costly and perhaps disastrous course of action.

The following list, derived from a report by the Faculty Land Use Problem Definition Seminar, "Data Needs and Data Manipulation," available from the Institute for Environmental Studies, University of Wisconsin and authored by Lewis, Loucks, Moore and others, is intended to serve as an example of the specific data elements that appear to be necessary to provide an adequate description of environmental systems and to provide a basis for planning and conflict-resolution are industrial siting, energy generation distribution and use. No attempt has been made here to determine scale, units of measure, or degree of specificity of the data. Such determinations can only be made in relation to specific problems and geographic areas.

A. Geological Data
1. Topographic maps
2. Soil family groupings
3. Soil reconnaissance maps
4. Soil associations maps
5. Land capable of supporting intensive farming
6. Engineering groupings of soils

   7. Surface geology map
   8. Bedrock geology map
   9. Bedrock surface contour map
10. Precambrian surface map
11. Overburden thickness map
12. Glacial deposits map
13. Unique geological features — mounds, natural bridges, caves, sinkholes, waterfalls, fossil sites
14. Mineral resources maps
15. Active and abandoned quarries, mines, and pits
16. Aeromagnetic map
17. Seismic activity maps
18. Gravity map

B. Climatic Data
  19. Precipitation
  20. Temperature
  21. Mean relative humidity
  22. Annual evaporation
  23. Annual transpiration
  24. Storm frequency and type
  25. Wind velocity (monthly)
  26. Prevailing wind rosettes (annual)
  27. Atmospheric inversion probability map
  28. Atmospheric particulates
  29. Atmospheric sulfur

C. Hydrologic Data
  Groundwater
  30. Phreatic water surface maps (with observation well network delineated)
  31. Potentiometric maps of major bedrock aquifer systems
     32. Depth to water table (secondary map)
     33. Quality — chemical (contoured TDS and/or separate maps of major cation and anion concentrations) — near-surface (phreatic)
  34. Quality — chemical (contoured TDS and/or separate maps of major cation and anion concentrations) — bedrock system(s)
  35. Quality — particulates, in near-surface water
  36. Quality — particulates, in bedrock water
  37. Map of estimated total water storage volume from average-annual water table position to Precambrian surface
  38. Groundwater recharge regions
  39. Groundwater discharge regions
  40. Map of estimated volumes of discharge
  Surface Water
  41. Lake inventory maps
  42. Flowing water inventory maps
  43. Lake volumes (acre feet)

44. Stream storage: mean low flow
45. Stream storage: mean annual flow
46. Stream storage: maximum flow at flood state
Lake Characteristics
47. Miles shoreline
48. Existing reservoirs
49. Potential reservoir sites (based on topographic, geological, and hydrologic considerations)
50. Quality — particulates (mg/1)
51. Quality — chemical (by ppm of TDS)
52. Quality — chemical (pH)
53. Quality — chemical (specific conductance)
54. Sedimentation rates
55. Major point pollution sources by type
56. Mean annual lake level fluctuation
57. Temperature of hypolimnion
Stream Characteristics
58. Quality — particulates (mg/l)
59. Quality — chemical (by ppm of TDS)
60. Quality — chemical (pH)
61. Quality — chemical (specific conductance)
62. Existing wild rivers
63. Wild rivers potential
64. Maximum depth
65. Bottom Characteristics

D. Biological Data
Terrestrial Elements
66. Native vegetation maps (including forest)
67. Native plant community types
68. Forest inventory maps (timber types by species, density, and age class)
69. Cultural vegetation type maps (man-induced)
70. Animal community types (insects and mammals)
Aquatic Elements
71. Wetlands inventory types (including emergent aquatic plants)
72. Submerged aquatic macrophyte communities
73. Fertility and productivity (stage of eutrophication)
74. Aquatic invertebrate communities
75. Fishery resource types

E. Social Data
(historical data should be assembled whenever appropriate)
76. Land ownership
77. Owner activity
78. Special interest areas or uses
79. Existing intensive land use maps
80. Population density
81. Population composition

Additional information recommended for storage and periodic updating, includes data with practical value as indices of land use trends.

82. Indices of land use trends
    a. land value trends
    b. land speculation trends
    c. public facility planning
    d. semi-public facility planning

§4.6.5. Simulation and Optimization. Once a model has been developed which accurately describes the behavior of a complex system, it can be used in simulation studies to show how the system can be managed in real life for optimal benefit.

A review of the preceding sections should suggest that the concern with accurate and efficient sampling, complex analysis of variance, regression problems and simulation studies, with statistical tests of the various models is leading ecologists into complex areas of mathematics. Model-building has lead to an increased concern with differential and difference equations, partial differential equations and related topics of mathematical analysis. The concern with optimization and multistage decision processes has brought an operations research viewpoint into ecology leading to a concern with the literature in engineering and applied mathematics on gradient techniques, systematic search procedures and dynamic programming.

The sophisticated techniques of modern engineering have become a part of field ecology through telemetry, remote-sensing, and the more frequent use of electronic devices.

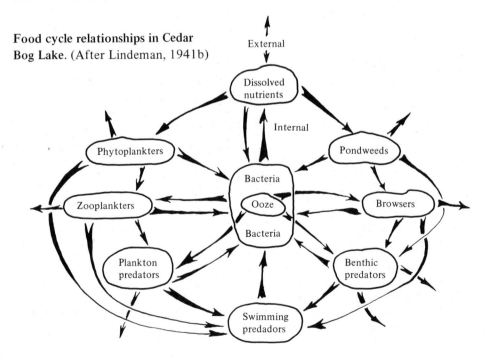

Food cycle relationships in Cedar Bog Lake. (After Lindeman, 1941b)

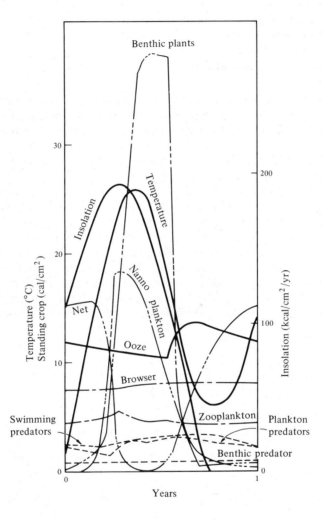

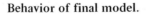

Behavior of final model.

As ecology becomes more concerned with the simulation of resource systems and the interaction of natural resource systems with man, a new link is being forged with the social scientist, the political scientist, the lawyer, the legislator, and the geographer. Eventually concerned citizens will have to become part of the team, hopefully before extinction.

§4.7. Impact on the Physical Environment. The effects of an industrial plant on its physical surroundings may include physical changes to the plant site, diversion of local resources, and changes in the off-site environment resulting from releases of materials, noise, and heat. There may also be indirect impacts from other development activities stimulated by industry.

The physical and biological portions of an ecosystem are so closely interdependent that they are often difficult to separate.

The severity of environmental impacts often vary widely, may often be reduced by appropriate changes in industrial practices.

While many industrial waste materials are foreign to natural systems, many others are normal constituents of the natural environment but may become "pollutants" if released in quantities sufficient to cause system changes considered detrimental. Even such common environmental components as water vapor or heat may be considered pollutants if introduced into the atmosphere or waterways in quantities sufficient to induce fog or icing conditions or to disrupt the fish and plant communities of a stream by raising water temperature.

§4.8. Environmental Toxicants. The environment is now contaminated with a myriad of potentially toxic substances, many of which have now become constituents of nearly everything that man uses. In trace amounts certain substances are essential to life; yet in larger quantities these same substances may be toxic. The balance between these two levels is often unknown. Because of man's own activities, substances not formerly present are now found in the human body. Many new substances are being formulated and new commercial applications for well known substances are being found almost daily.

While approximately 2 million chemical substances are already known and several thousand new chemicals are discovered each year, most of these compounds are laboratory curiosities that will never be produced commercially. However, several hundred of these new chemicals are introduced into commercial use annually. Of particular concern because of their rapidly increasing use are the metals, metallic compounds and synthetic organic compounds.

United States consumption of metals with known toxic effects has increased greatly in the last twenty years, and much of the data on their use underestimate the increasing pervasiveness of metals in our environment as the result of the many new metallic compounds being formulated and their use in ever widening product lines.

Similarly, the use of synthetic organic chemicals is growing rapidly, and although many of these substances are not toxic themselves the sheer number of such substances, their increasing diversity and use, and the environmental problems already encountered from some, such as DDT, the polychlorinated biphenyls and mercury, indicate the existence of a problem of worldwide magnitude.

Toxic substances enter the environment through complex and interrelated pathways. Among the key processes for which man himself is responsible are manufacture, consumption and disposal.

Raw materials are extracted from the environment in a crude or natural form and then successively refined, processed and manufactured into more diverse and complex forms. The processes associated with manufacture may produce wastes to which man may be exposed at each intermediate step. Such wastes often contain not only the original substance but also modified, and perhaps more toxic, substances. The end products of manufacture are consumed by man as food or are used as durable and non-durable goods. Such consumption and use can result in further exposure to toxic substances.

Consumption, however, is not the end of most consumer products, because after they have served their often limited purpose, they must be disposed of. Except for direct recycling, many current disposal methods return material to the environment, but almost always to a different place and often in a different chemical form. The disposal process alters the patterns of distribution and concentration of substances which naturally occur in the environment and may involve new chemical substances which may be more toxic than the original substance. Disposal processes often facilitate assimilation by living organisms, interaction with other chemicals and enhancement of inherent toxicity (synergism).

Both metals and synthetic organic chemicals are potential environmental toxicants, however, significant differences exist in the ways in which the two classes of materials enter the environment and ultimately affect man.

Metals are recovered from ore deposits either directly or as by-products in the course of refining other metals. Pure cadmium, for example, is not found uncombined in nature in commercially usable quantities. Commercial amounts are obtained as a by-product of smelting zinc. During the mining and refining processes, dusts and gases enter the atmosphere. Metallic salts formed during these recovery and refining processes can escape as waste products to surface and ground water. Undesirable concentrations of metals and metallic salts in the environment have been reported from such sources.

Metals, unlike synthetic organic compounds, have always been present in the environment, and living organisms, including man, have evolved in their presence. Blood and other tissues are composed of a complex mixture of chemical elements including the metals. Some metals are essential to life at low concentrations but are toxic at higher concentrations. In addition the form in which the metal occurs— whether as a pure metallic element, an inorganic metallic compound or an organic metallic compound—strongly influences its toxicity. The danger to mankind from metals in the environment depends on the chemical form of the metal and its concentration.

Most synthetic organic substances are not essential to life, though many share with metals the characteristic of toxicity. As with metals, the concentration and type of exposure to a particular synthetic organic substance are key factors in determining its effects.

The total effect of all toxic substances on a single species, even man, is difficult to quantify with accuracy because of the lack of knowledge about the sub-lethal effects of toxic substances. Although many environmental toxicants can cause death or injury if man is exposed to them in sufficiently high concentrations, the effects of long-term exposure to such toxicants at low levels, singly or in combination, are generally little known.

There are many difficulties inherent in the testing of chemical compounds for adverse effects. Extrapolation of data on dose effects obtained from animal studies to man must consider species variations in response. Toxic substances rarely occur in the environment in isolation, so that possible synergism or antagonism of two or more substances adds to the difficulty of adequate testing in the laboratory and interpretation of field data. Carcinogenesis, mutagenesis, and teratogenesis are often recognized as the sole significant biological effects.

Carcinogenesis is the ability of a substance to cause cancer. *Chemical mutagenesis* is the induction of genetic mutations — permanent and transmissible changes in the genes of an offspring from those of the parents of earlier generations.

Teratogenesis is the production of physical or biochemical defects in an offspring during gestation; it is limited to a particular offspring. The many deformed infants born of women who had ingested the drug thalidomide during pregnancy furnish a particularly horrifying recent example of teratogenesis.

Further complicating matters is the selective resistance of individual members of particular species. The effects of any given substance may vary among individuals of a single species as well as among different species. Differences in effects are often functions of age, sex, health condition and history, stress, different metabolic patterns and other less understood factors.

After a contaminant enters the environment, it may be diluted or concentrated by physical forces, or may undergo chemical changes including combination with other chemicals that affect its toxicity. A contaminant may be picked up by a living organism which may further change and either store or eliminate it.

While the results of the interaction between living organisms and chemical substances are often unpredictable, such interactions may produce materials that are more dangerous than the initial contaminant. A now classic example is the inorganic mercury which was thought to settle safely to the bottom sediments when discharged into water. Now it is known that certain bacteria convert inorganic mercury into toxic, soluble, organic mercury compounds, such as methylmercury, which pass throughout foodwebs eventually reaching man.

DDT, a synthetic organic chemical, is nearly insoluble in water, yet it is found in high concentrations among some fish-eating birds as a result of two factors: DDT is much more soluble in lipids (animal and plant fatty materials) than water so it selectively transfers from water to living organisms such as phytoplankton (microscopic plants that form the base of most aquatic food chains), zooplankton (microscopic animals that generally "graze" on phytoplankton), and this concentration process is enhanced by the process of biological or trophic level magnification, whereby successively higher concentrations of the toxicant are passed along toward the top of each food chain.

Synergism is another complicating phenomenon. Two or more compounds acting together or successively on an organism may affect that organism in a way greater than the sum of the separate effects of such compounds on the organism.

For example, the toxic effects of mercuric salts are accentuated by the presence of trace amounts of copper, cadmium acts as a synergist with zinc and cyanide in the aquatic environment to increase toxicity. Conversely, sometimes the presence of one substance lessens the effect of another substance of an organism. Arsenic, a toxic substance itself, counteracts the toxicity of selenium and has been added to poultry and cattle feed in areas where animal feeds are naturally high in selenium.

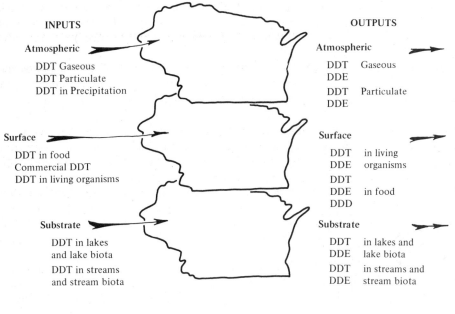

INPUTS

Atmospheric

DDT Gaseous
DDT Particulate
DDT in Precipitation

Surface

DDT in food
Commercial DDT
DDT in living organisms

Substrate

DDT in lakes
and lake biota

DDT in streams
and stream biota

OUTPUTS

Atmospheric

DDT Gaseous
DDE

DDT Particulate
DDE

Surface

DDT in living
DDE organisms

DDT
DDE in food
DDD

Substrate

DDT in lakes and
DDE lake biota

DDT in streams and
DDE stream biota

STORAGE

Surface

DDT in living organisms
DDT in dead tissue

Substrate

DDT in soil
DDT in deep water bodies
DDT in deep organic sediments

DDT exchanges: Wisconsin regional ecosystem.

An additional problem arises because many environmental toxicants are not exclusively air and water pollutants, but can be found in varying quantities at various times in air, water, soil, food and industrial, commercial and consumer products. The multiplicity of ways by which man can be exposed to these substances makes it difficult for agencies administering particular pollution laws to consider the total exposure of an individual to a given toxicant, a consideration essential for the establishment of adequate environmental standards. Also, no agency has considered itself completely responsible for all environmental toxicants present in and cycled throughout ecological systems. The agency most concerned with the total environmental distribution of its principle pollutant has been the Atomic Energy Commission where much of the significant work on material cycling and biological mangification has been done. Yet, until the recent *Calvert Cliffs* decision (1973) by the U.S. Court of Appeals, the AEC has not considered environmental concerns part of its statutory mission. Fortunately, now that the courts have directed the AEC to become concerned about the environmental effects of atomic energy among other things, it appears the AEC has the scientific capability to comply, albeit reluctantly.

§4.8.1. Environmental Material Cycling. Once released into an ecological system, materials may be transported by the atmospheric circulation and by water movement. They may in turn be absorbed and transported by living organisms . Chemical changes may also give the substance properties quite different from those of the form in which it was released. Thus, inorganic mercury released into waterways, once thought to remain in lake and river sediments as an environmentally inert substance, is transformed biologically into methylmercury. This much more toxic form is found in fish flesh and other organisms in water bodies receiving inorganic mercury discharges. This physical, biological and chemical movement and transformation is often referred to as biogeochemical cycling.

§4.8.2. Geographic Extent of Impact. While the effects of waste releases are often local, atmospheric and hydrologic transport permit effects far from the source of the waste material. An example of this phenomenon is atmospheric transport of sulfur and nitrogen oxides. High acid levels are being observed in rainfall in areas far from industrial centers. The high acidity of the rainfall is attributed to the transformation of sulfur and nitrogen oxides into sulfuric and nitric acid. While the exact sources of the sulfur and nitrogen oxide are not known in this case, research in Scandinavia suggests that these gases may be transported large distances. (Sulfur dioxide originating in England and in the Ruhr Valley may be transported an average of more than 620 miles before reaching Scandinavian soil in rainfall) While acid rain poses no apparent threat to health, at this time, it can damage structures and may cause serious ecological damage.

§4.8.3. Effects Over Time. Environmental impacts may be separated in time as well as space from the introduction of waste materials. The time lags involved vary widely and depend upon the properties of the substance and organisms involved. An example of long time lags is provided by DDT. A modeling study of DDT movement in food chains indicates that some organisms in a food chain are not expected to attain their maximum DDT concentrations in response to a constant level of DDT application for approximately four times the average lifetime of the longest lived organisms in the chain. The magnitude of the time delays involved here become apparent when one considers that some predatory birds have life spans in excess of 50 years. Another modeling study of DDT shows that even with complete cessation of DDT application in 1972, DDT and its principle environmental metabolite, DDE are expected to be present in human tissues in measurable concentrations 50 years later.

These time lags point to a clear need to consider the consequences of waste releases in advance, since the problems which arise may be with us for a long time.

§4.8.4. Biological Concentration of Pollutants. Many systems are capable of concentrating certain substances. This may occur simply by accumulation, such as the gradual buildup of sludge deposits on a stream or lake bottom, or such concentration may involve biological processes. Selective uptake and retention by organisms is a particularly important concentration mechanism. The degree of concentration varies widely and depends upon the organisms and substance involved, but some chemical elements have been found in aquatic plants and animals at concentrations up to 100,000 to 200,000 times higher than the concentration of the same element in the surrounding water. The effects of this concentration may be far reaching. Examples are the concentration of DDT and its metabolic breakdown products in certain food chains, and the concentration of mercury in fish food chains until certain fish attain methylmercury levels considered unsafe for human consumption.

§4.9. Direct Environmental Impacts. Direct environmental impacts will be considered here as those effects upon the physical environment resulting from the actual generation and distribution of electricity rather than from development or other human activity stimulated by the power plant or transmission lines. Effects in water, air and soil will be considered separately.

§4.10. Water Related Impacts. Industrial activities may include both manipulation of local hydrology, such as use of groundwater and creation of reservoirs, and release of waste heat and materials to local waterways.

§4.10.1. Manipulation of Local Hydrology. Large scale pumping of groundwater may result in both a lowering of the local groundwater table and a diversion of water needed elsewhere. Such effects are generally predictable using standard hydrologic techniques (non-equilibrium well hydraulic methods in conjunction with flow net analysis).

Construction of a dam generally raises the local groundwater table and may cause effects beyond the confines of the intended reservoir. Again, standard hydrologic techniques exist for the prediction of such effects.

Dam construction may significantly alter the quality of water released to the stream below the dam, and may also reduce streamflow. Loss of water in reservoirs to evaporation is likely to be significant mainly in areas with warm climates. Water quality changes in reservoirs are often significant, and may include a rise in water temperature and undesirable changes in the chemical content of the water.

Manipulations of hydrologic conditions may have effects beyond those intended and these effects may often be predicted by use of standard hydrogeologic methods.

§4.10.2. Thermal effluent. Temperature is an environmental parameter of great importance to biological systems. Long term temperature shifts may cause profound changes in natural community structure, and short term temperature changes may radically affect the behavior or survival of individuals of many species. An extensive literature exists on the biological effects of heated water discharges.

A particularly large source of thermal discharges is the steam electric power generation industry. This source is becoming increasingly important with the trend toward construction of larger generating facilities and the increasingly frequent construction of nuclear facilities which release more heat into cooling water per unit of electrical energy than do fossil fuel burning plants. (See §1.16.1)

The temperature change caused in a stream or lake by a given effluent release rate depends upon effluent temperature, the quantity of receiving water available for mixing with effluent, and mixing efficiency. The mixing efficiency is dependent upon the technique used in effluent release and the hydrodyamic properties of the receiving water body.

River flow rates often vary widely on a seasonal basis and the effluent may represent a considerable percentage of total flow during low flow periods, thus increasing the temperature rise. Many hydrologic models exist for the prediction of water availability, and many models also exist for simulation of dispersion of thermal discharges in lakes and streams. The quality of predictions obtained from these models is variable, but many are sufficiently well developed that the range of thermal effects to be expected can be estimated.

Alternatives to thermal release directly into a stream or lake include the use of cooling towers, cooling ponds, or canals which allow some reduction in effluent temperature before release. Tradeoffs to be considered include the economic and aesthetic costs of large and relatively expensive cooling towers, meteorological and ecological factors related to the introduction of large amounts of water vapor into the atmosphere by some cooling towers which may cause fog and icing conditions, and the ecological, aesthetic and economic value of the relatively large amounts of land required by cooling ponds.

§4.10.4. Material Releases. Industrial wastes include a broad spectrum of substrances too numerous to list here individually. Many of these compounds are known to be toxic, and the environmental and human toxicity of many others is unknown. Further complicating the picture is the fact that many wastes may be released by a large variety of industrial processes.

Materials released into aquatic systems may be either in dissolved or particulate form. Particulates will generally settle to the bottom, with larger particles lost close to the source and finer particles remaining in suspension longer, especially under turbulent conditions. Dissolved materials will usually be transported longer distances unless withdrawn from the water by organisms.

Suspended particulates generally decrease the aesthetic appeal of water, decrease light transmission, thereby reducing the amount of light available to vegetation for photosynthesis and may be directly detrimental to fish. Settling of particulates onto stream and lake bottoms may effect habitat suitability for desirable species of aquatic organisms.

Some organic compounds introduced into an aquatic system may be toxic, either directly or after concentration by food chains. A more common problem is depletion of the dissolved oxygen content of the water as the organic substances are broken down by oxidative reactions. Lowering of aquatic oxygen levels generally changes the species distribution of fish and other organisms present in a body of water, usually discriminating against game fish species. Periodic changes in dissolved oxygen content can result in large fish kills. The accuracy with which transport of materials by aquatic systems can be predicted is highly variable, depending in large part upon knowledge of the environmental interactions of the substance involved and the complexity of the interactions. No generalizations can be made except to say that much success has been achieved already and that this is an area of active research. An example of such a model is the Hydrologic Transport Model developed in 1970 by Huff and Kruger, which has been successfully used to simulate movement through a watershed of strontium-90 and cesium-137 originating in fallout from nuclear explosions. Models describing the effect of organic waste releases on dissolved oxygen content in rivers have long been in use. Mathematical techniques and data for prediction of movement of many substances through both aquatic and terrestrial food chains are also available. While data gaps exist and movement of all substances cannot be predicted with confidence, much is known about the hydrologic transport and food chain concentration of many substances.

§4.10.5. Impacts on Soils. Wastes released by industrial plants may eventually find their way to soils. In general, less attention has been given to the consequences of waste deposition on soils than to impacts on air and water.

One area of increasing public concern is the fate of potentially toxic materials deposited on soils. Some of these can accumulate in the soil to toxic levels, and some may be taken-up and concentrated by plants, thus finding their way into food and animal feed. There are a number of unanswered questions regarding the pathways, retention, and toxicity of many elements in soils, plants and animals.

§4.10.6. Subsurface Geology. Groundwater hydrology is especially significant where extensive pumping of groundwater is planned. Full knowledge of the characteristics of the aquifer is essential. The mapping of specific geologic formations is also critical in areas where seismic activity is probable. This consideration is especially important for installations in which rupture of containers would be disastrous, for example, power plants, whether nuclear or fossil-fueled.

§4.11. Indirect Impacts. Generating, distributing and utilizing electricity and siting an industrial plant may have physical environmental impacts other than those attributable directly to such activities. Some industries, in particular, electric power plants and high voltage transmission lines, encourage the location of other industries nearby. These industries may in turn cause environmental impacts such as those just discussed. The economic effects of the plant may also encourage local commercial and residential development.

§4.12. Impact on Biological Systems. This chapter considers the possible biological effects of generating, distributing and using electricity on the environment. It is sketchy, but should serve as a quick guide to things the biologist knows that the concerned citizen should know.

A natural system is a complex interaction of plants and animals, air, water and soil and geology which maintains its existence as a system by using energy to cycle materials. Disturbing any of the parts of the system disturbs the entire system. Disturbing a system faster that it can absorb the disruption will destroy it.

The most commonly used method for evaluating environmental impact is a matrix such as that developed by Luna B. Leopold, Frank E. Clarke, Bruce B. Hanshaw, and James R. Balsley of the United States Geological Survey and published in USGS Circular 645, *Procedure for Evaluating Environmental Impact.*

Obviously, any industrial development will change the natural system in its vicinity. It is the determination of which changes constitute harm that gives trouble, since the dynamic equilibrium of the ecosystem is not only complicated, and only partially understood but also subject to fluctuations, sometimes wide fluctuations. For instance, power plants are often built near water sources such as rivers, and the flood plain of a river is a system which is subject to natural disturbance at regular intervals. This fact creates a necessity to assess changes wrought by human intervention against time. If a river floods and washes all the vegetation off a given area of bank, there is a predictable succession of plants and animals that will return and grow there, until the next disturbance, which may be years, or even centuries later. If that area is built upon, or otherwise appropriated by man, the natural succession will not take place, and even though the building was put on what was empty land at the time, man has unquestionably altered the natural system. There are areas which are naturally disturbed by fire, like the chaparral areas of Southern California, which are changed by man's suppression of fire. Equally real, although less obvious, changes occur all the time in any natural system.

§4.13. **Assessment of Damage to Biological Systems.** Ordinarily damage is calculated in human terms—monetary loss, or hazard to human health, however, these terms have broadened to include damage to (human) aesthetic or recreational values (See Chapters 11 and 12).

Even in human terms, harm is sometimes hard to assess if it is incremental and over a long time span. Statistically, city dwellers get emphysema more often than country dwellers, but which air pollutant emitted by which industrial operation shortened which person's life and by how much? Some sorts of environmental disruption have consequences which are not perceived for years. DDT, for example, concentrates in certain food chains and may demonstrate effects miles from the source and years after the application.

§4.13.1. **Commercial and Game Species.** Immediate losses of wildlife are one sort of harm arising from industrial development. If these losses include timber or game species, they are calculable in money, and are usually afforded legal notice. If industrial development is likely to injure a trout fishery, or a commercial oyster bed, or a stand of merchantable timber, courts have been convinced that such development entails harm to the environment. Commercially valuable species and game have received much study, and about them many trustworthy predictions can be made. There are governmental agency experts available to make such predictions.

It should be borne in mind that wildlife management techniques are directed toward increasing supplies of some species at the expense of others. What is the best thing to do for deer may not be the best thing to do for the forest, or the wolf, and, like any sort of human interference in a natural system, game management practices have far reaching consequences.

§4.13.2. **Species Without Commercial Value.** Species which have no value as game, food, timber or fur have, ordinarily, no immediately calculable monetary value. There are a few exceptions—species of extraordinary beauty or rarity which people may pay to see. The boat which carries people to view the whooping cranes at Aransas Wildlife Refuge makes a substantial profit each winter—a substantial but unusual compensation. Some well-publicized rarities have a considerable public following, but most do not. With the drying of the Everglades, the beautiful Everglades kite is threatened by the loss of its food supply. Many birdlovers are vigorously trying to protect the kite, but there doesn't seem to be much concern for the passing of the snail that it eats. "Save the Everglades snail!" is just not an effective rallying cry.

The same is true of many species of small fish, insects, plants, and amphibians. No one knows "what they're *good* for." They are unlikely to have been studied intensively, and their ecological contributions are likely to be unknown.

There is also a possibility that industrial development, especially in undeveloped areas may harm or destroy species which are not even known to be present, since many areas of this sort have never been assessed biologically.

§4.13.3. Loss of Diversity. For some years ecologists assumed intuitively that the complex interactions among species imparted to the community several characteristics that were more than those which could been attributable to the activities of each species taken separately. The whole seemed more than the sum of its parts. For the last seven or eight years the studies of diversity in ecosystems have legitimized this intuition. Diversity can be thought of as a web through which energy and materials flow and maintain the ecosystem. Intersections in the web represent species. Obviously the more species the system has the more options it has. A high number of options increases the probability that the system may recover if some part is damaged. The loss of a species may reduce the capability of the system to withstand sudden changes. The chestnut blight did not significantly affect the diverse forests of eastern North America but the spruce budworm seriously reduced the amount of forest in the undiverse forests of Canada. This is the typical plant ecologist's definition of diversity, simply the number of species per number of individuals in the sample. An individual of a rare species has the same value as one of an abundant species.

However, among some animal ecologists and some plant ecologists, individuals of rare species are thought to play more important roles than their more abundant neighbors. Evidence indicates that species may be rare because they use or convert rare materials under unusual conditions. If these materials are limiting, the dynamics of the whole ecosystem may depend on these rare species to an extent much greater than their numbers would indicate. These key species need not be spectacular to humans.

Several ecologists have suggested that diversity and thus stability increases geometrically with increase in contiguous area. Doubling the contiguous area quadruples the diversity. All else being equal the community might be considerably more harmed by taking a strip out of its middle rather than off one edge.

§4.13.4. Disruption of Natural Succession. Ecosystems are not fixed, they are constantly changing. If there is little human interference these changes are predictable. This predictable series of changes called succession ends in a climax plant community. The climax differs with the climate. One of the difficulties we see is an inability to measure harm to the community which would ordinarily come had the system not been altered.

§4.14. Effects of Industrial Construction. An industry may affect an ecosystem either directly, as a result of any structures or paving or indirectly through its effluents. The construction itself displaces the plant community at the site.

The movement of air, water, and wildlife through an industrial site is usually disturbed. Any large building creates updraughts that affect migratory bird patterns, while millions of birds collide with towers, smokestacks and tall buildings each year. Residential buildings across the floor of San Bernadino Pass in Southern California have stopped the seasonal movement of desert animals through the pass. Dams, channels and intake systems have entrapped fish of a wide variety of species.

§4.15. Effects of Industrial Production. Industries have been known to introduce toxic materials and heat into air and water and to modify local weather. Toxic materials may be born in air or water. There are many reports of plants being killed by airborne sulfur oxides and ozone. Fish and aquatic plants have been killed by chemicals, particulate matter and organic material in water. The decaying organic material may produce materials toxic to other life but the greater effect is that organic materials use up the oxygen needed by other organisms and deprives them of oxygen, nutrients and light.

Although toxicity of materials was historically defined in terms of human health, toxic effects on fish and wildlife now must be considered. In some cases toxic materials are hard to identify because of synergistic interactions. Even when the material is a well known poison such as cyanide, the evidence linking the discharge of cyanide to a particular fish kill in the Houston Ship Channel was circumstantial. For environmental toxicants such as DDT and the PCB's which are very long lived and widely used and which accumulate throughout biological food chains, individual responsibility is particularly difficult to establish.

§4.15.1. Heat. The introduction of heat into a lake or stream has varied results. In some cases it has been shown to improve a warm water game fishery, while being only weakly linked with early and severe algal blooms in others. In some areas, waste heat keeps parts of lakes open all winter attracting Mallard, Black Duck and Coot to overwinter. However, since there is scant winter forage, these birds must be fed. In some cases the heated water encourages fish to feed and be active, while at the same time killing the potential food supply. In Lake Michigan local thermal patterns in the water may cause 15°F. changes in a few minutes, making measurements of the ambient temperature and the impact of effluent uncertain. When, as in the case of Lake Michigan, the amount of effluent is very small compared to the volume of the lake, the impact seems small, however, a substantial increase in thermal effluent, could materially affect the quality of the lake fishery. Here, as in many cases involving multiple use of public resources, no one use causes the ultimate degradation of the resource which really represents the sum of each incremental addition.

In southern states summer ambient water temperatures can be quite high in shallow waters. Shrimp, oysters and important fishes breed in the shallow, brackish lagoons along the intercoastal waterways. The young of these species are very sensitive to small changes in heat. A plant dumping its heated water into access channels can reduce hatching success and deprive the fish of large breeding areas.

These same brackish areas, are very dependent on seasonal floods of fresh water from the rivers and streams that feed them. The young of many species can only survive if the water is sufficiently brackish, which requires the appropriate dilution of salt by fresh water during the spring. Fish, such as salmon, and shrimp have adapted their life histories to come in on the spring flood when the water is high and properly diluted. Many industries divert water or create impoundments upstream in order to store water in the spring for use during the dry season. This deprives the spring breeders of both enough water and water of sufficient dilution. Again as with heat, no one particular diversion may interfere with the spawning cycle, but the total of each incremental loss certainly does.

§4.15.2. Weather. Local weather results from the interaction of moisture and heat, both of which may either originate locally or be brought in by moving air or water. Huge areas of concrete do not store water and tend to retain heat. As a result cities are usually warmer than the surrounding countryside. The increased heat further aggravates the dryness and dustiness by increasing the evaporation rate. Small factories have an affect on local air currents. Summer thermal uprisings can be detected by watching soaring birds.

Moisture regimes are modified both by drying and by the addition of moisture via cooling towers and cooling impoundments. Desert plants (Saguaro) have been reported to die as a result of increased humidity caused by the large impoundments along the Colorado River.

§4.16. Irreparability of Biological Systems. Once the possibility of damage has been established, the next problem is to determine whether the harm is irreparable. Some sorts of biological losses are renewable. For instance, along the northwest coast selectively cut conifers will be replaced within a reasonable time. Some sorts of losses *seem* renewable, but require new energy inputs into the system, such as timber replacement in a slower-growing forest which requires the addition of fertilizer. The substitution of faster growing trees for slower reduces diversity, like replacing a mixed southern hardwood forest with a slash pine plantation, so that measuring "renewal" in terms of board-feet of lumber may be misleading.

Other environmental impacts may reduce animal or plant species for a time, and then the question of renewability revolves around whether conditions have been so changed as to preclude restocking, and if conditions favor restocking, whether there is a sufficient population of the species elsewhere to provide the breeding stock.

If an industrial plant dumps a toxin into a stream once, and kills all the fish, then stops dumping the toxin, the fish will probably return. Unless an entire species is completely destroyed in the kill, in which case, the loss of that species would be irreparable. If the effluent permanently changes the character of the stream so as to make it uninhabitable by some species, the condition would also be essentially irreparable.

Changes, such as soil erosion, timber cutting, water pollution or air pollution, can make a relatively stable system, unstable; while other changes, such as fire and flood control and breakwaters, can artificially stabilize a relatively unstable. Either may constitute irreparable damage to the system as a whole.

§4.16.1. Cost of Repair.

§4.16.1. **Cost of Repair.** A number of industrial sites have planned and built natural areas as part of the landscaping around their buildings. Usually the dominant plant species are introduced, and often exotic species are used. These plantings, while aesthetically pleasing and of educational value, do not immediately become the ecosystems after which they are modeled. They may come close after 20 years or so with grasslands and four times that long with forests. These small projects are initially very expensive, so that we can see that it would take considerable time and money to replace a natural community that cost so little to destroy. Replacement cost would be a useful method to determine the value of lost ecosystems. The same could be used for species, however, an extinct species is impossible to replace and therefore priceless or valueless depending on your point-of-view.

§4.16.2. **Rate of Repair.** Natural ecosystems vary in the rates at which they may be repaired. A trampled sand beach can be substantially restored on the next high tide; but a single footstep on alpine tundra can kill a forget-me-not the size of a quarter that has grown for 75 years. Chop down an aspen grove and it can return in 10-20 years, but climax redwood or high altitude pines may take 500-1,000 years. If the land is plowed or graded, the pines and forget-me-nots will never come back even if left alone. The redwoods might recover in several hundred years and the aspens in 30-50. The rate of repair is an indication of sensitivity of the system and magnitude of harm done to it.

§4.17. **Impact on Air.** Air pollutants are conveniently divided into two classes: gases and particulates. Particulates are transported distances which are dependent on their size as well as meteorological conditions, with the largest particles being deposited closest to the source. Deposition processes include *fallout,* or simple earthward movement by gravitational acceleration, *washout,* the sweeping of materials out of the atmosphere by falling precipitation, and *rainout*—the formation of water droplets or ice crystals around particulates which serve as condensation nuclei, and subsequent removal in precipitation.

Gaseous wastes are transported by the atmosphere until removed, usually after sorption onto water droplets or other particulates. Physical, chemical and photochemical processes may transform such wastes during transport. An example is the chemical transformation of sulfur dioxide to sulfur trioxide, which may then combine with water to form sulfuric acid, resulting in acid rain.

Impact of atmospheric particulates ranges from soiling of surfaces to human illness arising from their inhalation. Atmospheric transport provides a mechanism for accumulation of toxic materials in surrounding ecosystems to levels at which they may become damaging.

Impacts of gaseous wastes range from human illness to corrosion of man-made structures.

Technology exists for removal of up to 99% of particulates from atmospheric discharges, (See §3.19.1), however removal of gases from atmospheric discharges is usually more difficult.

Movement of wastes discharged into the atmosphere depends upon the height of the stack used, local topography and meteorological conditions. A large variety of models exist for simulating atmospheric transport of materials, including those originating in single point sources, line sources, (such as a roadway), multiple point sources, and area sources. Multiple point source and area source models lend themselves well to urban situations. Area source models in particular lend themselves well to air quality management through land use regulation. A good summary of model uses and effectiveness in urban situations is the *Air Management Research Group Report on Models for Prediction of Air Pollution* (Organization for Economic Cooperation and Development, 1970).

Model precision and accuracy varies depending upon the model used, input data available, and the type of prediction sought. In general, predictions of concentrations averaged over time and space are more reliable than predictions for a particular time or for a small area. Predictions for gas transport tend to be more reliable than those for particulates.

§4.18. Problems in Determining Causes of Environmental Damage. We have just reviewed the immediate environmental damage which might result from industrial impact. These are the most obvious effects to assign causes for, but they may not be the most important.

§4.18.1. Space. Some effects are separated from their cause by intervening space. For instance, an industry may require electricity, and result in further impact on the area where the electricity is generated. If water is impounded, not only is the area which is flooded affected, but the character of the ecosystem downstream may be radically altered. Smoke may change weather downwind. It is necessary to consider a wide range of effects in order to fully understand the impact of generating, distributing and using electrical energy.

§4.18.2. Time. Other effects can be separated from their causes by time. Persistent toxins may take years to concentrate in the food chain sufficiently to do measurable harm. Reduction of breeding space may endanger a species but not be apparent for as much as twenty years.

§4.18.3. Synergism. To complicate the problem still further, two or more causes can combine to result in an impact greater than either would have had alone.

§4.19. Social and Economic Effects of Industrial Development. Determination of the environmental impact of industrial plant siting must consider the effects of industry on the socio-economic environment of a region. The social and economic effects of industrial development vary widely with the type of industry and the characteristics of the region in which the industry is locating. This section delineates some of the regional characteristics which determine the extent of impact.

The social and economic effects of an industry locating in a suburb of a major metropolitan center will differ drastically from those of an industry locating in or near a small agriculturally oriented rural town.

§4.20. The Metropolitan Suburban Setting.

The impact of an industrial plant in a metropolitan suburban setting is extremely complex. A metropolitan suburban community whose principle growth has been in population rather than economic activity (the residents work outside of the community) usually finds itself in a disadvantageous tax position. Extensive and costly systems of services and facilities have been provided from a property tax base which has a relatively low value per unit of land area (residences). When an industry with a relatively high value per unit of land area enters the community, the tax position improves and more development occurs. A suburban community with extensive facilities and services usually incurs far more benefits in tax dollars from new industry than the additional cost of facilities and services required. Favorable labor force characteristics, extensive public services and facilities, and the presence of new industry may induce related industries to locate in the community. More services are then needed to support the new industrial firms and their employees and more jobs are created. This continuing cycle is often called a *multiplier effect*.

The effects of an industrial plant siting in a metropolitan suburban community will very likely include associated or support industry, industrial and business services and increased retail and service businesses to support the growing work force. Sewage and water facilities, highways, and community services expanded to accommodate the new growth may be attractive to other industries totally unrelated to those already in the community. Residential areas will also expand to house the growing work force.

The relationship between the indirect social and economic effects (through land use) and the natural environmental systems becomes immediately apparent. Additional industries may increase existing pollution levels. All development displaces natural communities and may drastically alter the hydrologic system—especially quantities and patterns of storm-water run-off.

Multiplier effects caused by industrial plant siting in a rural setting are not as great.

§4.20.1. Changes in age composition.

Social demographers have long been aware of a surplus of old people in small towns and rural counties. As early as 1942 Smith labeled the small town as "America's old folk's home". Two demographic trends: (1) the steady out-migration of young adults and (2) the in-migration of retired urbanites and agricultural people, have produced a top heavy age distribution in these areas. This abnormal age distribution is an important economic determinant in many rural areas and small towns.

One of the major arguments for bringing industry to depressed areas is that the expanded employment opportunities will stifle the out-migration of young adults and thereby re-establish a more or less normal age structure, and research has generally substantiated this argument.

§4.20.2. **Increase in Income.** It is well known that many small towns and rural counties are lagging in terms of economic growth. One of the major reasons these communities are attempting to attract industry is to establish a stable economic base and thereby raise the general level of living of their citizens.

Research along these lines has given solid support to the assumption that industrial development produces an increase in per capita income.

§4.20.3. **Increased employment.** Many residents of small towns and rural counties are underemployed. Advocates of industrial siting in these areas argue that the expanded employment opportunities will decrease unemployment and underemployment, and, increase employment not only in jobs at the new plant but also result in a multiplier effect—generating employment in other areas, e.g. filling stations, restaurants, construction, and the service industries.

Income and employment changes take two forms—direct and indirect. The direct change is the income (payroll) and employment at the new industrial plant. The indirect change is the income and employment caused by the new jobs and new income at the plant. An example of indirect employment change is the hiring of more clerks by merchants to handle the increased business caused by the new plant and its employees.

The indirect impacts are commonly measured by a multiplier. The employment and income multipliers, while similar, differ in magnitude because the jobs generated in different sectors of the economy pay different wages. The multipliers can measure both direct and indirect change or indirect change only.

Several techniques exist for analyzing the economic and spatial relationships among industries. Inter-industry analyses are premised on the concept that the ties or linkages between businesses transmit changes in economic activity from one business to another. For example, an increase in steel production can lead to increased activity in the transportation industry increased coal production for coke, and additional activity for businesses servicing the iron mining industry. The export-base and inter-industry approaches both make comparable estimates of the indirect economic impacts. The inter-industry analysis provides a more detailed description of the changes, but this comes at the expense of increased data costs.

§4.21. **Limitation of Analysis Techniques.** While the techniques to estimate the multipliers are generally accepted, there are limitations to the techniques that must be considered. While the techniques are generally accepted, the results are not general. The multiplier for a manufacturing firm locating in northern Wisconsin can vary greatly from the multiplier for that firm if it located in metropolitan Chicago. The reason is that the necessary supporting and linked services for the firm are likely to be available locally in the Chicago area but not in northern Wisconsin. The indirect effects from the new plant still occur but they occur somewhere other than in northern Wisconsin.

The reverberations from a change caused by manufacturing can vary significantly from a change caused by wholesale and retail trade.

The geographic spill-overs are important considerations when evaluating economic and environmental impact of industrial development. But geographic spill-overs are only recently being empirically examined by the techniques of export-base and inter-industry analysis.

Inter-industry analysis provides a clue to the need for supporting business services by an industry but fails to incorporate fully the spatial aspects, i.e., where the supporting firm locates. Another technique—industrial complex analysis—is an attempt to combine the spatial aspects with inter-industry analysis. This technique attempts to minimize the costs to a subsystem of activities (not a single industrial sector) which are subject to important production, marketing or other interrelations. While this technique is not designed for the prediction of ancillary businesses, it appears to be useful for that purpose.

§4.22. Negative Effects. While the positive social and economic effects of industrial development are fairly straightforward and widely recognized at both professional and lay levels, the negative consequences of such siting have received much less attention. Several have argued that, since most Americans equate industrial development with progress, there has been little concern over the possible liabilities of such development. The severe social, economic and environmental problems that plague our over-industrialized metropolitan areas, however, clearly indicate that industrialization is not the panacea for all problems. Some of the most important potential problems of locating large industry in small towns are:
1. income inequility,
2. "leakage" of benefits, and
3. size and density of population.

§4.22.1. Income Inequality. Although industrial development generates an increase in per capita income these increases are not proportionately distributed throughout the population. more specifically, some very sizable segments of the population may actually be adversely affected by industrial development.

Old people comprise a large portion of the population of small towns and rural counties, and many of these elderly individuals settle in such areas to take advantage of the fact that the cost of living is lower than that of large cities. When a large industry moves into the area, however, the entire economic structure of the region is altered. Old people, already at the lower levels of the economic hierarchy, are likely to be relegated to an even lower economic status as the income of younger individuals sharply increases.

While the absolute income of the elderly may not decline, their economic status relative to younger residents of the area can be drastically reduced. Given the fact that the elderly people are already concentrated in small towns and rural counties this problem should be recognized by those who would encourage the industrial development of these areas.

A second segment of the population which may be adversely affected by industry consists of the female heads of households in the area. Many of these female headed families are located in small towns and rural counties which are actively seeking industry. And the type of industry which communities are most eager to attract are heavy manufacturing facilities. These plants have predominantly male payrolls. Thus female household heads in the area may find themselves in the same position as the elderly—unable to take advantage of expanded employment opportunities. While not all industries have such a high proportion of males on their payroll it should be re-emphasized that the types of industries which most communities are seeking to attract have a high male specific labor demand.

§4.22.2. "Leakage" of Benefits to Surrounding Areas. Large commuter fields often develop when a plant is located in a rural area or a small town. In many instances people commute to the plant from more than 50 miles away. This indicates that the benefits of the plant may be spread over such a wide geographical area as to have a dilute effect. While this phenomenon may not at first appear to be a problem it can easily become one if leakage exists to such an extent that the community in which the plant is located receives only a small share of the benefits but almost all the burdens of the new industry.

Many communities, eager to attract industry, will develop industrial parks, promote tax subsidies, expand the school system, and revamp the transportation system, all with the idea of recouping these expenses when industry locates in the community. But when the siting actually takes place the benefits may be distributed over such a wide area that the community cannot regain its expenses let alone attain economic stability.

§4.23. Electricity and Nuclear Energy. In the United States most electricity is generated via the steam cycle. (Figure 1.26) The steam source is fossil fuel, a fire box and a boiler which provides a means to effect the transition from stored chemical energy in the fuel to mechanical energy of the expanding steam which is utilized in the turbine to turn an electric generator wherein the energy of motion is con-

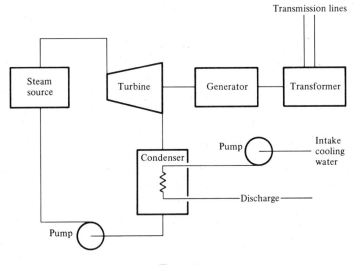

Fig   4.1

verted into electrical energy which is then distributed to the customers. In the nuclear electric power plant the fossil fuel, fire box and boiler are replaced by a nuclear pertinent features of the nuclear processes (see also §§2.20-2.21).

§4.24. Nuclear Fission. The fission process, in which the nucleus of an atom splits apart into two or more fragments, generally occurs in very heavy nuclei. Although fission occurs spontaneously in only an infinitesimal fraction of U^{238} nuclei, it can be easily induced by neutron absorption. In the case of U^{235} slow neutrons suffice; whereas U^{238} fission only occurs when the neutron energy exceeds 1 Mev. In natural uranium $<1\%$ of the atoms are the easily fissionable U^{235}, whereas the remainder are U^{238}.

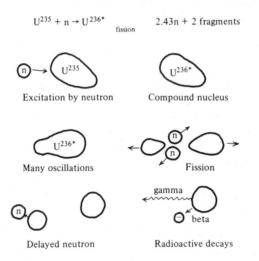

$$U^{235} + n \rightarrow U^{236*} \xrightarrow[\text{fission}]{} 2.43n + 2 \text{ fragments}$$

Excitation by neutron   Compound nucleus

Many oscillations   Fission

Delayed neutron   Radioactive decays

Fig   4.2

It is not U^{235} which undergoes fission, but actually an excited state (*) of U^{236} which is produced by absorption of the neutron. This occurs in the following sequence:

$$_{92}U^{235} + {_0}n^1 \rightarrow {_{92}}U^{236*} \rightarrow \text{(fission or other decay made)}. \tag{4.1}$$

The U^{236*} has a finite life-time ($T_{\frac{1}{2}} \approx 10^{-14}$ seconds) before the decay event occurs. During this time the nucleus behaves much as a drop of fluid undergoing a variety of oscillatory motions until it breaks apart. A typical fission event would be as follows (see Figure 2)

$$_{92}U^{236*} \rightarrow {_{36}}Kr^{90*} + {_{56}}Ba^{143*} + 3{_0}n^1 + \sim 200 \text{ MeV} \tag{4.2}$$

The two primary fragments, Kr^{90*} and Ba^{143*}, are usually formed in highly excited states, which quickly decay to their lowest state of excitation by prompt emission of one or more γ rays, and then both products subsequently undergo a series of radioactive decays. In addition to the fission fragments, a few neutrons are produced which may be absorbed by other U^{235} nuclei. This gives rise to a chain reaction under appropriate conditions. Practically all of the neutrons are emitted in the initial fission event (prompt neutrons); but in the case of slow neutron induced fission of U^{235} a fraction of 1% of the neutrons are emitted by highly excited fission fragments a finite time after the formation of the fragment. These *delayed neutrons* may be emitted as long as a second or two after fission, and are extremely useful in controlling the chain reaction.

The total energy release from a fission event is very large, ~ 210 MeV, and appears as kinetic energy of the fragments and neutrons, γ-rays emitted promptly after fission, and the kinetic energy of particles and γ-rays emitted from the radionuclides.

The kinetic energy of various fragments and particles is quickly converted to heat by means of collisions with surrounding atoms in the fuel and other structures.

Energy from radioactive decay, however, is released at later times in accordance with the half-lives of the various isotopes. Neutrinos and the energy they carry are lost from earth since they have negligible probability of interacting with matter.

The fission event has hundreds of variations and that shown in Equation 4.2 is only one. Radioactive decay of the fragments follow decay chains such as:

$$Kr^{90}(32.3s) \rightarrow Rb^{90}(43m) \rightarrow Sr^{90}(28.9y) \rightarrow Y^{90}(64h) \rightarrow Zr^{90}(stable) \quad (4.3)$$

and,

$$Ba^{143}(12s) \rightarrow La^{143}(14m) \rightarrow Ce^{143}(33.0h) \rightarrow Pr^{143}(13.6d) \rightarrow Nd^{143}(stable) \quad (4.4)$$

The half-lives are shown in parentheses (s = seconds, m = minutes, h = hours, d = days, and y = years). It is seen that this one fission event yields two fragments each of which decay in four successive steps until they convert themselves into stable isotopes of zirconium and neodymium. Note that one of the decay products in the Krypton chain is the strontium isotope, Sr^{90}, which has a 28.9 year half-life, whereas the other seven isotopes all have relatively short half-lives.

More than 120 radioactive isotopes have been identified among fission fragments. The half-lives of these isotopes range from fractions of a second to millions of years. The mass distribution of the fission products from U^{235} + n is shown in Figure 4.3. Note that the largest yields occur in the vicinity of mass numbers 90 and 140 indicating that the nucleus tends to break up into one small and one large fragment when slow neutrons are absorbed.

It is the radioactive isotopes among products shown in Figure 4.3 which create most of the public health and safety problems associated with nuclear energy. Only limited quantities of radionuclides can be tolerated in the biosphere.

There are several other isotopes which exhibit neutron induced fission similar to U^{235}. Among these, U^{233} and Pu^{241}, are of practical interest for nuclear energy production. Such nuclides are called *fissile*. All except U^{235} are generally considered to be "man-made" isotopes since their half-lives are short compared with the age of the earth. However, detectable amounts of U^{233} and Pu^{239} have been found in

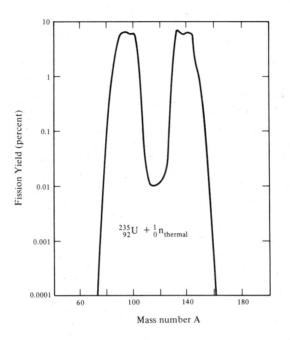

$$^{235}_{92}U + ^{1}_{0}n_{thermal}$$

Fig   4.3

natural ore deposits. In the discussion of chain reactions, the concepts apply equally to U^{233}, U^{235}, Pu^{239} and Pu^{241} except that the values of various constants are slightly different.

§4.24.1. The Fission Chain Reaction. From the example cited in equations 4.1 and 4.2 it is seen that a fission event initiated by a neutron produces additional neutrons. When U^{235} undergoes slow neutron fission, slightly more than two neutrons are emitted on the average; however, since some of the absorbed neutrons lead to decay of the U^{236*} by means of γ-ray emission rather than fission, the actual number of neutrons produced and available per neutron absorbed in U^{235} is slightly less. In any practical system some of the neutrons are lost from the chain reaction by leaking out of the system or by being captured in materials other than the U^{235} which might be present.

The kinetic energy of the neutrons released by the fission events can vary from <0.1 to >10 Mev; however most of those produced lie in the range of 1 to 2 MeV. These are commonly referred to as *fast neutrons*. Not all these neutrons are absorbed and lead to fission with equal probability. The fission probability of the fissile isotopes is much larger for neutrons of thermal energy . The fast neutrons can be degraded in energy to the thermal region, which permits a chain reaction to be sustained with relatively dilute fuel. This is done by providing a medium, called a *moderator,* which consists of nuclei with small mass and low absorption probability for neutrons. Since in an elastic collision with hydrogen the neutron can lose significant fractions of its energy (on the average ½), a hydrogenous medium such as water is an effective moderator. Heavy water (D_2O) and graphite are also suitable.

Because water also is a good heat transfer medium, a whole class of reactors, called *light water reactors (LWR)*, have been developed in the United States which use thermal neutron chain reactions with water as the neutron moderator. The BWR and PWR types mentioned earlier are in this class. The HTGR employs graphite as the moderator and helium as the heat transfer medium. When a reactor uses un-moderated neutrons in the chain reaction, it is called a *fast reactor*.

§4.25. Breeding. Power reactors of the present era—Boiling Water Reactors, (BWR's) and Pressurized Water Reactors, (PWR's)) are fueled with uranium slightly enriched in U^{235} (see Table 4.1). The bulk of the fuel consists of U^{238}. During operation some neutrons are captured by the U^{238} producing Pu^{239} in the following reactions:

$$U^{238} + n \rightarrow U^{239} (23.5m) \xrightarrow{\beta^-} Np^{239} (2.35d) \xrightarrow{\beta^-} pU(2.44 \times 10^4 y). \qquad (4.5)$$

Since Pu^{239} is fissile, it participates in the chain reaction in a manner analogous to the U^{235}. In a typical light water reactor approximately 0.55 atoms of Pu^{239} are created for each U^{235} atom consumed. The 0.55 is called the *conversion coefficient*, K. The production of Pu^{239} enhances the amount of the fuel. When $K < 1$, the reactor is called a converter (or burner).

In the High Temperature Gas Reactor (HTGR), thorium plays an analogous role, through the reaction:

$$Th^{232} + n \rightarrow Th^{233}.2m) \xrightarrow{\beta^-} Pa^{233} (27d) \xrightarrow{\beta^-} U^{233} 6 \times 10^5 y). \qquad (4.6)$$

The conversion coefficient for HTGR is somewhat higher. Eventually U^{233} will replace U^{235} in HTGR fuels.

If $K > 1$, the reactor is called a *breeder reactor*. This condition cannot be achieved in *thermal reactors* employing the U^{238} - Pu^{239} cycle, although it appears possible for the Th^{232} - U^{233} cycle. In order to obtain breeding with U^{238} it is necessary to use fast reactors. Breeder reactors are a desirable development because most of the world's potential fission fuel is in the form of Th^{232} and U^{238} (see §2.21.2). These materials are said to be fertile since they can be made fissile by the reactions cited in Equations 4.5 and 4.6. They can extend the fission power resources by orders of magnitude.

Table 4.1: Typical Characteristics and Operating Parameters of Commercial Power Reactors.[1] The properties listed are for a 100 MWe capacity.

PROPERTY	BWR	PWR	HTGR[4]
FUEL			
Compound	UO_2	UO_2	UC_2-ThC_2
U^{235} Enrichment	~2%	~3.2%	(Fully Enriched)
Pellet Size: height × diameter (in.)	0.58 × 0.48	0.60 × 0.37	(See text)
Length of Fuel Pin (ft.)	12	12	(See text)
Spacing Between Fuel Pins (in.)	0.74	0.56	(See text)
Number of Pins Per Element	49	204	(see text)

Total Number of Elements	764	193	3841
Total Fuel Weight (Tons)	186.1	108.3	$2.3(U^{235})$
Core Size: height × diameter (ft.)	12 × 15.6	12 × 11.1	15.6 × 31.1
Conversion Ratio	~0.5	~0.5	~0.75
THERMODYNAMIC			
Coolant Inlet Temp (°F)	376	545	758
Coolant Exit Temp (°F)	546	610	1449
Reactor Vessel Pressure (psia)	1050	2235	700
Net Plant Efficiency	32.8	32.5	40.7

§4.26. Reactor Types. The boiling water reactor (BWR) is shown schematically in Figure 4.4. In this system the coolant (ordinary water) is allowed to boil in the reac-

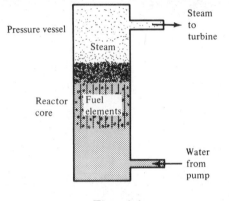

Fig   4.4

tor core. The steam is collected in the dome of the pressure vessel, and goes directly to the turbine. The fuel consists of uranium oxide (UO_2) pellets in the form of ceramic-like cylinders in which the uranium is enriched to ~2% U^{235}.

The UO_2 pellets are sealed in zircalloy tubing to form a fuel pin typically 12 feet long. Arrays of 49 pins are assembled with a spacing of ~ ¾ inches between centers in a zircalloy housing to form a fuel element. Each fuel element contains ~ 500 lbs. of UO_2. The core of a 1000 MWe reactor would contain ~ 760 such elements totaling ~ 186 tons of UO_2. The equivalent core diameter would be 15.6 feet.

Plants of this type which are in operation or nearing completion are shown in Table 4.2.[5]

Table 4.2. BWR-Type Commercial Nuclear Power Stations in the United States[5]. Only those plants which are operating or expected to be in operation before the end of 1972 are listed. Experimental and "demonstration" plants which have fulfilled their function and have been shutdown are not listed.

Name	Location	Electric Power (MWe)	Thermal Power (Mwt)	Year of Start-Up
Dresden #1	Morris, Ill.	200	700	1959
Big Rock Point	Big Rock Point, Mich.	70.3	240	1962

Humbolt Bay #3	Eureka, Calif.	68.5	240	1963
LaCrosse	Genoa, Wis.	50	165	1967
Oyster Creek #1	Toms River, N.J.	650	1930	1969
Nine Mile Point	Scriba, N.Y.	625	1850	1969
Dresden #2	Morris, Ill.	809	2527	1970
Millstone #1	Waterford, Conn.	652	2011	1970
Monticello	Monticello, Minn.	545	1670	1970
Dresden #3	Morris, Ill.	809	2527	1971
Browns Ferry #1	Decatur, Ala.	1065	3293	1972
Quad Cities #1	Cordova, Ill.	809	2511	1972
Vermont Ynakee	Vernon, Vt.	514	1593	1972
Quad Cities #2	Cordova, Ill.	809	2511	1972
Pilgrim	Plymouth, Mass.	655	1998	1972
Cooper	Brownville, Neb.	778	2381	1972
Peach Bottom #2	Peach Bottom, Pa.	1065	3294	1972
Edwin I. Hatch #1	Baxley, Ga.	786	2436	1972

The pressurized water reactor (PWR) is shown schematically in Figure 4.5. In the PWR the primary loop is pressurized to prevent boiling. The primary coolant delivers heat to a secondary loop at the steam generator. Steam formed in the sec-

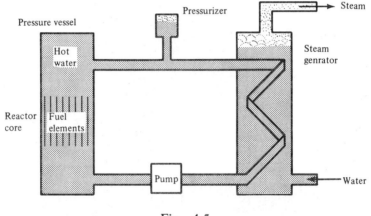

Fig 4.5

ondary loop drives the turbine. The fuel is similar but not identical to the BWR type. Dimensional and operating parameters for a PWR are shown in Table 4.1. Plants of this type which are in operation or nearing completion are shown in Table 4.3.[5] Only the identifying words in the formal name have been given. Most are called, units are located at the same site and are assigned unit numbers also, e.g. Turkey Point Station, Units 1 and 2 are fossil whereas 3 and 4 are nuclear. Since most plants are sited in remote areas, the location is sometimes given as the nearest population center or in terms of some prominent geographical feature. Locations cited here are those listed in the AEC licensing dockets.

Yannacone–The Energy Crisis—17

Table 4.3. PWR-Type Commercial Nuclear Power Stations in the United States[5]. Only those plants which are operating or expected to be in operation before the end of 1972 are listed. Experimental and "demonstration" plants which have fulfilled their function and been shutdown are not listed.

Name	Location	Electric Power (MWe)	Thermal Power (MWt)	Year of Start-Up
Shippingport	Shippingport, Pa.	90	505	1957
Yankee	Rowe, Mass.	175	600	1960
Indian Point #1	Buchanan, N.Y.	265	615	1962
San Onofre #1	San Clemente, Calif.	430	1347	1967
Connecticut Yankee	Haddam Neck, Conn.	575	1825	1967
Robert Emmett Genna #1	Ontario, N.Y.	420	1300	1969
H. B. Robinson #2	Hartsville, S.C.	700	2200	1970
Point Beach #1	Two Creeks, Wis.	297	1518	1970
Palisades #1	South Haven, Mich.	700	2212	1972
Oconee #1	Seneca, S.C.	841	2452	1972
Indian Point #2	Buchanan, N.Y.	873	2758	1972
Surry #1	Gravel Neck, Va.	780	2441	1972
Turkey Point #3	Turkey Point, Fla.	693	2200	1972
Point Beach #2	Two Creeks, Wis.	497	1518	1972
Zion #1	Zion, Ill.	1050	3250	1972
Oconee #2	Seneca, S.C.	886	2568	1972
Three Mile Island #1	Goldsbora, Pa.	831	2452	1972
Ft. Calhoun #1	Ft. Calhoun, Neb.	457	1420	1972
Surry #2	Gravel Neck, Va.	780	2441	1972
Salem #1	Salem, N.J.	1050	3250	1972
Turkey Point #4	Turkey Point, Fla.	693	2200	1972
Parie Island #1	Red Wing, Minn.	530	1650	1972
Maine Yankee	Wiscasset, Me.	790	2440	1972
Kewaunee	Carlton, Wis.	540	1650	1972
Crystal River #3	Red Level, Fla.	858	2452	1972
Rancho Seco #1	Clay Station, Calif.	804	2452	1972
Calvert Cliffs #1	Lusby, Md.	845	2570	1972
Donald C. Cook #1	Bridgman, Mich.	1054	3250	1972
Beaver Valley #1	Midland, Pa.	847	2660	1972
Arkansas #1	London, Ark.	820	2452	1972

High temperature gas-cooled reactors (HTGR) have been used extensively in England for the past 15 years. The English reactors are cooled with carbon dioxide. In the United States, the HTGR is just entering commercial competition using helium as the reactor coolant. Steam to drive the turbine is produced in a secondary loop by the hot helium gas. In principal the schematic diagram is similar to Figure 4.5 except that high pressure helium gas replaces the pressurized water in the primary loop. In a HTGR, graphite serves as the neutron moderator (in both the BWR and PWR the water serves the dual role as coolant and moderator). The latest version of an HTGR is the Fort St. Vrain Nuclear Generating Station.

The fuel for the reactor consists of a mixture of uranium and thorium—the fissile and fertile atoms. Early in the lifetime of this plant the fuel will consist of fully enriched ($\sim$ 93%) U^{235}. In future years the U^{235} will be replaced by U^{233} bred from the thorium.

The fuel particles consist of tiny spheroids which contain a mixture of uranium- and thorium-carbide (UC_2 and ThC_2) ranging from 100-300 microns (μ) in diameter. Fertile particles containing only ThC_2 are slightly larger, 300-600 μ. These "kernels" are coated with successive layers of porous pyrolytic carbon, high density pyrolytic carbon, silicon carbide (SiC), and finally high density pyrolytic carbon. These successive layers trap and retain most of the fission products. The fuel and fertile particles are mixed with graphite filler and placed in holes in larger graphite fuel blocks. The reactor is constructed by stacking the hexagonal fuel blocks into an array six blocks high by 247 columns. The Ft. St. Vrain Station is designed to produce 330 MWe at a net plant efficiency of 39.4%. Table 4.2 cites some properties of a 1000 MWe "scaled-up" version of the Fort St. Vrain unit.

The HTGR is attracting considerable interest because of its relatively high net efficiency ($\sim$ 40%) which means less waste heat (see §1.16.1), and because it can operate partially on the thorium-U^{233} cycle at a high conversion ratio ($\sim$ 0.8) thus reducing the requirements for uranium. The two existing American HTGR's are listed in Table 4.5. Again only the identifying words in the formal name have geen given and the location is stated as the nearest population center.

Table 4.4. Miscellaneous Commerical Nuclear Power Stations in the United States. The table includes only those operating or scheduled to begin operating in 1972, and omits several small demonstration plants which have served their function and been shut-down and dismantled.

Name	Location	Type	Electric Power MWe	Thermal Power MWt	Year of Start-Up
Enrico Fermi #1	Lagoona Beach, Mich.	Sodium Cooled Fast	60.9	200	1963
N Reactor	Richland, Wash.	Graphite dual purpose	790	4000	1963
Peach Bottom #1	Peach Bottom, Pa.	HTGR	40	115	1966
Fort St. Vrain	Platteville, Col.	HTGR	330	842	1972

Another type of reactor which needs mention is the sodium-cooled fast breeder reactor of which several small prototypes have been built (Enrico Fermi #1, Table 4.4). This type of system may evolve into the first United States breeder reactor—the liquid metal-cooled fast breeder reactor (LMFBR). The fuel of a sodium cooled fast reactor consists of UO_2 or PuO pellets. The core is cooled by a primary loop of molten sodium which transfers the heat to a secondary sodium loop which in turn produces steam in a tertiary water-steam loop. Such systems can operate at a much higher temperatures than the BWR or PWR and consequently promise relatively high net plant efficiencies of about 41%.

§4.27. **The Nuclear Fuel Cycle.** To understand many aspects of the nuclear controversy it is well to be aware of the total nuclear fuel cycle. This includes not only the history of the fuel within the reactor, but also the many steps of the process by which a given uranium atom is taken out of the earth and ultimately returned as radioactive waste, bearing entirely different chemical and physical properties. The steps of the cycle include uranium mining, milling, fluoriding, isotopic enrichment, fabrication into fuel elements, reprocessing or irradiated fuel, recycling of recovered uranium and plutonium, radioactive waste management, and transportation of the materials between each of these steps. At each stage there are possibilities for environmental contamination or degradation which must be controlled. In this cahpter we can only sketch a few of the main points which bear on radiological safety and resource use.

Radon (Rn) gas is a health hazard in underground mining, since uranium occurs naturally in association with radium and the radon is formed by natural decay of the radium. Currently, slightly less than half of the mines are underground and the rest are open pits. The accumulation of radon is not a hazard in the open pit mines; however underground mines must be flushed with large volumes of uncontaminated air to reduce the radon concentrations to safe levels for the miners. Breathing radon and uranium dust is a serious occupational hazard, and worker exposure during underground mining must be very strictly controlled. Monitoring of mine water for contamination is also necessary for the associated settling basins and run-off. The dust from open pit mines and tailing piles from milling operations must also be monitored and controlled.

Each pound of uranium oxide (U_3O_8, called yellowcake) that is derived by milling ore of 0.2% grade requires about 500 lb. of raw material. A 1000 MWe nuclear power plant takes about 182 metric tons/yr. of yellowcake, and this comes from disturbing roughly 17 acres/yr. of mining land and 2.4 acres/yr. for a tailings pond. At this stage radioactive materials to be monitored include the natural uranium, Th^{230}, Ra^{226}, and Rn^{222}.

In the fuel fabrication cycle, before enrichment of U^{235} begins, the U_3O_8 concentrate is converted to a gaseous compound UF. In the fluorination process small amounts of uranium are emitted to the air and process water. During the isotopic enrichment process in a gaseous diffusion plant the U^{235} content of natural uranium is increased. The second uranium stream, depleted in U^{235} is stockpiled for future use in breeder reactors. The plants used in this process are among the largest industrial establishments in the world, and large amounts of energy are needed to perform the enrichment, however, the electrical energy which can ultimately be provided by the enriched uranium is about 22 times the total energy (mainly coal at present) needed to produce the fuel.

After enrichment in U^{235} the UF_6 is converted to UO_2 which is then sintered and formed into fuel pellets. These are sealed in zircalloy tubes and form the fuel rods which are used in the reactor. When the U^{235} undergoes fission and the inventory of fission products builds up within a fuel element, it becomes intensely radioactive. Concurrently the plutonium content grows from zero to a value which makes recovery desirable. This plutonium must be chemically separated from the fission products, and requires special facilities to carry out the hazardous operations. The fuel reprocessing plant where this is done may emit somewhat higher levels of certain radioisotopes (e.g. Kr^{85} and H^3) than would occur at the original reactor site. The radioactive fission products must then be controlled and permanently sequestered from the biosphere in a high-level waste repository.

§4.28. Reactor Safety and the ECCS. With the Atomic Energy Act of 1954, as amended, Congress fixed upon the Atomic Energy Commission the responsibility for regulation of the production and utilization of atomic energy and related facilities, in order "to assure the common defense and security, and to protect the health and safety of the public." To accomplish these purposes the AEC after review of proposed designs and public hearings, issued separate licenses to construct and to operate each nuclear power plant. Until 1970 the AEC interpreted its responsibility as only in the fields of plant safety and radiological safety, but as a consequence of the National Environmental Policy Act of 1969, as interpreted by the U.S. Court of Appeals in the *Calvert Cliffs* case, it is now required also to consider the total environmental impact of a plant before it may issue licenses.

It has been central to the philosophy of the AEC that the primary responsibility for safe and reliable operation of a power plant or other facilities used in the nuclear cycle rests with the owners and operators of the facilities. The AEC, through its regulatory staff, reviews matters having to do with construction and operation of a nuclear plant throughout its entire lifetime from initial siting to ultimate disposal. In this regulatory function the primary aim is to insure that the radioactivity produced by the fissile material throughout the entire nuclear fuel cycle be safely managed without harm to people or the environment.

The guiding philosophy of the regulators seems to be to provide multiple barriers between the radioactive materials and the biosphere. In the reactor itself this involves designing and building the system so that the products of fission cannot escape from the sintered fuel and pass successively through the fuel cladding, the reactor pressure vessel, the biological shield, and finally the reinforced containment building. However, the reactor and its associated condenser-turbine systems are sufficiently complex that possibilities exist for breaching the multiple barriers. These can have consequences which range from the trivial and inconvenient to levels approaching natural disasters such as floods, hurricanes and the collapse of dams. The design goal is to guarantee that the more serious the accident the less probable its occurrence. To this end there has been introduced into the design process the concept of the most serious plausible accident, called the *design basis accident* (DBA). For a water cooled reactor the DBA would be the most severe *loss of coolant accident* (LOCA), (postulated to occur precipitously by reason of the complete and sudden break of a large pipe at the most critical location in the primary coolant system) Such a LOCA would initiate a rapid increase in fuel temperature since, even though the chain reaction were stopped, there would still be a very large amount of heat stored in the fuel and generated on a continuous basis, but diminishing in intensity, by the radioactive fission products. The fission product decay heat must be continuously removed, otherwise the core of the reactor will rise in temperature and eventually melt. The melt-down of a large fraction of the fuel would destroy the reactor as a power source, and could result in a breach of the containment, allowing the escape of radionuclides and ionizing radiation. To guard against such an accident, each watercooled reactor is provided with several independent emergency core cooling systems (ECCS) which are designed to provide adequate reactor core cooling even though the normal water flow is stopped. These systems must act reliably within a few seconds if they are to protect against core damage, and they must operate for several days to extract the fission product heat as the shorter half-life radionuclides decay away. The fuel must then be kept submerged in water for several months thereafter. The reliability of the ECSS has been challenged by many critics who raise many technical questions which have not yet been clearly resolved even in the protracted series of public hearings held by the AEC during 1972-73.

During routine operation of facilities it is anticipated that there will be controlled releases of radionuclides into the environment. The AEC has set allowable limits, acting within constraints from other governmental agencies, for the maximum exposures which human beings may receive from the ionizing radiation which result from the nuclear industry, and They are set forth in the code of Federal Regulations (10 CFR Part 20). In addition to the total dose limits, there are also limits on the concentration of each individual radioactive isotope in water and air, or in cases of mixtures of isotopes, the maximum activity of the total mixture.

§4.29. The Public Controversy Over Nuclear Power.

§4.29. The Public Controversy Over Nuclear Power. Other sections of this Chapter considered the principles of operation of nuclear electric generating plants, the nuclear fuel cycle, and some aspects of nuclear safety, however, there is still continuing public debate over the wisdom of continuing development of nuclear energy as a means of generating electricity.

Issues have been raised for which purely scientific proofs or solutions cannot be found, since they involve consideration of personal value systems, and therefore lie outside the boundaries of pure natural science. Still, each question has strong scientific components!

If the public is to make any wise decisions regarding the use of nuclear energy, the scientific community must furnish the technical information and advice. At the present time the information coming from the scientific community is inconsistent, confusing, and frequently emotional rather than objective. It is important for the concerned citizen to be able to distinguish between the scientific aspects of the controversy and the emotional aspects. The remaining sections of this Chapter shall consider many of these arguments and consider opposing sides of the controversy. The conclusions are for you, the reader, to draw, hopefully after further study.

§4.29.1. The rush to nuclear capacity.

§4.29.1. The rush to nuclear capacity. In 1972, utilities placed orders for 33 domestic nuclear stations having a total capacity of 36,000 megawatts of electricity. The capital investment required to build just 22 of these new plants will amount to approximately $8.8 billion. At the end of 1972, 29 plants were in operation, 55 were under construction, and 77 were committed but not yet begun.

In view of much outspoken opposition to nuclear power, and the long delays which such projects have encountered during licensing and construction, why do utilities persist in selecting *nuclear* energy as a fuel? The utilities' answer to this question has several components:

1. long-term supplies of conventional fuels have an uncertain future;
2. coal can no longer be burned in many regional airsheds because the new air quality standards cannot be met with present-day coal combustion technology;
3. economic considerations appear to favor nuclear over fossil fueled plants throughout most of the country;
4. nuclear plants are the least damaging from the environmental point of view; and
5. no other viable short-term alternatives exist at the present time.

The utilities also point out that in recent years electrical demand has been increasing at a rate between 5.5 and 7% per year, (See Chapter 9) and since the companies are legally required to satisfy the demand in their franchise area they must double their generating capacity in the next decade if the growth of demand continues. Failure to keep pace with the demand can affect the reliability of the electrical system necessitating voltage reductions and/or load shedding at times when demand exceeds capacity. Such measures endanger the public safety because sudden failure of the electrical distribution system can trap people in dangerous situations, such as elevators, as well as halting critical services. They also extract economic penalties from the community because of lost working hours and damage to equipment and property. If a utility is to meet its obligation to the community it serves, it must have reliable and adequate supplies of electricity.

§4.29.2. The growth of opposition to nuclear power. Utilities have been experiencing increasing opposition to their efforts to construct new facilities: opposition to unrestrained growth in the use of electrical energy, and opposition to the use of nuclear energy. Individual opinions vary widely, with some advocating abolition of growth and others tolerating growth with some constraints.

Although the question of growth in the use of electrical energy is not the subject of this chapter, several aspects of this question are relevant and will be mentioned. Obviously we cannot indefinitely continue the consumption of energy (whether electrical or other) at an exponentially increasing rate. Sooner or later the growth must cease, perhaps abruptly crashing. Some suggest that we are still far from the physical limits to growth imposed by nature; however even a simplistic model demonstrates that continued exponential growth at the present rate would lead to disaster in a few generations. Other scientists believe that the evidence indicates that population trends and economic factors might produce a levelling of electrical demand in the next two decades.

Critics argue that promotional advertising stimulates an increased demand for electricity. Furthermore, when new capacity is constructed the utility must find a market for the electricity it provides in order to earn a profit on its investment. Many utilities have abandoned promotional advertising completely and now engage only in institutional-type advertising which attempts to explain their predicaments to the public. Some have conducted campaigns to educate the public on energy conservation matters. (See Chapter 9) However, even with such efforts they still anticipate a rising demand aggravated by the need to retire overage units, and thus claim the need for new capacity, further arguing that the majority of the new plants must be nuclear.

§4.29.3. Public health controversy and the peaceful atom. The controversy over the potential hazard of the peaceful atom began in the late 40's when Leo Goodman, a labor union official located at Oak Ridge, Tennessee, began to complain that the AEC was not taking proper cognizance of what he claimed were radiation-induced injuries to workers in its facilities. The origins of public concern over nuclear energy actually are deep-rooted, and probably associated with the use of atomic weapons upon Hiroshima and Nagasaki. Several others joined in those early and still relatively low key criticisms of AEC recognition of radiation hazards during the early attempts of New York Consolidated Edison Company to construct a nuclear plant at Ravenswood in densely populated New York City and the bitterly criticized Enrico Fermi plant near Detroit. These initial complaints were essentially all instigated by individuals, with citizens groups involved only to a very limited extent. Additional individuals later joined the opposition, most notable among them David Lilienthal, who served as the first chairman of the AEC from 1946 to 1950.

The problem of the atmospheric testing of nuclear weapons played a large role in developing the overall hostile public attitude towards radiation, and led Nobel Prize winning Linus Pauling to write his first anti-nuclear book, *Life or Death in the Nuclear Age*. Thus the opposition to nuclear power was intertwined with opposition to nuclear weapons and their use, and was complicated by a growing distrust of politicians, the government and big business.

Although public opposition did lead to action with respect to above-ground nuclear testing and its associated fallout, it had little effect on the initial development of nuclear power.

The first large-scale opposition to nuclear power took place during 1966 in connection with the second Indian Point reactor in New York.

By this time a number of organized groups had emerged, and additional books, most notably *The Careless Atom* by Sheldon Novick, and *Perils of the Peaceful Atom* by Elizabeth Hogan, increased public concern and unrest.

The current controversy over the public health hazards from ionizing radiation may have started when Dr. Ernest Sternglass, Dr John Gofman and Dr. Arthur Tamplin appeared on the scene in the late 60's. The principal reason for the significant impact of their statements, as opposed to those of most previous critics, was that they were scientists who had personal experience with radiation. In addition, they are excellent public speakers who, with apparently inexhaustible energy, debated and testified frequently and at length, throughout the entire country.

It was at about this time that potential hazards associated with nuclear energy other than release of ionizing radiation or radionuclides received some attention, although in vague and nonspecific terms. Thermal pollution, reactor safety and accidents, transportation of radioactive materials, fuel reprocessing plants, and the long-term storage of high level radioactive waste, have now become significant concerns in any discussion about the use of nuclear energy for electrical power generation.

A significant milestone, covered widely in the press, was the controversy with respect to the licensing of the Monticello plant of Northern States Power Company, Minnesota. The issue involved serious questions of States Rights: whether the federal government had the exclusive right to set standards and regulate emissions from reactors, and whether an individual state could impose more restrictive regulations than Federal agencies. Litigation finally ended in a ruling favorable to the federal government, but organized opposition groups appeared at the hearings on the Monticello reactor, and such groups proliferated all over the country. Often two or more such groups were active at the same time in a particular case. A National Coalition of opposition groups has appeared, which combines the resources and talents of individual groups.

More recently, following the AEC proposal to reduce reactor effluent limits, attention has focused mainly on reactor safety. Perhaps the key article in this respect was that by Ian A. Forbes, et. al, of the Union of Concerned Scientists (UCS), which appeared in the magazine *Nuclear News*. This article stressed the extreme dangers to which the authors consider the American Public may be exposed as the result of a hypothetical "catastrophic" accident in a nuclear power plant. The report was written by a group of Boston area professors, which included individuals with impressive credentials, and has been followed by additional articles by the same group.

At this time routine releases of low-level radiation from reactors appears to have become almost a "dead issue" as far as the media and the public are concerned. Now the question of nuclear safety, particularly the efficiency of the emergency core cooling system (ECCS) is the primary issue.

Very few of the anti-nuclear groups advocate abolition of nuclear power. Most seek mainly to ensure that the potential hazards of nuclear power have in fact received adequate attention and that the community at large is being protected to the maximum extent possible. For example, the Sierra Club, a large and influential conservation group which has been very vocal in its opposition to some of the practices associated with current nuclear reactors, has stated that nuclear power plants, for most situations, appear to offer the best source of electrical power for the immediate future, and that minimal environmental damage is associated with this form of energy. They do not oppose nuclear power plants in principle.

The "nuclear controversy" without question has had an enormous impact, not only on the power industry in the United States but on the development of all projects that might have an impact on the environment. It has resulted in a new general level of public awareness, court decisions with far-reaching consequences, and new laws. To varying degrees, government and industry have become very concerned over the impact of their operations on the environment, and for the most part have taken positive steps to reduce the undesirable effects of their activities.

The frenetic tempo of the environmental movement has been slowed, but it has not disappeared. Because of the broad level of public awareness that it has promoted, the basic concerns will persist and undoubtedly grow. The overall movement, although counter-productive in certain instances, has probably had a net positive effect. Safety problems that did require more attention than they had received are now receiving that attention. Most important of all, the movement spurred an interest on the part of the public and the government not only in nuclear energy, but in the energy problems of the country in general. Consequently, in the next few years, an overall energy policy should emerge which will take into account not only a realistic assessment of the energy needs of the country now and in the future, but the best ways of meeting these needs under a variety of circumstances with the least damaging impact on the environment. It seems inevitable that nuclear energy will play a very significant and safe role in satisfying the immense energy needs of the country.

§4.30. Nuclear Power? The basic issues. During the past few years the opposition to nuclear plants has increased in intensity and in sophistication. Nuclear plants must be licensed (by the AEC among other agencies), and public hearings are required as part of this process. Private organizations and members of the public can participate in the hearings as intervenors with the right to present their own witnesses and to cross-examine utility company witnesses. The opposition has been based on several arguments:

1. the AEC cannot adequately protect the public because of a conflict of interest stemming from its statutory mandate to both regulate and develop (or promote) the use of atomic energy.
2. the existing designs of water cooled reactors (LWRs) have a finite potential for catastrophic accident and adequate safety research to prevent such accidents has not been completed.
3. During routine operation, nuclear electric generating plants emit radioactivity into the air and water which represents a public health hazard.
4. There is hazard in the transportation of radioactive wastes.
5. The problem of disposal of high level radioactive wastes has not been satisfactorily resolved.
6. Sufficient uranium to operate existing plants at reasonable prices will be available only for a few decades.
7. The widespread use of nuclear energy affords the opportunity for the diversion of fissile materials into nuclear weapons, thus increasing the likelihood of nuclear war or nuclear blackmail.

Other arguments are also presented but the above points appear to be those most often raised.

The proponents respond that the regulatory performance of the AEC does adequately safeguard the public, that the present LWR designs achieve a standard of safety far superior to other equally potentially hazardous activities of society, and that the procedures for transporting, storing, and safeguarding nuclear materials are adequately developed. Thus, the large-scale exploitation of nuclear energy for the benefit of mankind is justified and far outweighs the risks.

In the following sections these arguments will be examined in more detail.

§4.30.1. The Institutional Status of the AEC. The Atomic Energy Act of 1954, as amended, required the Atomic Energy Commission to develop and control the peaceful application of atomic energy. To meet these and other statutory responsibilities the Commission established independent branches to separate the regulatory functions from the operational and research functions. The expressed intention of Congress was to provide federal research and development support to make new peaceful applications of atomic energy a reality; but this was to be done so as to "strengthen free competition in private enterprise". Thus, when a new product or process was developed to the point at which private industry could enter the business with some prospect of success, federal support would be suspended. However, responsibility for safety could not be delegated to industry but remains vested in the AEC.

Reactors for commercial power plants were developed according to the general plan. The preliminary research was conducted by AEC contractors; the early prototypes were built at AEC operated national laboratories. As a final step, the AEC entered into various partnership arrangements with utilities and manufacturers to build several small demonstration plants to test the performance of various concepts. In a few of the commercial projects started in the late 1950's some further federal assistance was given in the form of fuel leases and plutonium repurchase agreements. For the contemporary commercial plants no federal subsidies remain. The utility must pay full costs for all services rendered by the AEC or other federal agency. However, the development of the LMFBR (breeder) is following the same format as the early development of BWR and PWR reactors and the first AEC-industry demonstration breeder project is under way.

§4.30.2. The AEC regulatory record. Proponents of further nuclear power development argue that:

no accidents have occurred in commercial nuclear power plants which have resulted in injuries, excess exposure to radiation or inconvenience to the public;
no employees have been injured or killed in nuclear related accidents;
no injuries, excess exposure to radiation, or inconvenience to the public have occurred as a consequence of accidents involving transportation, reprocessing, and storage of wastes from fuel for commercial power plants. (Dozens of minor accidents do occur each year in the transport and handling of radioisotopes for *medical* applications, however.

Nevertheless, critics point to the lack of tight AEC control over uranium mine tailings which have been used in Colorado as cheap land fill, unnecessarily exposing people to higher concentrations of radon than normally would occur. The AEC has maintained that this is the responsibility of the States. There is continued concern about lung cancer in uranium miners, a problem which is well recognized and under better administrative control than the equally dangerous "black lung" disease (pneumoconiosis) in coal miners.

The AEC has not fared so well in overseeing storage of wastes from weapon development activities and has been criticised for laxness in monitoring the liquid waste storage facilities in the states of Washington and Idaho. Also there has been much public acrimony over atmospheric testing of nuclear weaposn, and about leaks from underground tests. These activities, while not part of the civilian power program, has created in some observers a lack of confidence in the credibility of the AEC, which has certainly not encouraged the rational consideration of nuclear power in the United States.

§4.30.3. The conflict of interest problem. Many people believe that the AEC has a built-in conflict of interest since it is required both to *develop* and to *regulate* the peaceful use of atomic energy. It is argued that with such an inherent internal conflict between promotion and regulation, the AEC cannot be trusted to be an effective policeman of the nuclear industry. From the point of view of establishing credibility of AEC information, this is a burden which nuclear power now bears whether justified or not.

Others believe that the dual responsibility is an advantage in that it gives the regulatory staff easy access to the technical expertise of the developmental branches. They would argue that the system seems to be working because of the excellent safety record that exists.

§4.31. Loss-of-Coolant Accident (LOCA). Operation of a reactor leads to buildup of fission product radioactivity. When control rods are inserted and the fission process stops, large amounts of heat continue to be generated while the radionuclides decay. At the instant of shutdown the afterheat from fission products would be about 1/16 full operating power, about 1/30 full power 100 seconds after shutdown, and about 1/200 after one day. If the heat transfer medium is removed, the temperature of the fuel elements will rise rapidly, they will be damaged, and ultimately they may melt.

Obviously it it necessary to guard against this by insuring that there will always be sufficient cooling even though the primary cooling system is not operating. In the safety design of Liquid Water Reactors, (LWR's) the applicant for the AEC license must postulate that failure of the primary coolant supply has occurred by an abrupt guillotine-like break of a major coolant pipe and then establish that the sequence leading to a fuel element melt down cannot happen.

To prevent fuel meltdown the reactors are provided with an *emergency core cooling system (ECCS)* which is designed to remove the heat stored in the fuel at the moment of coolant-supply failure and the heat which is subsequently generated by fission-product decay. The ECCS must continue to function without failure over a period of several days until this heat has reached a level where the reactor structure can safely dissipate it. Afterwards, the core must remain immersed for several months while the remaining heat generated by radioactive decay subsides. At no time should the temperature of the core be allowed to rise to the point where melting of the fuel cladding can occur, or where an exothermic chemical reaction may occur between the water and the zirconium-alloy fuel cladding. The AEC has set four basic Interm Acceptance Criteria which must be met by the ECCS of a power reactor in order for it to be licensed to operate. During a loss of coolant accident:

1. the fuel elements shall not attain or exceed a certain calculated temperature;
2. the amount of cladding that can react chemically with water or steam may not exceed 1% of the cladding in the reactor;
3. the cladding temperature transient is terminated at a time when the core geometry is still amenable to cooling and before the cladding is embrittled; and
4. the core temperature is reduced and decay heat is removed for an extended period of time as required by the amount of long-lived radionuclides in the core.

Engineered systems for attaining these goals are quite different for the BWR and PWR reactors, but both provide for redundant systems which do not have common modes of failure. Controversy has arisen because full scale physical tests have not yet been conducted which prove that all systems function properly.

In addition to the challenge of whether the AEC criteria are satisfied there is also controversy about the suitability of the AEC criteria themselves.

Critics further protest because the AEC has ruled that a catastrophic failure of the pressure vessel itself is an incredible event which need not be considered in the safety analysis, while a recent AEC report (WASH 1250) gives a rate of 10^{-6}/yr (one chance in 1,000,000 per year) for catastrophic failure of the pressure vessel itself. A rate such as this can not be estimated from direct reactor experience; but, rather, it is based on industrial experience with all sorts of pressure vessels.

The matter is further complicated in that the probabilities quoted consider failures related to manufacturing, operational, and inspectional causes; but do not consider happenings such as airplane crashes or sabotage which occur with probability P_X. Unfortunately, even though the probability of failure associated with a particular technology can be vanishingly small, the critics who focus on P_X cannot be nearly so neatly answered.

If there is both a loss of primary coolant and a failure of the emergency core cooling system, the meltdown of the reactor core may breech the barriers, including melting the steel pressure vessel and the concrete biological shield. This has been called the *China Syndrome* since the molten core material may melt itself many feet into the ground. The probability of rupture for the primary coolant pipe has been variously estimated in the AEC report WASH 1250 to range from 10^{-3} to 10^{-5}/yr, (one chance in 100,000) decreasing as the severity of the break increases from minor cracking to outright severance. In the latter case operation of the ECCS is absolutely imperative, and the probability of failure of this system to start up when needed was given as 10^{-2}/demand (one chance in 100 per demand). This would then imply that the probability of a serious event would be of the order of $10^{-5} \times 10^{-2}$ or 10^{-7}/yr. (one chance in 10 million). However, critics point to the absence of unequivocal tests of a full scale ECCS, and would discount the 10^{-2} failure estimate. Doing this raises the probability of the core melting rate to an unacceptable level when a large number of reactors requiring that type of ECCS are in use.

The AEC has held an extended series of hearings over a period of two years on the questions raised by various critics of the ECCS, and a formal statement of its findings has recently been issued giving no estimate of the combined probability of loss of coolant and ECCS failure.

§4.32. Transport of Radioactive Materials. Large quantities of radioactive materials must be shipped around the country with a potential for accidental spillage. Considering the whole nuclear industry fuel cycle, there is the transportation of ores and concentrates, unirradiated nuclear fuel materials, fabricated fuel elements, irradiated spent fuel elements, radioisotopes, and radioactive wastes. Containers must be provided which will shield the biota from the radioactivity of such materials. For high-level wastes derived from spent reactor fuel, provision also must be made for in-transit cooling.

Most of the material transported is only weakly radioactive, however, depending on how long the material has been allowed to decay, the activity in a high-level shipping cask from present reactors might contain radionuclides with an activity of 7 megacuries which produce 30 kW of heat. This, of course, is sufficient radioactivity to represent a substantial potential hazard to the public at large should it all be released in a single catastrophic event. By AEC regulations, however, the shipping containers must be leakproof, capable of withstanding severe impacts and fire, and even if breached, limit the maximum dose released to certain prescribed and relatively safe levels. Although at present most wastes are carried by truck, high-level wastes are to be shipped in massive containers on specially designed railcars. Comparable shipping systems may be designed later for truck transport. It should be noted that many other toxic and hazardous materials such as oxygen, hydrogen, other liquified gases, petroleum products, petrochemicals of known carcinogenic potential, concentrated acids and alkalis, and high explosives are transported every day with considerably less care and greater potential for public disaster.

Because a strategy has not been decided as to how nuclear plants will be sited and distributed around the country, one cannot make really firm estimates of transportation hazards far into the future.

Proponents point the fact that shipments of both high-level and low-level radioactive materials have been made routinely and safely for 25 years without a single case of radiation exposure to the general public. Opponents warn that this record can not continue.

§4.33. Storage of Radioactive Wastes.

One of the most burdensome problems posed by nuclear technology is that of keeping the radioactivity which it spawns from contaminating the biosphere during succeeding generations. This involves question of permanent storage as well as transportation of the radioactive wastes.

The fissile isotopes are long-lived alpha particle emitters. Storage and handling methods for articles contaminated by such isotopes are essentially uncomplicated because penetrating radiation is not involved, and the concentration of such contaminants is small. These are so-called low-level wastes, and the amount of heat generated by them is at most a few hundredths of a watt per cubic foot. By contrast the products of fission in the fuel elements involve beta and gamma rays which are penetrating, and may have long as well as short half-lives. Such fission products also represent a large source of heat. These high-level wastes must be handled with great care requiring substantial radiation shields and provision for dissipation of the heat generated as a result of radioactive decay.

In normal nuclear power plant operation today, roughly one-third of the fuel is replaced every year. After a holding period of some months the spent fuel is sent to a chemical reprocessing plant to recover unspent uranium and plutonium, which are separated from the radioactive fission products. The fission products are then concentrated and stored in liquid form. The shorter-lived radionuclides decay with the generation of large amounts of heat as the radiation is absorbed within the liquid of the storage vats. This wet storage is only an interim matter, however, since the metal-walled vats corrode over the years and must be replaced. Technology has been developed for converting the partially aged liquid wastes into solidified glass, calcined, or ceramic concentrates, and the evolution of heat is now more manageable because only the longer-lived radionuclides remain. High-level wastes in the solid form, when provided with suitable encapsulation and shielding, may then be shipped to a repository with much greater safety.

Such solid wastes from fuel processing average about 3 cubic feet per ton of fuel, and initially have an activity of 4.4 megacuries and are a 20 kW heat source.

This activity decreases to 0.31 megacurie and 1.1 kW in 10 years and to about 0.034 megacurie and 0.11 kW in 100 years.[00]

These figures are for fuel that has been enriched to 3.3% U^{235} and that has been exposed in the reactor for 33,000 megawatt-days thermal per ton, with an assumed delay of 150 days before solidification. A PWR of 1,000 MWe capacity has 108 tons of fuel. When the entire fuel is used, such a 3,000 MWt reactor would give rise to 300 ft^3 of solid waste after three years of operation.

Because only 100 ft^3 solid waste each year is involved, which corresponds to a cube 4.8 ft. on each edge, the size of a permanent repository for such waste need not be very large.

Many possibilities have been suggested for permanent storage of these high-level wastes, including rocketing them to the sun, depositing them in deep wells, rock caverns, the Antartic ice cap, bedded salt layers, or by storage above ground in concrete vaults. Of these, the most carefully investigated method has been that of storage in deep bedded-salt strata, which appears in principle to solve the problem. (See §2.21) This is because such beds of salt, (for example, those in central Kansas) have been isolated from ground water for hundreds of millions of years, and the properties of salt are favorable in terms of thermal conductivity, compressional strength, shielding, and plastic flow.

On the other hand, in an attempt to bring the problem of radioactive waste disposal into focus, it has been claimed[00] that the stone of the Great Pyramid of Cheops could be rearranged into a series of smaller vaults and safely contain all the nuclear wastes from 5,000 years of generating electricity from nuclear energy in the United States.

At a site chosen for minimal earthquake activity, any cracks in the salt bed would heat themselves.

The Federal German Republic is pursing the salt bed method; however, the proposed repository at Lyons, Kansas has become the subject of controversy because of the existence of old oil well prospecting bore holes in the salt bed which might allow ground water infiltration. The AEC has announced[00] as an interim measure that wastes will be stored above ground in concrete repositories while it is investigating other suitable bedded salt areas in the United Stated.

Because there are isotopes with very long half-lives (Pu^{239} has a 24,400 year half-life) the ultimate storage site must remain "safe" for many tens of thousands of years. This imposes a constraint which is relatively unique in terms of man's experience, and represents a source of misgiving to many people who argue that fission reactor development should wait until permanent waste storage facilities are demonstrably safe. There is an alternative to storage, and that is consumption of portions of this nuclear fuel in other reactors.

§4.34. Uranium Resources and the Economics of Supply. Some critics of nuclear power point out that the uranium resources of the United States, if used in water cooled reactors can provide power for only a few decades, and that it is unwise to commit our technology to such a short-term future. This argument, of course, also can be made for oil and natural-gas-fired steam plants. It is partly an economic one since the discovery and acquisition of raw materials is strongly influenced by prices they command. Notwithstanding this, it is apparent that light-water reactors alone cannot supply power indefinitely into the future because of the finite size of uranium ore deposits, (See §2.22), however, it appears that sufficient uranium ore is available within the U.S. to fuel reactors of the present kind into the next century at a cost which need not be excessive. Furthermore, breeder reactors would assure that the fuel content of $U_{33} O_{38}$ would effectively be amplified by a large factor, somewhere between 30 and 130 times. Thus the fission process could provide us with energy at least for 1,000 years and more probably longer (See §2.23).

Opponents of the breeder reactor argue that a uranium supply shortage is not indicated, and should not be used as a reason for pushing the breeder project to avoid the economic penalties of increasing LWR fuel costs in the near future.

5

FUELS FOR POWER: COSTS, BENEFITS, AND RISKS IN PERSPECTIVE

Victor P. Bond

Editor's Introduction.   Dr. Victor P. Bond examines the so-called "nuclear controversy" in the light of the American experience, where it has been manifest in the protracted debates among scientists, laymen and the press over the pros and cons of this relatively new source of energy. Dr. Bond reviews this experience and indicates the direction in which the overall course seems to be moving. He recounts briefly, in approximately chronological order, the events as they have occurred and are occurring. Although the specific issues have changed, the "nuclear controversy" is by no means over and with the coming confrontation over the allocation of limited fossil fuel resources, it is difficult at this point to predict with any degree of confidence what course will eventually be pursued.

When someone writes on a controversial subject, the reader is entitled to know something of the background and qualifications of the author so that they can make an assessment of overall attitudes and perhaps biases and as an aid in evaluating what the author has to say. Dr. Bond is primarily a physician (MD) with a substantial (Ph.D.) background also in physics and medical physics. His scientific career has been concerned largely with the biomedical effects of radiation, and he has been closely associated with studies on the effects of the radiation associated with nuclear processes. He has participated in the "nuclear controversy" principally in the context of the potential effects of radiation from power reactors, and has endeavored to counteract some of the more extreme statements that have been made by some critics of nuclear power. He is employed by the Brookhaven National Laboratory, which is operated for the U. S. Atomic Energy Commission by a private corporation made up of nine of the leading universities located in the Eastern United States. This paper represents his own personal viewpoint, and does not necessarily represent the position of the Brookhaven National Laboratory, The Associated Universities, Inc., the Atomic Energy Commission, or any other governmental agency or organized group.

When Dr. Bond speaks of radiation effects on biological systems, including man, he speaks from extensive direct knowledge and experience in this area. However, in his discussions of questions basically related to engineering (e.g., possible severe accidents and waste storage) he states that his knowledge is derived principally from extensive reading, and discussions with those he considers highly experienced and knowledgeable in these fields.

By way of introduction, Dr. Bond says:

> "It is obvious that the ultimate choice as to how much *if any* additional power will be provided to the people of the world; and in what form this additional power will be provided, whether fossil, nuclear or otherwise, cannot be determined solely by the government, by power companies, by scientists or by any other groups. Public acceptance of any choice is necessary for it is ultimately the people as a whole who will determine public policy and programs.
>
> I believe that nuclear power should and must play a markedly increased role in providing electrical power. However, I am not attempting to 'sell you' on any particular solution to the overall problem. My objective is to furnish the background information to aid you in understanding the debate and to help you to arrive at an informed opinion with respect to the direction you would like power development to proceed."

The American experience with respect to electrical power production has undergone and is undergoing considerable evolution. Early and frequently strident, debates tended to focus on the risks and benefits of only one possible power source *nuclear*, and on one possible hazard *ionizing radiation from routine releases*. Discussions have tended to become more moderate, and deal with all feasible power sources, as well as the costs and benefits associated with the fuel production and transportation, the operation of such power sources, and waste disposal. Radiation from routine releases is no longer a major issue.

A number of attempts have been made to assess the total benefits and risks of nuclear versus fossil fuel plants, in order to allow a rational basis for deciding on what type of fuel source is best for a given situation. More data and more refined analyses are needed; however, most of those completed to date have concluded that the overall cost in terms of possible damage to health and the environment is least for nuclear power, next for oil and the greatest for coal-fired plants. One very recent and extensive analytical report has concluded that the routine operation of a nuclear plant presents a smaller public health risk than routine operation of an oil-fired plant; and that the public health risk due to accidental releases from nuclear plants is less than that from oil-fired. It is increasingly realized that there is and will continue to be a role for many energy sources in meeting the needs of society to advance civilization. The net effect of the "nuclear controversy" has probably been positive, in that most of the partisans have found it necessary to moderate their positions and converge toward positions that are more defensible and that are in the best interest of meeting the energy needs of society at the least potential risk to public and the hazard to the environment.

§5.1 The Development of the Nuclear Controversy. The "nuclear controversy" has evolved considerably in the United States over the past several years. Initially, both critics and defenders of American power policies focused on one form of power generation, nuclear—almost totally excluding either the benefits or the risks of other types of power from consideration. The critics centered almost exclusively on the potential hazards of radiation from these nuclear devices, and the proponents responded accordingly. The tone of the presentations and debate was often strident, with extreme positions being adopted on both sides of the argument. Consequently, this led to the production of more rhetorical heat than public enlightenment.

With increasing awareness of an impending energy crisis, the arguments have assumed a larger frame of reference. Positions have moderated and consideration is being given to other sources of power, and their possible advantages, disadvantages and hazards associated with these sources. At the same time the Atomic Energy Commission (AEC) has made some changes that have tended to allay the fears of some of the critics of nuclear power. Former Chairman Schlesinger made it clear that the AEC is not a partner of the power companies and is not attempting to promote nuclear power (with the exception of the breeder reactor, a promotion which is the direct result of a Presidential directive). Also, much more restrictive routine release rates for radioactivity have been proposed by the AEC. Although one can justify these more restrictive release rates on the basis that they are technologically possible and that all exposures should be kept as low as practicable, many believe, on the basis of cost-benefit analyses, that the proposed release rates are excessively and unnecessarily restrictive unless similar constraints are imposed upon competing sources of power generation   and that the change came about in large measure as a result of extreme pressure from the critics.

As a result of the famous *Calvert Cliffs* decision in the courts, the AEC was forced to alter its regulatory practices to achieve compliance with the National Environmental Protection Agency (NEPA), and, as a result, extensive "environmental impact statements" are required for each new reactor.

With the fading of radiation risks from routine reactor emissions as a prime focus for critics, other issues, principally that of reactor safety and particularly the adequacy of the emergency core cooling system (ECCS) have come to the fore. A number of articles pro and con have appeared on the subject, and public hearings are currently being held.

On the other hand, the press and the public have become increasingly aware of the extent of air pollution and the resultant deleterious effects resulting from the burning of fossil fuels, particularly coal and oil. Articles on the adverse effects of air pollution on individuals, particularly those with lung disease have appeared. Damage to plant life is obvious in the Los Angeles area, as in other countries including Sweden and Germany. Gas masks have been ordered by the Labor Office in Venice.[26] Severe damage to the Coliseum in Rome, to the Cologne Cathedral[27] and buildings in Sweden[28] is ascribed to sulphuric acid produced when sulphur oxides are released into the air with the burning of fossil fuels.

Also, serious investigations of the possibility of "cleaning up" and making more efficient all sources of energy now available are being made. Gasification of coal is receiving increasing attention, having been identified as a priority goal by President Nixon in his Clean Energy Message to Congress, 4 June 1971. Such approaches as combined gas and steam turbines which may appreciably increase the efficiency of fuel consumption, are also under consideration. Increasingly we realize that no one source of power is adequate and that there is a proper role for many of the available sources and that all need additional developmental work.

The public also is increasingly aware of the limited availability of fossil fuel resources, not only because of innate scarcity but also because some abundant fuels, like coal, simply are not produced in adequate amounts in this country. Natural gas is in very short supply. Some 90% of the oil required to generate electric power on the eastern seaboard is imported, and the implications of dependence for these vital supplies on international political considerations is increasingly apparent, as is the severe public safety hazard associated with a significant lack of power for even a brief interval. Articles on real, not predicted energy shortages appear almost daily in the press.

§5.2 The Need for More Power.   Figure 5.1 depicts the rate of increase in world population, which in itself implies a rapidly increasing need for power [30, 31]   Figure 5.2 shows the per capita rate of energy usage in countries as a function of per capita income. Figure 5.3 shows rate of increase of *per capita* use of electrical power in the USA. These factors lead to the rather startling projections of the increased need for electricity in the USA shown in Figure 5.3; with the requirement doubling in less than 10 years. The demand for additional power in developing countries is real and urgent. It is of interest that the *daily* energy need of the average American household requires

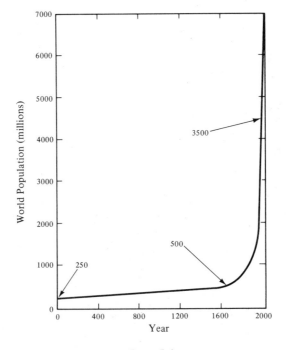

Fig   5.1

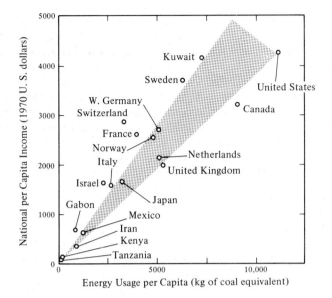

Fig 5.2

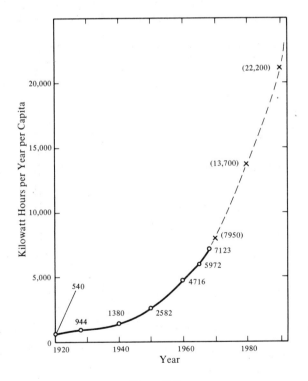

Fig 5.3

the use of the equivalent of approximately 21 kg of coal, 4 liters oil-equivalent of hydropower, 9 liters of oil products, 25 liters of natural gas and 1/2 liter oil-equivalent of nuclear power.

§5.3 Availability of Natural Energy Resources.

Most natural energy resources will be almost completely depleted, at the present rates of consumption, within limited periods of time. Figure 5.4 shows schematically the rate of production after the time at which the resource is first used for energy production. The inverse of the curve, particularly at later times, probably reflects rather well the relative cost per unit amount.

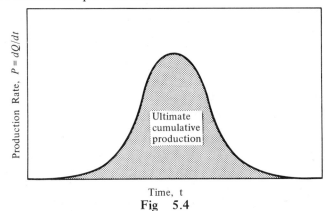

Time, t

Fig   5.4

It is difficult to predict accurately how much longer certain fuels will last, since this depends on a number of factors including the amount of time and money expended in searching for new resources, environmental considerations, the degree to which foreign imports are utilized, and the amount of money the public is willing to spend for the fuel as availability declines and/or one is forced to use poorer grade ores or other materials. Approximations, however, can be made. Natural gas in the United States is on the declining part of the curve in Figure 5.4, (see Chapter 2) and the available resources will last only a few more years, probably no more than a decade. Oil similarly is on the declining portion and will probably last no more than one or two decades (oil resources off the eastern seacoast can be tapped if the objections of environmentalists are ignored; however, the available total is estimated at only some 5.5 billion barrels, less than one year's supply for the entire USA). More costly shale oil could last on the order of 30 to over 100 years, depending on the cost the public will accept. Coal will last for several centuries, perhaps four or five, depending again on the tolerated cost in terms of dollars and environmental damage. The availability of uranium suitable for use in thermal (non-breeder) reactors is measured in decades to perhaps a century, depending again on cost considerations; with breeder reactors these fuels will last several centuries.

The following obvious point should be stressed: Almost all of the above fuel materials are useful for many purposes other than for "central station" heating plants, e.g., oil and coal are used for motor, rail and air transport, important chemical products, etc. Thus the urgency to develop non-fossil fuel alternatives for central station power plants is greater than indicated by the avove time figures. (See Chapter 3.)

§5.4 Different Approaches to the Problem of Power. In this paper, power, refers to central station generating plants that provide electrical power for every day work and living. Excluded are the myriad additional requirements and sources of power such as gasoline for automobiles, fuel for aircraft and trains, etc., although these have to be taken into account in any overall assessment of the energy requirements of a society. Many possible approaches to providing additional power are considered by Dr. Hubbert in Chapter 2 of this book.

In discussing these possible sources of additional power, I should like to first eliminate wind, tidal action and harnessing ocean currents on the grounds that their practicability is either highly questionable or undemonstrated, and that the total recoverable energy would be insignificant in the overall picture. Also, while I certainly do not wish to dismiss controlled thermonuclear reaction plants, we have no assurance if and when this approach will be feasible. Thus, while a great deal of attention must be given to the possible use of fusion reactors, it does not appear sensible to orient our present plans toward this source of energy. Similar comments apply to solar energy. Geothermal energy in its present form is available in limited regions, and certainly should be developed. However, its application on a wide scale would require new technologies just now in the early stages of speculation. Additional hydroelectric power is availabile to a limited extent but is opposed by environmental groups even more strongly than nuclear power.

The approaches that can be considered feasible at the present time, include use of little or no additional power, or reliance upon the available sources: natural gas, oil, coal and nuclear.

§5.5 Severe Restriction of Power Consumption. Power consumption has been increasing at a high rate in the United States, with a doubling time of ten years or less, (Figure 3)[30, 31] and most projections of power needs for the future have been obtained by extrapolation from this base. Critics of nuclear energy have pointed out that power consumption is rising at a much faster rate than is the population and have stated that this is due in some measure to advertising by public utility companies to increase power consumption and thus because of the unique rate-structure of public utilities, profit. Some advocate zero power growth, arguing that additional power is only used for such non-essentials as aluminum beer cans and aluminum trays for TV dinners. Serious attention has been given to reducing the waste in power consumption, and thereby reducing the total need for energy.[33] Much energy is now consumed in an unnecessary or wasteful manner, and the demand for energy growth is partly a matter of choice. One must examine just what standard of living one wishes to enjoy, and then must evaluate its potential impact on the natural environment. There is much merit in such analyses and they deserve serious consideration.

However, even strong advocates of reducing the rate of energy consumption recognize that the rate of power consumption will continue to increase at an appreciable rate, although hopefully at a slower rate than predicted earlier. This is true of many countries that now enjoy relatively high standards of living, and certainly is and will be true for developing countries.

Therefore, one must work from the assumption that most of the predicted increased power needs have to be dealt with on both a short term and a long term basis. It must also be recognized that any source of additional power, no matter what this may be, carries with it hazards to human beings and to the environment.

The task before society is to look objectively and dispassionately at the energy sources that are now available or will become available in the near future and to make as good an assessment as possible of the benefits and costs associated with each method of power production in a given locality and situation. No one potential power source should be considered or examined to the exclusion of others. Only when society is armed with such information is it in a position to make a rational decision with respect to what is best for it.

It would be sad indeed to have any given method of power production rejected and its potential benefits ignored because of irrational fears about real or supposed risks associated with that particular method.

§5.6 Nuclear Power.

All available means to fill additional power needs should be used and developed since no one form is likely to solve all of the problems associated with power generation. Nuclear power should and will be employed to supply a large portion of these needs.

Nuclear power has certain advantages. It is clean. It is economically competitive. It is as safe if not safer than most other forms of power production. The supply of Nuclear fuel if breeder reactors are used, exceeds available supplies of other fuels. There are certain disadvantages also. At the same power rating for some contemporary (thermal) Nuclear reactors, the amount of thermal waste is greater than with fossil fuel. However, with present and future high temperature gas reactors (HTGR) and breeder designs, nuclear is and will be at least as efficient as most fossil fuel cycles in this respect.

The second potential disadvantage of nuclear power has to do with the possible hazards associated with fission products. These must be examined in the context of several circumstances related to the overall cycle of obtaining, using and disposing of nuclear fuel. These include 1) mining, 2) routine releases, 3) severe accidents, 4) transport of radioactive material, 5) fuel reprocessing, and 6) long term storage of high level radioactive wastes. I shall deal with these in turn.

All mining operations are notoriously hazardous with an accident rate higher than for most other major industry. Also, workers in mines containing appreciable amounts of uranium and its daughters (many different types of mines; not only uranium mines), are exposed to a risk of lung cancer, particularly from radon and its daughter products. Although there is no question that there is an increase in lung cancer rate among these miners, the quantitative relationship is not clear. The analysis is extremely complicated by the fact that many miners are, or have been, heavy cigarette smokers, which in itself is capable of producing lung cancer.

One can say, however, that the total social cost of mining nuclear fuel is considerably less than deep coal mining in which "black lung disease" (pneumoconiosis) is highly prevalent, mainly because the total amount of mining required for a given level of energy production is considerably less for nuclear power than for coal-fired plants. The annual operation of a 1,000 megawatt (MWe) coal-fired plant consumes some 2,300,000 tons of coal while a nuclear plant of similar capacity requires only 175 tons of uranium (U_3O_8) which can be produced from about 80,000 tons of 0.2% ore (assuming a 65% plant factor).

Routine releases of radioactivity contained in the air and liquid effluents from nuclear power plants were the subject of bitter controversy a couple of years ago; however, it is rarely heard of now. One of the principal arguments advanced earlier was that, while the dose to the public from routine reactor operations has been extremely low and has constituted only a very small fraction of the so-called "standards" for such releases, the "as low as practicable" philosophy had not been explicitly written into the AEC regulations. Critics maintained therefore that even though releases were much lower than the "standards," industry had been notorious in releasing up to the absolute limit allowed by law so that this would eventually happen with nuclear reactors. This was proclaimed in the face of the fact that releases from nuclear reactors had been maintained at an extremely small percentage of the "standards" long before the radiation controversy became a popular subject in the press. Nonetheless, the AEC did propose, in effect, to lower the standards by a factor of close to 100, and to codify what always had been practiced, namely, that releases of radiation should be held to "as low as practicable." Even such vocal critics as Dr. Gofman now have stated flatly that routine releases represent no problem.

§ 5.7 Radiation Effects; Introduction. An enormous literature exists on the spectrum of possible effects of ionizing radiation on a large variety of species and biological systems, including man. Thus it would require the equivalent of several textbooks to cover the subject adequately. In this presentation an effort will be made to set forth some illustrative data and concepts in a relatively straightforward manner. In so doing one runs the risk of oversimplifying, and it must be kept in mind that the subject is complicated. Each situation is likely to have particular aspects that may require more detailed and extensive consideration than is provided here, and thus each must be considered specifically. An effort will be made to separate that which is now generally regarded as established from that which is as yet unsubstantiated or subject to different interpretations. Although several effects of radiation will be dealt with, the primary emphasis will be on those areas I believe to be of principal interest in the context of this meeting, i.e., possible late effects, such as cancer induction (including leukemia) in the general population, associated with relatively low doses of principally low LET (linear energy transfer) radiations delivered at low dose rates.

Although specific citations are given throughout the text, a few documents useful for specific aspects of the overall problems are referred to here (see text for meaning of terms used). Early effects of radiation exposure are covered in detail in *Mammalian Radiation Lethality, A Disturbance in Cellular Kinetics*,[2] as are industrial radiation accident cases (Chapter 6) through 1965 (relatively few serious accidents have occurred since). Late effects in general are covered extensively in two recent reviews of available literature.[3,4] Possible late effects of low doses and dose rates are considered in detail in the same two documents, as are risk estimates for man from exposure to radiation. Both documents also give details on sources and amounts of exposure of human populations. The documents[3,4] are both interesting and useful, particularly with respect to possible effects of low-level exposure, since both committees worked with essentially the same data but came to different conclusions. Internal radiation, radiation in the environment and genetic effects are dealt with in both documents.

Philosophies of radiation protection and of radiation protection guides are dealt with in detail in Reference 1. Specific information on internal emitters and maximum permissible concentrations of isotopes are given in References 5 and 6.

§5.8 Early Effects of Radiation.   All substances and agents, including those that are usually considered to be non-toxic (even pure water), have the potential for serious injury and even death if the dose administered is high enough. Radiation is no exception. If massive doses of radiation, in the order of many thousands of rads, are applied over the entire body in a short period of time, severe symptoms and death can occur within minutes or less, to hours. If the total dose is in the 600-1500 rad range, death probably is inevitable and could be anticipated within a week or ten days, independent of treatment. In the range of perhaps 250 to 500 rads over the whole body in a short period of time, "radiation sickness," resulting primarily from destruction of the bone marrow is seen. The patient may show initial nausea, vomiting, and diarrhea lasting hours to a few days. Following this many "recover" at least temporarily to enter the so-called "latent" period. During this period the bone marrow and blood counts become increasingly depressed. When the blood counts reach low levels at two to four weeks following exposure, depending on radiation dose, the patient's temperature may rise due largely to neutrophile (white count) depletion and consequent infection. The patient may also show signs of bleeding due to a depression of the blood platelets which are instrumental in preventing loss of blood from the vessels.

A large body of data exists on the early effects of external radiation exposure in animals.[2] Also, sizeable populations of human beings who have been exposed to varying amounts of external "whole body" radiation have been observed, e.g., the Japanese exposed in Hiroshima and Nagasaki; the Marshalese exposed to fallout radiations in 1954.[2] Also, even though peacetime nuclear activities including potential radiation exposure (excluding mining activities) has an excellent overall safety record, second only to the communications industry, a total of approximately 50 individuals have been occupationally exposed to radiation at doses high enough to require hospitalization. Of this total approximately 10 individuals have died and their deaths have been attributed to their radiation exposure. The records of a total of 48 cases of hospitalized exposed individuals, including eight who died (record up to 1965) have been reviewed in detail.[2]

Thus the early effects of radiation are documented and well known.[2] Although in some incidents it has not been determined for some time after the incident that the illness under consideration was due to radiation exposure, and even though it has been difficult in many accidents to determine the dose received with a satisfactory degree of accuracy, there has usually been little difficulty in the physician making the clinical presumption that the clinical syndrome(s) seen were ascribable principally to the radiation exposure.

§5.9 Genetic Effects.   While direct evidence of genetic effects from the exposure of human beings is virtually nonexistent, no one seriously questions that radiation exposure at high doses and dose rates will lead to genetic effects in the descendents of the individual(s) exposed. Extensive studies in Hiroshima and Nagasaki designed to demonstrate genetic effects were negative; however, such effects are difficult to demonstrate and these negative findings are not taken as evidence that genetic effects in man do not occur. Such effects have been seen in a number of experimental species from bacteria to mammals, and almost surely occur in man.

Dose effect curves for genetic effects were studied in lower organisms.[7] These data were best represented by curve D of Figure 5.5, i.e., a linear, non-threshold situation appeared to obtain, Similar data were obtained in initial studies on mice[7,8,9] in which the effects of irradiation of the mature sperm were studied.

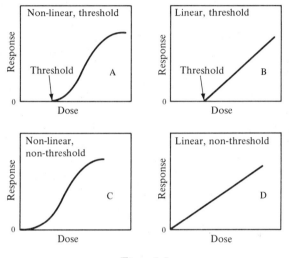

Fig 5.5

With such a curve, no repair or recovery, and no dose rate effect would normally be anticipated. Further studies were then done[8,9] on the genetic effects of radiation on earlier stages of cells that later develop into sperm. Here a significant dose rate effect was observed. It appears that the reason that sperm do not have the capacity for repair of radiation damage observed in their earlier cell stages is that they have essentially no cytoplasm. The effects of radiation on the female reproductive cells were then studied. There were effects at high doses and dose rates; however a very strong dose-rate effect was observed. In fact, when the dose rates were lowered far enough, no difference was observed between the exposed and unexposed, i.e., recovery of radiation damage apeared to have been complete.[4,5,7-9]

It should be stated here that the incidence of the effects observed was extremely small, and therefore, it was very difficult to obtain statistical validity. In some experiments, particularly at the lower doses, it required on the order of 15,000 mice or more to detect only a very few mutations. Under these cond9tions it is essentially impossible to determine accurately the shape of the dose-effect curve, or to detect effects at very low doses and dose rates. For genetic effects in mice, it has been stated that, "For example, if one assumes a linear dose-response for genetic effects, then to find with 95 percent confidence, the predicted 0.5 percent increase in genetic effect in mice at a dose of, say, 150 mrem would require 8 billion animals"[10] (same conclusions from Ref. 11). The 150 mrem is equal roughly to the average exposure from background radiation in the United States, or to the upper limit radiation protection guide of 170 mrem/yr for the general public. While it is this difficult to demonstrate a presumed effect at these low dose levels, it is even more difficult, in fact impossible by this approach, to prove the negative—that there is some threshold dose below which the effect does not occur. One might, using this approach, set an upper limit of risk at very low doses, but could not prove that it is zero below some dose level.

From additional experiments it was concluded that the overall situation, for mice, may be approximately that depicted schematically in Figure 5.6[2,4] With high doses and dose rates, either curve 5.6A (linear) or curve 5.6B (curvelinear) represents the data adequately, because of the wide limits of error represented by the dotted lines on either side of curves 5.6A and 5.6B. However, from dose rate and other studies, it appears that curve 5.6B probably most accurately represents the data obtained at high doses and dose rates. For the male mouse, curve 5.6C, many well represent the data olbtained at low dose rates. Note that though the curve is linear, the slope (effect per unit dose) is considerably lower than that for curve 5.6A or 5.6B. For female mice, in which it appeared that recovery was essentially complete, curve 5.6D (a threshold situation) may best represent the data.

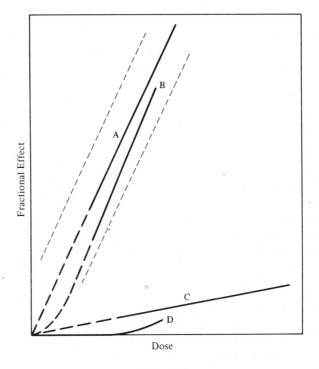

Fig   5.6

In summary, no one seriously doubts that genetic effects result from radiation exposure in man, even though they have not been observed directly. Data are available principally from experiments using bacteria, *drosophila* and mice. There are uncertainties and questions in attempting to translate these data to man. Genetic effects were clearly evident at high doses and dose rates in male and female mice. At low dose rates genetic effects were clearly evident in the male mouse but no effects were detected in the female mouse. I doubt that many would dispute that radiation does produce some degree of genetic effect in the population even at low doses. There are differing opinions, however, with respect to the extent of the possible effects at low doses and dose rates. Most, I believe, would agree that conservative assumptions should be made in attempting to evaluate the extent of such possible changes in human populations.

§ 5.10 Late Effefts, High Doses, and Dose Rates. It has been clearly evident, since shortly after Konrad Roentgen discovered X-rays in 1895, that high doses of radiation can lead, after a period of some years, to an eventual increased incidence of neoplastic diseases in animals and in man. These early observations have been confirmed in a number of situations. An enormous literature exists, which has been extensively reviewed and analyzed recently by both a scientific committee of the United Nations[3] and by a committee of the National Academy of Sciences.[4]

Somatic effects other than the development of neoplasia can develop. Lens opacification or "cataract" is known to occur in animals and in man. Evidence for "non-specific" life shortening in animals has been put forth; however, the National Academy of Sciences Committee[4] reviewed such data that are available for man and concluded that there is at present no evidence of such an effect in man.

§ 5.11 Dosimetric Concepts. An understanding of the basic concepts of radiation dose is essential to dealing with the different possible radiation effects that may be encountered under different exposure conditions. It is common to describe the amount or dose of administered drugs or toxic agents in terms of weight or volume, expressed in grams, cubic centimeters, etc. (or grams/unit weight). The analogous unit for the dose of radiation is the *rad*, which indicates the energy absorbed per gram of material or tissue and is equal to 100 ergs/gram.

Other quantities are also used to express "dose." Thus the *roentgen* (R) is sometimes used. Technically, this indicates "exposure," but is roughly equal to and can be converted into the absorbed dose in rads. Another quantity commonly used is the dose equivalent expressed in units of "roentgens equivalent man," or *rem*. This quantity arises because different radiations, e.g., X-rays, neutrons, and alpha particles, are not equally effective per unit dose (in rads) in producing biological effects. Neutrons and alpha particles deposit energy in larger "packages" on a microscopic scale than do X-rays or gamma radiation; technically they have a higher linear energy transfer, or LET. This higher effectiveness per rad is taken into account in radiation protection by assigning a "quality factor," or Q; the Q for X-rays or gamma radiation is unity. The dose equivalent in *rem* for any radiation is equal to the absorbed dose in *rads* multiplied by the appropriate Q.

There are real and important differences between *rads* and *rems*; however, recoginition of these differences is not essential for most of this presentation. Hence, I shall use the terms interchangeably unless otherwise specified. One also speaks for convenience of *millirads* or *millirem*, which are one thousandth of one *rad* or *rem*, respectively.

In experimentally or clinically controlled conditions, there is no significant problem in specifying adequately the dose received by the individual. It is measured in a location having a known relationship to the individual, which can be taken as the exposure or dose received, or from which one usually can if necessary (by calculations or measurement under simulated conditions) determine the dose to the various parts of the body. However, uncertainties arise under less well controlled conditions, including those associated with situations in which there may be a question about the exact location of the measuring device(s) at the time of presumed exposure, instrument accuracy, partial shielding of either the dosimeter or the body, the type and energy of radiation involved, and the geometry of the instrument versus that of the specimen or body.

The term radiation as used here includes only "ionizing" radiations such as X-rays, gamma rays, electrons, neutrons, and alpha particles. Radiation such as ultra violet, visible light, and microwaves are not considered. Radiation can be applied as "external radiation," or "internal radiation" from "internal emitters." With external radiation, the radiation source is external to the body, e.g., the situation of an X-ray machine used in diagnostic radiology. With internal radiation, radioactive materials gain acess to the body and thence usually to the blood stream. As with any administered substance, the distribution in the body is determined principally by physiologic factors. There is often some localization in one organ; (e.g., radioactive iodine goes largely to the thyroid) thus, for a given intake, different organs may receive different absorbed doses of radiation.

A dose of external radiation, however, can differ in an important way from a dose of a drug or of an internal emitter. The external beams can be directed by various means. Thus one may receive, as an example, 10 rads to one finger, 10 rads to the entire body (whole body exposure), or to essentially an intermediate volume and/or kind of tissue(s). The effects can be very different. One may deliver hundreds or thousands of rads to a localized area, e.g., a finger, and while local damage will be sustained, the injury is most unlikely to threaten life. Three hundred rads to the whole body, on the other hand, would almost certainly produce serious overt symptoms and signs of illness, and could be life-threatening. A given dose of radiation is most damaging to the individual if it is delivered over the entire body. For purposes of further orientation, single whole body doses of 25 or perhaps fewer rads may be detectable by means of chromosomal analyses or other specialized techniques.[1] However, one generally cannot detect changes in the individual at these exposure levels by means of more ordinary laboratory procedures such as blood counts. Radiation exposure does depress the bone marrow and blood cell counts, and the extent depends on dose. However, because of individual variation in such counts, it is unlikely that it can be established with assurance that such a change has occurred in a given individual below approximately 50 rads. Such changes at these dose levels might be detected statistically in a large group, however, by comparing the spectrum of, or median blood counts of the group presumed to be exposed with that of a suitable comparison group. Even mild early symptoms or illness are rarely if ever seen in the normal individual below about 75–125 rads[1]. The dose to the entire body that would probably be lethal to one-half of a given group of human beings within a period of several weeks, in the absence of treatment (the median lethal dose. LD_{50}) is about 300 rads to the midline of the body.[1,2] This LD_50 dose will vary somewhat with age, perhaps sex, physiological condition, and state of health. In radiotherapy, several thousand rads may be delivered to localized volumes of tumor and normal tissues over a period of days to weeks, since usually only a small fraction of the body or of organs are exposed over an extended time interval. The body can usually withstand such localized damage to normal tissues adjacent to the tumor.

Another consideration of considerable importance is the temporal pattern of dose delivery, the *dose rate*. A given dose of radiation is most damaging if delivered in a relatively short period of time (minutes). This subject is covered in more detail below.

§ 5.12 Categories of Effects in Toxicology. In toxicology (of which the evaluation of radiation effects is a part), a variety of effects categories must usually be considered. The broad categories include somatic effects (those in the individual to whom the agent was actually administered), which are in turn subdivided into early effects (within minutes to hours or weeks), late effects (appearing after a few years or more), and genetic effects which appear only in the descendents of the individual who received the agent.

Early clinical manifestations (used here to mean findings by history and physical examination, as popposed to findings by laboratory test) are seen essentially only after the administration of relatively large or massive doses of the toxic agent, although effects that are not associated with clinical manifestations may in some instances be detected by ordinary or special laboratory procedures. The kinds of clinical manifestations depend of course on the toxic agent. If the dose is high enough, there is a certainty that each individual exposed will show clinical manifestations and may die. There is usually little problem, if the individual is known to have received the toxic agent, in the physician making a clinical presumption that a cause and effect relationship exists between the administration of the agent and the effects seen.

Late effects must be considered to be possible either after the administration of high doses (sufficient to produce early clinical manifestations), or after lower doses which may cause no detected early manifestations. However, a difference frequently exists between early and late effects. For instance cancer, a late effect of some agents, occurs "naturally" in some fraction of the population. I know of no method of distinguishing unequivocally, on the basis of medical findings alone, between most late effects that may have been associated with the administration of a given toxic agent, and the corresponding "natural" disease. As an example, an increased probability of developing myelocytic (not lymphocytic) leukemia is associated in man with exposure to large doses of radiation. However, the myelocytic leukemia seen in irradiated populations present clinical, pathological, biochemical, and cytological changes that appear to be identical to those found in myelocytic leukemia arising in unexposed populations. Further, while early clinical manifestations can occur in one-hundred percent of exposed individuals if the dose is high enough, late effects usually appear only in a small and frequently in a very small percentage of those exposed. Thus while the exposed individual has an increased *probability*, usually small, of a given late effect, there is no certainty that the effect will ever develop. These considerations assume considerable importance when one deals with possible cause and effect relationships.

Genetic effects also can occur after high or low doses of many agents, and as with some late somatic effects, one must deal usually with probabilities and not certainties. The spectrum of genetic effects that can be produced with some toxic agents is of course very broad.

§ 5.13 Dose-Effect Relationships in Toxicology. In making toxicological evaluations, it is often necessary to deal with dose-effect relationships in which groups of individuals have been or are exposed to different doses of the toxic agent, following which the *percentage* of exposed individuals that respond by developing a given effect over some stated period of time is determined. Note Figure 1A, which depicts schematically the classical "S-shaped" dose-effect curve well known in pharmacology and toxicology. Consider a hypothetical experiment for illustrative purposes: such a curve might be obtained, for instance, by giving different but similar groups of individuals (e.g., 100 per group) graded or increasing numbers of aspirin pills and determining the percentage in each group that show a particular effect. If each individual in a group is given one pill, probably none will show the effect. As the dose is increased, perhaps to three pills to each individual, some small percentage out of a group of 100, say 5% may show the effect, reflecting individual variation in sensitivity. At some dose level, perhaps 10 pills to each in a group of 100, 50% may show the effect. A few in another group might not show the effect even if 20 pills are ingested. At very high doses "all" in a group of 100 would show the effect. If one plots the percentage of each group that manifest the effect versus the dose each individual in that group received, we obtain a dose-effect curve such as that depicted in Figure 5.5-A.

The shape of the lower part of this S-shaped curve is not well established, however, it has been commonly assumed that there is a "threshold" below which no individual would respond even if the group tested were increased greatly in size over the 100 in each of our hypothetical groups above. Curve B (Figure 5.5) is a variation of curve A, however, in this case the "threshold" is lower and it becomes much more difficult to say whether or not a threshold exists.

Before radiation toxicology was developed, it was generally assumed that curves A, B, or C (Figure 5.5) characterized the response to essentially all if not all toxic agents, i.e., either there existed a clear threshold, or the effect at very small doses was vanishingly small. It was in radiation toxicology that the possibility that curve D, Figure 5.5 might apply in some cases was focused on. Other branches of toxicology have now recognized that agents other than radiation may have to be considered in this manner.

Curve 5.5D represents the so-called "linear, non-threshold" situation, and its implications are far-reaching. If such a curve applies, there is of course no "safe" dose below which there can be presumed to be no effect on the population, i.e., there is a chance, however small it may be, that some exposed individual will show the effect. Such an interpretation, applied to the aspirin analogy, would lead to the expectation that in a very large population given even a very small fraction of one pill, a few individuals would show the particular effect. If a harmful or toxic agent is ubiquitous in the environment such that very large populations, e.g., that of the United States, are exposed, then even a very small percentage effect would mean a large absolute number of affected individuals. (Consider that the number of Americans killed per year in auto accidents amounts to approximately 0.025 per cent. Yet this percent of 200,000,000 Americans is the approximately 50,000 killed per year.)

§5.14 Dose Rate Effects. Biological systems are constantly subjected to toxic insults, and their large capacity to repair or to recover from such insults is widely known and appreciated. More specifically, certain quantities of toxic agents, i.e., strychnine or arsenic, given in a single dose may produce severe effects and even death. However, the same total quantities if given in smaller amounts over days, weeks or months, may produce no detectable effects. Some of the "recovery" with a "low dose rate" is of course associated with excretion of the toxic agent, but some of it has to do with recovery from small increments of damage that in themselves are insufficient to produce detectible manifestations.

In the instance of damage from exposure to low-LET ionizing radiation, cellular, and other mechanisms also appear to be able to effect repair and recovery. Reference is made here only to ionizing radiations such as x- or gamma rays, technically of low linear energy transfer, or LET, the type of radiation to which the public is exposed principally. Cellular repair of sublethal damage is diminished or essentially absent with high LET radiation effects.

Thus a marked dose-rate effect (the lower the dose rate the greater the dose of radiation required for a given degree of effect) has been demonstrated to apply almost universally.

The shape of the dose-effect curve is indicative of the ability of a biological system to repair or recover from damage. Consider curve D. Figure 5.5. This is a so-called "one-hit" type of curve, i.e., each lesion laid down by the toxic agent results in the given effect and there usually is little or no recovery. The effects usually are dose-rate independent, i.e., the effect of a given total dose is the same if the entire dose is delivered in a very short time interval or over a long period of time.

Curves such as A and C, Figure 5.5, however, indicate that repair or recovery does take place. These are "multi-hit" curves, i.e., damage can be laid down at low doses that by itself is incapable of producing the effect. It is assumed that only when two or more such damaging lesions are produced in close juxtaposition in the same biological specimen that the effect appears. Further, if there is a time interval between the laying down of the initial damage and later exposure (low dose-rate or fractionated exposure used), the organism may repair some or all of the initial increment of damage and in this respect it may again become pristine. The broader the "shoulder" or initial curved portion of these curves, generally the greater is the indicated capacity for repair or recovery.

§5.15 Leukemia and Cancer Induction. The types of cancer now generally interpreted to be increased in incidence following exposure at high dose and dose rates include: myclocytic leukemia, cancer of the female breast, lung cancer, thyroid cancer, cancer of the bowel, and possibly others. The data are sparse, but are sufficiently quantitative to allow rather accurate predictions of what would happen in an irradiated population exposed to high doses at high dose rates. For instance, it is possible to say that, in a large population exposed to, say, 100 rads, the incidence per year of these different types of cancers will be increased over the normal incidence by some calculated percentage (see below).

With respect to the incidence of cancer following exposure at low doses and dose rates, at the millirem to few rem per year level, with the possible exception of fetal exposure dealt with below, there are no clear cut direct observations. I know of no cases of leukemia or cancer in the adult human being that can be shown definitely to have resulted from exposure at these low doses and dose rates. This does not at all mean that the incidence may not be increased, or that such an increase is not important. The main reason for the lack of data is that the incidence is low, and therefore, an increase, if it exists, is extremely difficult to demonstrate. The problem of numbers with respect to the induction of malignancies by irradiation is similar to that outlined above for genetic effects.

In the absence of actual data at low doses and dose rates, the only current method of obtaining estimates of what the degree of effect might be is to extrapolate from data at high doses and dose rates (more accurately, to interpolate between data obtained at high doses and dose rates, and background effect at zero dose). The

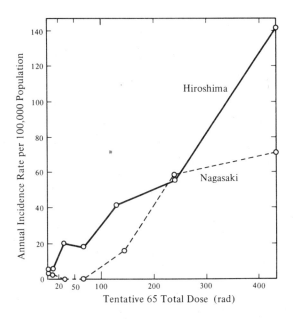

Fig 5.7. **Annual Incidence Rate of Definite and Probable Leukemia (All Form) per 100,000 Population of A-Bomb Survivors in** ABCC **Master Sample by Tentative 65 Total Dose and City, October 1950–September 1966**

question arises as to how one should extrapolate, i.e., linearly or using a curvilinear or "threshold" function.

To illustrate the problem, an example of some of the most complete data available on the induction of cancer in human beings by external radiation is shown in Figure 5.7. These data will be used to illustrate the problems involved in attempting to evaluate the cancer-induction potential for any type of malignancy following exposure particularly at low doses and dose rates. The figure shows the increased incidence in myelocytic leukemia in the individuals exposed in Hiroshima and Nagasaki.[4] To repeat, these data are given here because they are among the most complete that are available on any kind of cancer induction by external radiation in man; nonetheless the total number of leukemias in the individuals from Hiroshima (5 rads or greater) is 61 and the total number from Nagasaki is 15. With these small numbers, it is not possible to determine with satisfactory accuracy the shape of the curve for either Hiroshima or Nagasaki, and certainly the numbers of cases at low doses (in the range of a few rads) are inadequate to allow a firm statement with respect to whether these doses are leukemogenic or not.

Before expanding further on different approaches to extrapolation or interpolation one possible exception to the statements made above about radiation-induced cancer at low doses must be recognized. This has to do with a reported increase in the incidence of cancers of several types in children whose mothers (and therefore the fetuses) received diagnostic x-irradiation of pelvimetry while the mother was carrying the child. Such observations were first made by Alice Stewart in England,[12] and similar observations have been made by others since. Alice Stewart is a very careful worker, and essentially no one would doubt the reported increase in malignancy in her studies and in some other similar ones. On the other hand, extensive studies on children irradiated in Hiroshima and Nagasaki while they were *in utero*[13] and some other retrospective studies, did not show such an increase in malignancy. Thus the question of cause and effect in this situation is not generally agreed on. There is the possibility that whatever the reasons were that prompted the use of diagnostic x-ray for pelvimetry might also have been in part responsible for the development of the malignancy, and therefore that the irradiation may have been incidental to and not the sole cuase of the malignancy. Whatever the differing views may be, there is agreement that one should assume that a cause and effect relationship exists until or unless it can be demonstrated otherwise, and that every effort should be made to reduce the number of pelvimetry x-rays used during pregnancy.

With respect to the arguments for and against linear extrapolation of the data from Hiroshima and Nagasaki, it is necessary to point out that the radiations in the two cities were different.[3,4] The radiation in Hiroshima had a large component of fast neutrons; that in Nagasaki was essentially pure gamma radiation. Neutrons in general have a higher effectiveness per unit dose than do gamma rays, and the ratio is referred to as the relative biological effectiveness or the RBE of neutrons. The RBE for neutrons for leukemia induction in man is not known with satisfactory accuracy.

One can argue that the data from Hiroshima particularly are consistent with a linear relationship down to low doses, and that while no increase in leukemia was seen in Nagasaki below about 100 rads, the data are so sparse that one perhaps should combine them with the data from Hiroshima. In doing so, one implicitly assumes that the RBE for neutrons is unity or close to it.

If this approach is taken one can estimate from the linear no-threshold model what the incidence might be at low doses. This leads to a total yield of about 20 excess leukemias per million people exposed, over a period of about 20 to 25 years after exposure, or of the order of one to two excess leukemias per 10^6 (million) people per year per rem. The incidence of other types of cancer is of the order of seven or eight times for that of leukemia alone, for exposure to high doses and dose rates. It has been a rather common practice in recent years to estimate on this basis the total number of cancers that might occur in the 200,000,000 people in the United States if they were exposed to the radiation guidelines of 0.17 millirem per year. If this were to happen year after year, at equilibrium, on the linear no-threshold type hypothesis, it woul lead to approximately 5,000 to 7,000 additional cancer deaths per year,[4] assuming no dose rate effect.

Estimates considerably larger than this have been made principally by Dr. J. Gofman and Dr. A. Tamplin (see Ref. 4 for publications of these authors), and such numbers have appeared repeatedly in the press. A recent National Academy of Sciences Committee reviewed the evidence presented by these authors (see Appendix III of Ref. 4), and concluded that such high estimates were excessive, even using the linear no-threshold assumption. Among the reasons given are that their assumption of the same "doubling dose" for all radiation-induced malignancies is not supported by the data; that their assumption of an undiminished rate of cancer induction over an indefinitely long period following exposure is not reasonable; and that their assumed value of the ratio of dose at the body surface to that in deep seated organs, important in evaluating an increase in cancer incidence reported in patients treated radio-therapeutically, could not be substantiated by actual measurement of the ratio. Professor Ernest Sternglass has alleged repeatedly that very low level exposure from such sources as fallout from weapons testing, nuclear reactors and power plants, and fuel reprocessing plants has increased the incidence of infant mortality as a result of exposure from these sources. A number of scientists have reviewed his claims in detail, and have concluded that his claims are not supported by his evidence.[14]

The arguments against using linear extrapolation from data obtained at high doses and dose rates are more technical, and are as follows: Exposure to neutron radiation such as that in Hiroshima generally do not yield curves that are curvilinear with increasing slope as the dose increases; in general data obtained with gamma radiation lead to curvilinear relationships of this nature. Furthermore, in a number of radiobiological systems it has been shown that the relative biological effectiveness or RBE of neutrons is a strong function of dose, i.e., it changes rapidly with dose and becomes very large as the dose is decreased. If this applied in Hiroshima and Nagasaki, then both the Hiroshima and Nagasaki curves cannot be linear. This would argue that the true curves may in fact be very close to that shown in Figure 5.6 (curve A for Hiroshima, a neutron component; curve B for Nagasaki, gamma rays only), even though the data are not sufficiently extensive to allow one to state this with certainty. The curve for Hiroshima may well be linear, or near linear (or even convex upward at low doses). The curve for Nagasaki then might be expected to be curvilinear, (similar to curve C, Figure 5.5). The actual data from Nagasaki, taken at face value, indicate a curvilinear (increasing slope relationship for leukemias. Radiation to which the public is exposed is practically all similar to gamma radiation in that it is of low LET; not like neutron radiation. On this basis, one might expect that the increased incidence of leukemia in human populations exposed to very low doses of low-LET radiation would be vanishingly small and might even be zero.

This view is strengthened by studies on the effects of dose rate. It is not possible from available studies on human beings to determine if a dose rate effect for leukemogenesis does or does not apply. For induction of myelocytic leukemia in mice, however, a clear-cut and large dose-rate effect was observed.[15] As indicated above, a dose-rate effect is widely associated with curvilinearity, increasing slope.

It is possible that some dose-effect curves may be "hyperlinear" in the low-dose range, i.e., the curve may be convex upward or paraboloid. If this applied generally, then extrapolation from higher doses to low doses might often underestimate the risk at low doses. A recent report[16] has listed a number of dose-effect relationships for cancer induction in animals and man, indicating that several show such a parabolic relationship. However, most of these reported to show this have been obtained either with high-LET radiations, internal emitters (see § 5.25), or both. Thus the results have little necessary bearing on the shape of dose-effect curves for low-LET external radiation. I am confident that most scientists would agree with the thesis that the so-called linear, no-threshold hypotheses leads to an upper limit of risk for exposure to low doses and dose rates for most neoplasms, for external low-LET radiations.

An additional argument that invokes apparent differences between dose-effect curves for tumors that are easily induced by radiation and those that are not has been made.[3] Data consistent with linearity are reportedly seen in systems in which the target tissue is susceptible. It is further stated that radiation-induced human tumors usually involve resistant target tissues and thus one might expect curvilinearity with increasing slope to be the rule.

Some data on radiation effects, particularly those on human beings exposed to different amounts of radium,[17] have been interpreted as indicating that a "practical threshold" may obtain. As the doses and dose rates become smaller, the appearance time of malignancy appears to increase. If the dose is low enough, the appearance time of the potential tumor may exceed the remaining life span. Other dose-effect data have not suggested an inverse correlation between dose and appearance time, and there is not general agreement as to the extent of the applicability of the idea of a practical threshold.

Also, current theories of the action of ionizing radiation[18] indicate strongly that a curvilinear relationship rather than linear applies for x-ray or gamma ray dose-effect relationships in radiobiology in general.

Both recent extensive reports[3,4] discuss the data and views described above. However, one report[3] states, that, for the reasons given above, linear extrapolation from data at high dose and dose rates for low-LET radiation, would overestimate the effects at low doses and dose rates. Risk estimates for high doses and dose rates are given; however, it is emphasized that the estimates apply only for the high dose ranges and dose rates in which data exist, and that linear extrapolation should not be used to obtain estimates for low doses and dose rates, low LET radiation.

The other report,[4] while recognizing these views, did use linear extrapolation to derive risk estimates for low doses and dose rates.

From the above one can see why words such as "safe" or "unsafe" appear rarely if at all in the lexicon of guidelines-setting bodies. If the linear, no-threshold model is utilized, obviously one cannot say that doses below a given level such as the official exposure guidelines are "safe," nor conversely that exposure above these levels is "unsafe." "Safe" is a comparative word—safer than or less safe than what? Thus, in addition to guidelines for limiting exposure, one must invoke such considerations as "low as practicable," "no exposure without compensating benefits," "the possible hazards of accomplishing a task by methods involving some radiation exposure, versus the possible hazards of alternative approaches to accomplishing the task." The philosophies and problems involved in developing radiation guidelines are covered in some detail in Reference 1. The difficulties in determining what may be low as practicable, or low as reasonable have been discussed.[19]

In summary, the established facts are that radiation delivered at high doses and dose rates does increase the incidence of leukemia and several other types of cancers. Such effects have not been demonstrated at low doses and dose rates, with the possible exception of fetal exposure. If risk estimates are to be made, they must at present be derived by extrapolation (or interpolation) from data at high doses and rates. Extrapolation has been frequently carried out linearly, which leads to what most planners would regard as an upper limit of risk at low doses and dose rates for exposure to low-LET radiation. Many believe that such a procedure overestimates the risk by a considerable margin. Many would probably agree that there may well be no absolute threshold for tumor neoplasms, but that the effect at very low doses and dose rates is below that predicted by linear extrapolation. Essentially all would say that, in order to be prudent, one should assume that there is some effect on populations however small it may be, at even very low doses. Accordingly, all exposure of human beings should be kept to levels as low as practicable.

§ 5.16 Nuclear Accident. There is at present a great deal of confusion with respect to the probability of "severe" accidents and the extent of the damage to the public and to property that might result from such accidents. As of January 1973 there has been a total of 750 reactor-years experience throughout the world with commercial plants producing electricity, i.e. the equivalent of one plant operating for $7\frac{1}{2}$ centuries. While there have been "incidents" in these plants as in any industrial endeavor, nuclear or otherwise, there has never been an accident in which the release of radioactivity has been known to injure even plant personnel, let alone the general public. This is an enviable record and indicates that the probability of severe accident is quite low. On the other hand one can not conclude from this limited experience that accidents are not possible and that they will not occur in the future.

In assessing the potential impact of reactor accidents it is perhaps useful to approach it historically. One of the most widely cited documents on this subject is the WASH-740 Report, the findings of a committee established in the mid-50's by the USAEC to investigate the theoretical worst-imaginable consequences of a major accident in a large nuclear power plant.[39] Using the extremely improbable circumstances it was directed to assume, this committee postulated consequences of an accident that are truly horrendous, with thousands dead and multimillion dollar damages. Since the conclusions of the report are often quoted by nuclear critics; it is useful to review the conditions under which the report was developed, and the remarks of some of those who were on the Committee.

In 1956 the Joint Committee on Atomic Energy of the U.S. Congress took up the question of public liability insurance for power reactors. The insurance underwriters had put up a pool of 60 million dollars, and this was the maximum insurance a reactor operator could obtain privately. The questions in the minds of the utility executives and of the AEC was whether this sum was adequate, or whether some form of government backup should be provided. Thus, the study committee was asked to consider the cost in lives and property of a maximum "incredible" accident under the most adverse circumstances, namely the worst imaginable situation in which everything that could go wrong did in fact go wrong *simultaneously*. In essence, the committee report, WASH-740, was a scientifically conceived and produced work of science fiction, hypothesized to state an unimaginably impossible situation.

In talking with some members of the study committee, it was learned that they asked repeatedly for some model to explain how these extreme circumstances could possibly occur, since they could not imagine any under which such a series of events could happen at all, let alone simultaneously. They were told in essence to forget if and how such an accident could happen, and simply assess the consequences as if it did happen. Thus, rather than being an evaluation of reactor safety, the report is a statement as to what would happen if one-half of the total inventory of radioactive material in a reactor of a given size were dispersed over a highly populated area under the worst conceivable atmospheric and other conditions. As one member of the committee expressed it,[40] "...I believe it [the report] must rank as one of the most misused, and most misquoted documents in the whole nuclear energy field."

"At the outset, it is important to note that WASH-740 has nothing to say about the safety of power reactors. ...I repeat—the so-called 'Brookhaven Report' is not concerned with reactor safety. Nonetheless, the document has been widely quoted by critics as constituting positive proof that not only can such a phenomenon happen and the results postulated in the study occur, but that the likelihood is appreciable that they will happen."

The authors assumed that every safeguard of the hypothetical reactor failed simultaneously, that its containment was breached at once and completely, and that atmospheric dispersion conditions were the poorest imaginable at the time of release. It was assumed also that the entire one half of the inventory of fission products was released as an aerosol, which is simply incredible. It is extremely difficult to produce an aerosol of this material even in one wishes, let alone having it occur under the conditions of a reactor melt-down. The vast bulk of the material that might be hypothetically released from the fuel elements, even if it got through the fuel cladding, reactor vessel, biological shield, and other safeguards, would condense almost at once and could not be dispersed. In a real melt-down situation, the reactor containment would not be breached for many minutes to an hour. Even under the worst conditions, it is highly probable that only a small part of it would be destroyed. Thus, there would be time to use foam or other materials to retard releases from the containment. Not only was the postulated chain of events truly incredible, but the off-site consequences were vastly overestimated in WASH-740. The overall report, is, as expressed by Dr. Kuper,[40] more of a "mathematical exercise" rather than a scientific prediction. Parenthetically it is interesting to note that, because of the perfect safety record of commercial power reactors, the insurance pool has increased because of the lack of any substantial loss.

As I have previously indicated, more recently a group in Boston, the "Union of Concerned Scientists," headed by Ian Forbes,[29,26] has attacked the AEC and its regulation of reactor safety, particularly in regard to the reliability of the Emergency Core Cooling System or ECCS. On the basis that the probability of a catastrophic accident is too high and that a great deal more research and analyses are required, they have called for a moratorium on all nuclear power plants. They have raised a number of questions, and the fact that some of the group are scientists and engineers with impressive credentials has given the overall report an air of credibility. It is clear that they have raised a number of questions on reactor safety that must be answered, and that some may have validity. It is equally clear, however, that a large number of the people knowledgeable in this field do not agree at all with the Union's overall conclusions. However, there have been only spotty rebuttals and no detailed overall reply to the charges made.

Nonetheless, rebuttals have been put forth which indicate some of the defects in the Union of Concerned Scientists' reports.[42,43] Bray[42] stated that the authors of the report should address themselves to certain questions which to my knowledge they have not yet done. These include: (1) they dwelt only on negative aspects of the situation, with no apparent knowledge and certainly no reference to sources identifying positive aspects. (2) They quoted only from an appendix of a U.S. Atomic Energy Commission report on ECCS, in which a particular situation was divorced from overall reactor safeguards and its consequences analyzed. The results are represented as a conclusion of the report. The fundamental conclusion of the overall report gave implications opposite to those from the appendix used by Forbes, et al.[29] (3) The authors stated and implied that there is a serious lack of knowledge in a number of areas, when in fact there are numerous documents available which were not referred to or quoted by Forbes. Perhaps most serious is the fact that the authors used the consequences of WASH-740, which was not relevant, and used the consequences of the hypothetical accident put forth in their overall assessment of the result of ECCS's failure. As indicated above WASH-740 is not an authoritative reference to establish either the probability or the consequences of a core melt-down in a modern power reactor, with its many safeguard systems. (4) The authors implied that only one test was made of the ECCS, when in fact considerable test data are available. (5) The authors referred to an old report by C. G. Lawson, which flagged many technical issues that have been resolved in other more recent public documents readily available but not referred to by the authors. Bray concluded that "…as a result, the contents of the article, under scrutiny, raise serious doubts as to its credibility."[42]

Rasmussen[43] has also expressed objections to the approaches and conclusions of Forbes, et al., and concludes that "Unfortunately, in this case, I believe that the critics have become so enmeshed in the details of one specific problem that they have lost the perspective of its overall implications. As a result, they have reached some conclusions that I believe are not in the best interest of the public they are trying to serve. I believe that the risks to the public from the operation of nuclear power stations of the type being built today are not significant when compared to the risks imposed by other activities of our highly technical society."

Certainly the UCS members can and should express their opinions; however, one should be aware of the background of any individual or group who speaks or writes on these controversial issues. The authors of the critical reports indicated above are members of the Union of Concerned Scientists, a Boston area coalition of scientists, engineers and other professionals. Dr. Forbes is a nuclear negineer, Dr. Ford is an economist, Dr. Kendall is a nuclear and high energy physicist, and Dr. McKenzie is a nuclear physicist. None have had direct experience with reactor safety nor with the ECCS systems. The Union of Concerned Scientists was formed in March 1969, and has been most active in the areas of arms control and environmental pollution. It is an advocate organization concerned with environmental pollution from various sources. Most members of the UCS are also members of the Federation of American Scientists (FAS), which has been concerned over a number of years with the implications of scientific and technical developments on society. The UCS has made statements that imply an endorsement of their position by the FAS The FAS has studied their position and has declined to endorse fully their views on the adequacy of reactor safety.

A number of estimates have been made relative to the actual probability of various reactor components failing and of the probability of the so-called ' 'catastrophic accident." While there is some basis for such estimates of probability from a general knowledge of engineered facilities and their component parts, there is no "scientific" manner of arriving at such estimates with any degree of confidence. Weinberg[10] has referred to estimates of such extremely unlikely events as "trans-science." Thus such estimates are essentially a matter of judgment and educated guess. True estimates of the probability of an occurrence can be obtained only when one has accumulated a large library of such events, to which one can apply statistical methods for predictive purposes. Since the number of significant power reactor incidents is zero, such an approach is not possible. All such educated guesses indicate that the probability of serious malfunction and consequences is very low indeed. One estimate[44] is that the probability that the ECCS of a reactor will be called upon is perhaps less than once every 10,000 years of reactor operation, or of the order of 10^{-4} to 10^{-6} per reactor year. The probability that it will fail if called upon is an additional 10^{-2} to 10^{-4}. This gives a net probability of some release of activity into the containment, or even beyond it, to the range of 10^{-6} to 10^{-10} per reactor year.

In order to give some perspective as to the meaning of such probabilities, compare the risk figures given in Table 5.1. Here are given the probabilities of various events, or the probability of an individual being seriously injured or killed in various situations. Notice that these range from about one in a hundred (10^{-2}) per year for serious injury in an automobile accident, to the order of one in ten million (10^{-7}) per year or so from living in the vicinity of a nuclear reactor. Further perspective can be gained by considering the reactions of society to events with different probabilities of occurrence.[44] With the exception of injury in a motor vehicle accident, it is difficult to find serious hazards from human activities with a probability of the order of 10^{-3} per person year or less. When a risk is this great, the individual or society usually takes action to reduce the hazard. This level of risk appears to be essentially unacceptable to everyone.

Table 5.1. Chance of Serious Injury or Death-Per Year

Auto accident (disability)	1	chance in	100
Cancer, all types and causes	1	" "	700
Cancer from smoking	1	" "	2,000
Auto death	1	" "	4,000
Fire death	1	" "	25,000
The "Pill," death	1	" "	25,000
Drowning	1	" "	30,000
Electrocution	1	" "	200,000
Airplane trip, New York City-San Francisco and return	1	" "	1,000,000
Reactor Emanations; site boundary (5 to 10 mrem/yr.)	Less than 1	" "	1,000,000
Average for population within 50 miles of reactor	Less than 1	" "	10,000,000

At the level of risk of 10^{-4} per person year, people are willing to spend money, especially public money, to control a hazard. For instance, traffic signs and controls are provided, and fire departments are maintained. Campaigns are mounted to make people more aware of the risk, and there is an element of fear, for example, "the life you save may be your own."

At risk levels of the order of 10^{-5} per person per year, the hazards are still recognized and people are warned of the hazards, e.g., drowning, firearms, poisoning, etc. Some people may accept some inconvenience to avoid the risk, such as avoiding air travel.

Accidents that occur with the probability of the order of 10^{-6} per person per year do not appear to be of great concern to the average person who feels that they can't happen to him. Accidents with a probability of less than 10^{-6} appear to most people to be in the "never or can't happen" category.

The public is willing to accept one level of risk with one type of hazard, but very different levels with another type of hazard. For instance, automobile accidents represent a real killer in the United States as in other developed countries. Nonetheless, this very substantial risk is pretty much shrugged off and forgotten in trade for the convenience of the automobile. Reports of airplane accidents in which even more than 100 persons per accident have been killed are newsworthy, usually, for no more than one day, and are remarkable for their lack of impact on the public in general or their habits with respect to flying. The public appears to take a very different view with respect to accidents that they can at least imagine are under their control, versus those over which they feel they have no control. Thus, while the average person has really very little control over whether or not he is going to be in an automobile accident, he likes to think that he does. On the other hand, radiation emanating from any community source is regarded as something inflicted on him by someone else. Also, radiation is particularly susceptible to being used to arouse public fears. It is something that the public in general does not understand, and such phrases as "no amount is safe" and "produces leukemia and cancer" can be used, however much out of context, to instill fear if not terror. Thus many of the public, particularly after being exposed to such treatments of the problem, appear quite unwilling to even examine, much less accept, an objective and rational appraisal of radiation risks in comparison to some greater risks that they encounter in everyday life. Radiation doses received by the public from different sources are given in Table 5.2.

Table 5.2.   U.S. Population Exposure Year 1970

	Ave. Dose mrem/yr.
Natural background	100–150
Diagnostic X-ray	50–150
The "Standards"	170
weapons testing	3
jet travel, watches, color TV, etc.	1
Nuclear power plants	less than 0.001

Of course, statistics can be misleading and are easily misunderstood or misused. There is the story of the man who was told that the probability of one person having a bomb aboard an airplane is 10^{-3}, and of two persons having a bomb, 10^{-6}. He then always carried a bomb with him to reduce the chances of someone else having such a device to 10^{-6}. This, of course, makes no sense. And, the fact that we are exposed to one risk in no way justifies a cavalier approach to other risks. However, it is obvious from Table 5.1,   that it makes no sense, if one's interest is in saving lives, to attempt to deny oneself or society the benefits of nuclear power on the basis of possible hazards from radioactivity releases in the vicinity. To be logical, one should concentrate on the much greater risks associated with such items as auto accidents, fires, smoking, or even the radiation exposure received in diagnostic x-ray which is orders of magnitude greater than that received from reactor emanations.

While the last word has not been said, an accident in which a core melt-down occurs is an extremely improbable accident, in the range of probabilities far less than the "act of God" or natural disaster type of event generally contemplated in insurance contracts. Even in the most unlikely event that a core melt-down would occur, the probability of other backup safeguards failing is very remote. It appears doubtful that the bulk of the molten core would be broadly disseminated, other than for radioiodine and volatiles. Thus the probability of release of significant amounts of radioactivity into the environment is small indeed, and the risk to the public accordingly small to the point of almost vanishing, as a result of extensive and redundant safety systems, rigid quality control, and the continuing process of evaluation and reassessment. The damage to the public from reactor accidents is most likely less than that which would be encountered from using alternative energy sources, or from having significant energy shortage.

§5.17 **Waste Heat.** The possible effects of waste heat discharged into bodies of water must be looked at carefully, whether it derives from fossil fuel or nuclear plants. If the plant is located on a large body of water such as a very big lake or the ocean, then, with reasonable precautions, it would appear that the impact, if any, may be minimal. Under such circumstances it would make little sense to insist on costly and potentially unsightly cooling towers. However, if power plants are to be located on rivers or bodies of water such as relatively small lakes, then the problem must be examined carefully, particularly if several reactors are built or are scheduled to be built on the same river or lake. Extensive studies have been done on the effect of one reactor on a river (the Haddam Neck reactor on the Connecticut River). These detailed studies have revealed no significant damage from the waste heat. Fish do migrate up the river past the plant with apparently no interference, and there have been no significant changes in the aquatic life in the river. Under circumstances where a number of plants are to be built, the impact must be assessed very carefully. The impact of sustained significant temperature changes, or of intermittant temperature changes, has not been fully assessed with respect to aquatic organisms. If it appears that there will be significant damage, there is no alternative at present to cooling towers. Approaches to utilizing waste heat for agricultural or other purposes are being investigated.

The question of esturies requires special considerations since these are the breeding grounds for a wide variety of pelagic and other fish. If serious interference with the breeding habits of the fish are brought about either by thermal changes or by actual entrainment of small fish in the cooling systems of plants, the overall impact on commercial and sport fishing could be appreciable. It appears that no significant damage to estuaries of the United States has resulted as yet. However, a sufficient number of studies must be carried out to insure that the aquatic life is not adversely affected to any appreciable extent.

§5.17.1 **Thermal Effluent.** There is little doubt that continued public controversy will attend the discharge of thermal effluent from electric power generation facilities whether the source of such heat is nuclear energy or fossil fuels. During hearings held by the Atomic Energy Commission concerning the siting and operation of nuclear power reactors along the shores of Lake Michigan, Dr. John S. Bardach submitted the following affidavit which effectively delineates the issues raised by the proposed discharge of thermal effluent into public waters.

STATE OF MICHIGAN
CITY OF ANN ARBOR

John S. Bardach, being first duly sworn on oath, deposes and says:

1. My name is John S. Bardach and I reside in Ypsilanti, Michigan. By profession I am an aquatic ecologist with more than 20 years of teaching and research experience.

2. I am currently a full professor in the Department of Wild Life and Fisheries in the School of Natural Resources and a full professor of Zoology in the Department of Literature, Science and the Arts at the University of Michigan in Ann Arbor, Michigan. I was born on March 6, 1915 in Vienna, Austria and I became a naturalized American in 1953. I was graduated from the Realgymnasium in Vienna, Austria in 1933. I then

attended the University of Berlin, Germany, for two years and thereafter Queens University, Kingston, Ontario, where I received a Bachelor of Arts Degree in 1946. I received a Master of Science degree in 1948 and a Doctor of Philosophy degree in 1949, both from the University of Wisconsin.

3.   My present teaching duties at the University of Michigan include a course in Functional Ichthyology concerning itself with anatomy, physiology and behavior of fishes, and the supervision of graduate students in physiology, ecology and behavior of aquatic organisms. Moreover, I am involved in teaching Natural Resources Ecology, a course dealing with man's influence on and management of his natural environment.

*   *   *

6.   The following statements in this affidavit comprise my opinion based upon my 20 years of experience, teaching and research...

7.   Based upon my experience I have a scientific concern about the added heat load to the shore waters of Lake Michigan which would occur if nuclear generating plants of the type of Palisades were to discharge into the lake the waste heat in their cooling water in an unabated fashion.

My opinion that heat loading of the shore waters of Lake Michigan would contribute, in a long range fashion, to the deterioration of its ecology is predicated on two grounds:

A.   Present knowledge of hydrographic and meteorological parameters is insufficient to make proper prognosis of the cumulative ecological effects of several nuclear generating plants which would emit heated effluents into the lake. Only substantial further research will funish the basis for judgment whether thermal changes will have moderate or pronounced effects and whether these will be more or less gradual. Experience with other man-induced environmental changes (deforestation and resultant stream warming, for instance) that affected complex aquatic ecological systems make one confident in predicting that there will be changes, many of them not beneficial. I agree with the statement of David Ehrenfeld in "Biological Conservation," Holt, Rinehart and Winston, Inc., 1970 that "Changes in water temperature affect both the activity and energy requirements of aquatic organisms. Oxygen requirements also change. If the temperature rises, oxygen consumption increases but oxygen solubility in water declines. Many organisms have a narrow range of temperature tolerance. At some point lethal temperature is reached; this varies according to the rate of change of temperature, species of animal or plant and physiological condition of the individual. Since a rise of 10° C. is sufficient to double the rate of many chemical reactions, it can readily be understood why even a small amount of thermal pollution is sufficient to disrupt the organization of aquatic communities. Thermal pollution also damages ecosystems indirectly. Most important, it aggravates the effects of poisons and accelerates deoxygenation processes."

One nuclear generating plant will have some adverse effect and several of them would exacerbate conditions in a more than additive manner, due to the prevailing hydrographic conditions set forth below.

Some scientists believe that heated water remains on the surface and quickly loses heat to the atmosphere rather than to the water. However, present knowledge of water-air heat exchange and heat exchange between water masses in the regions of Lake Michigan to be affected is incomplete as there is not available information on all possible weather conditions such as different wind strength and directions and different current patterns along the shore under which these exchanges would take place. Nevertheless, and especially if there are a dozen nuclear generating plants along the shoreline and if the currents flow along this shore as they are indicated to do, long term adverse effects of heating the shallow water are likely to occur and eutrophication is likely to be accelerated.

Some of the likely changes to be enumerated below are not solely or not even predominantly caused by temperature increases but a temperature rise contributes to them; others are directly related to temperature rises. Like many biological phenomena they proceed slowly at first, soon to increase in geometric proportion and to become ever more difficult if not impossible to reverse.

B.   It is my opinion that even the incomplete knowledge about Lake Michigan which we now possess, coupled with general experience of ecological phenomena permit certain scientific conclusions and opinions, as set forth below, and that make the occurrence of long term deleterious effects of heat loading of Lake Michigan's shore waters highly probable indeed. I am particularly concerned about the following long range effects, covering two or more decades, of heated effluents from several nuclear generating plants being voided into Lake Michigan:

(1)   An increase in the rate of eutrophication of the shore waters.

(2)   A worsening of conditions favorable for, if not a threat to, the survival of a very valuable and unique fish fauna.

(3)   A change in conditions so as to favor less desirable fish species such as carp and alewife.

Discharges from nuclear generating plants will cause Lake Michigan shore waters to be threatened by increasing eutrophication. The cooling water discharges enter and are restricted to coastal waters, out to but a few miles from the shore. These lake areas also receive a substantial and increasing load of nutrients in the form of nitrogen and phosphorus compounds from domestic effluents and from agricultural runoff and, according to our present knowledge, the water in them does not mix substantially with the water in the lake at large during the fall, winter and spring; during the summer more mixing occurs but even then it is not continuous. Consequently fertilizing effects first and foremost occur in near-shore waters, proceeding faster at higher than at lower temperatures.

8.   It is my opinion that the shore regions of Lake Michigan contain relatively discrete water masses which do not mix with the waters of Lake Michigan during the year at large. This opinion is based upon my experience as well as upon a recent report of the Federal Water Pollution Control Administration, Great Lakes Region, Chicago, Illinois, entitled "Lake Currents" (Lake Michigan Basin), November 1967.

"Temperature records taken during the winter of 1961 through the summer of 1964 indicate that the following conditions occur.... The existence of thermoclines and thermal barriers during extended periods of the year greatly reduce mixing of the shallower shore waters and the waters of the hypolimnion with the main body of the lake. Such conditions promote the build-up of persistent pollutants discharged into isolated waters. Because of the prolonged periods during which such conditions can continue such build-ups can impair the uses of the water adjacent to the discharge points." Id. at p. 233.

"Thermal barrier conditions during the fall, winter and spring limit the outward extent of effective mixing volume.... The late spring storms and lake overturn break up the zonation due to the thermal bar and create conditions for effective mixing with the lake proper. However, during the summer when the thermal bar no longer exists, similar build-up occurs. Boundary effects, friction and the Southern gyre are probably responsible for the lateral transfer of water along the shore." Id. at p. 353.

"In general the shore currents move northward on both sides of the lake, except for periods during the late fall, winter and early spring.... Average current speeds on the western side of the lake range from 5 to 10 cm/sec. while those on the eastern side range from 12 to 14 cm/sec.... Inshore and offshore currents are quite separate from one another." Id. at p. 179.

Indicators of the changes generally subsumed under the term "eutrophication," and even now noticeable along certain portions of the shores of lower Lake Michigan, are the prevalence of algae, plankton and bottom organisms and eventually of fish tolerant of, by virtue of their evolution, and therefore adapted to warm oxygen-poor turbid water instead of those adapted to cool clear oxygen-rich water. The latter conditions were those of the Lake Michigan's geologic history and the organisms that evolved in the Lake, or in lakes like it, are therefore genetically fitted to them rather than to new man-induced ones. The biology of these fish and organisms do not permit rapid adaptation to man-induced changes.

Specific indicators of eutrophication are: blooms of bluegreen algae, the colonization of suitable substrates by the profusely growing green alga *Cladophora* the filaments which have known nuisance value. If effluents from nuclear generating plants generate stationary warm water masses for variable periods of time it is likely that point heat sources will accelerate localized conditions of chemical and organic pollution and that commercially valuable cold water fishes will disappear and be replaced by carp, suckers, alewifes and the like. Each of these factors singly, but more so in aggregate, decrease the value of shore porperty as well as certain recreational opportunities.

9.   The biological changes are gradual and cannot be properly ascertained by a one, two or even three year study following a local change such as a newly installed heated water discharge like Palisades Plant as overall climatic fluctuations may mask their effects. However, the biological changes gain momentum once they have begun unless the conditions favoring them are reversed.

The deterioration of Lakes Erie and Ontario since the turn of the century provides reliable and relevant analogies to the danger Lake Michigan faces from heated water. The average water temperature of Lake Erie has risen by 2° F. since 1920, due to heat loading, even without massive spot heat inputs such as occur through nuclear generating plants, and due to alterations in land use. Such a rise is tantamount to displacing the entire lake to a location 50 miles to the South. Small as the temperature change is, it is considered a contributory factor to the disappearance from Lake Erie of the Lake Herring, formerly the most abundant and valuable species in the Lake. It also favored the growth of undesirable algae in Lake Erie. Such changes in Lake Michigan as are presently being observed suggest that a comparable deterioration process may be already under way in Lake Michigan.

10.   The Great Lakes, but especially Lakes Michigan, Huron and Superior are the home of a unique, commercially valuable species complex of fishes, *i.e.*, the trout and salmon, related whitefishes and the lake trout. The numerically most abundant and commercially and recreationally most valuable among them spawn in in-shore or near-shore waters.

Experiments and observations at the Great Lakes Fishery Laboratory of the Department of Interior have shown them to be very sensitive to increases in water temperature, especially during their larval and juvenile periods. Their eggs are deposited on the bottom and their larvae must rise to the surface to gulp air for initiating the filling of their swimbladders. Without this one gulp they can never adjust their buoyancy. At that time they are but an inch or so in length and even if they sensed deleterious surface temperature, they are instinctively driven to the top where they will not be able to avoid warm water that could kill them. Such kills might be sporadic but could increase in frequency as patches of heated water multiply with several instead of one or two nuclear generating plants voiding heated water. The kills might also not be noticed due to the small size and semi-transparent nature of the fish larvae. Given such conditions, the effects of the kills would become apparent through a gradual irreversible reduction in the number of adult fish and eventually likely lead to the complete disappearance from Lake Michigan of a yet unpredictable number of their species.

It is ironic, in this context, that millions of dollars were and are spent to save these same species from the depradations of the sea lamprey and to rehabilitate them to their former abundance when the danger to them may now be shifted to thermal loading of the Lake from nuclear plants. The main difference between the two dangers is that lamprey control necessitated costly research before it could be implemented while the method of heat abatement of cooling waters is well within technological research today.

The Pacific Salmon, recently introduced into Lake Michigan is in far less direct danger from thermal change than the species native to the Lake, inasmuch as the numbers of the former will be replenished by artificial propagation. Only where spawning streams or shore areas near them become heated may direct temperature effects threaten some of their numbers.

While zones of heated water near the shore may not harm the salmon directly, they adversely affect the fishes' recreational potential. State of Michigan biologists believe that warm water near the shore such as prevailed due to climatic conditions in 1968 and 1969 kept the fish away from the shore and therefore out of the reach of the sport fishermen's craft. Heat loading of shore waters could well make these conditions that are adverse to recreational salmon fishing a much more permanent occurrence.

Some salmon which have established spawning runs in Lake Michigan streams will enter the shore waters near them on an instinct driven journey to their spawning grounds. It has been the experience of Michigan fishermen that they are far less likely to strike the fisherman's lure in warm than in cold water.

11. Alewives and other undesirable fish species will be favored through lake temperature increases. Alewives are shore spawning warm water fishes which have entered the Great Lakes inadvertently, through a man-made channel. In Lake Michigan they live at the lower edge of their temperature range and are very vulnerable to such cooling as occurs occasionally. If warm water becomes available to them they seek it out, entering into existing heated effluents in great numbers. A warming of shore waters is likely to favor them as are large streams of warm water from nuclear generating plants likely to attract them. They are, however, also delicate fish, prone to mass mortalities. Temperature conditions favoring their numbers could well be accompanied by far greater die-offs than have been experienced until now. Clean-up operations of millions of alewives are indeed costly to society. Over a twenty-year period their cost may well reach a significant portion of that involved in installing, initially, cooling devices for the effluents of nuclear generating plants. If ancillary losses in the tourist industry were to be included, the total loss may well equal or exceed the costs of installing cooling devices. In 1967 such total direct and indirect losses due to alewife die-off was estimated at 50 million dollars on the State of Michigan shoreline alone.

12. I am aware that there well could be possible beneficial consequences of heated nuclear generating plant effluents, especially in the first few years of a plant's operation, before the deleterious effects mentioned here would have time to build up. Fishing may improve in or near the effluent cones due to the attraction by the warm water of certain shore fishes such as perch, bass, pike and bluegill. These same water areas may, incidentally, also afford ice free fishing lagoons in the water. Water might become warmer in certain places and make them more attractive for swimming, and in the same places the swimming season might be prolonged, at least before the build-up of algae detracts from water contact sports. These beneficial consequences, given continued eutrophication influences from other sources will almost certainly be replaced over the

years with eutrophication, disappearance of native fishes and alewife nuisance. Any assumed beneficial effects would represent a poor interest indeed on a continuously devaluating large environmental capital. Cooling devices for the effluents should be installed on all present and future nuclear generating plants discharging heated water into Lake Michigan to prevent them for contributing to the devaluation of this capital. Later alleviation of thermal loading through subsequent modification of existing plants could well be more costly not only to the power industry, but, by virtue of the nature and time course of the changes indicated, it would also put an ever increasing economic burden on society at large.

§5.18 Radioactive Waste Management. Spent fuel rods, on reprocessing, yield large quantities of radioactive material that cannot be destroyed and therefore must be stored in a manner such that it cannot gain access to the biosphere. The material contains nuclides with half-lives of various lengths. Thus some of the radioactivity decays away essentially before it is stored, some will decay away in a matter of a few years to a few decades, and some will remain radioactive for thousands of years. Thus the effort is to provide assurance that the material either will "never" gain access to the biosphere; or, if and when it does, it will have decayed sufficiently or be sufficiently dilute so that the activity per unit mass of soil is comparable to that encountered from natural radioactivity.

Several approaches have been discussed and tried. Initially some radioactive wastes were deposited in the ocean; however, it appears, this practice has been, abandoned completely. The two principal approaches discussed in the United States are storage in salt beds or salt domes under the earth, or "engineered storage" on the surface of the earth. Extensive studies were done at Lyons, Kansas, in salt mines, and this initially was received favorably by most if not all. The obvious advantage is that these salt mines are dry on a geologic time scale, and therefore the probability of dispersal of radioactive materials stored there is small indeed. Some critics objected, however, on the basis that there were a number of wells drilled in the area, and that by this means water might gain access to the stored material. The intent had been to fill these abandoned wells to prevent the access of water. Although study continues on the use of such salt mines as storage areas; this approach to the actual storage of wastes in the United States has been abandoned, at least temporarily.[45]

The current approach is to use "engineered storage" at or near the surface. With this approach, radioactivity is imbedded in plastic material or concrete and stored in facilities on or near the surface. The advantage of this approach is that the material can be observed continuously and retrieved. Thus, at a later date, the material can be stored in some other fashion if newer or better approaches are devised.

The question arises as to the amount of land required for storage. While the amount of radioactive waste is large in terms of curies, the actual volume and the amount of land required is relatively small. Thus the land area that must be committed to such storage in the foreseeable future is of the order of a few thousand acres; a small area compared to that available and committed for other purposes.

The point is frequently raised, that such long term storage represents a strong commitment not only to the present society and its descendents, but perhaps even a different society at some date in the future. This is certainly true; and, therefore, the problem must be approached carefully.

§5.19 Transportation of Radioactive Wastes. Sizeable quantities of radioactive materials have been transported over American highways for a number of decades, without known harm to the public. Spills have of course occurred; however, these have represented relatively small amounts of radioactivity that constitute more of a nuisance and a cleanup problem than any threat to public health. Requirements for the strength of transportation casks have been made tightened stringently, and now radioactive wastes are stored in containers that are designed to withstand the impact involved even in severe accidents. It must be remembered that shipment of such materials are very clearly marked; there is no chance of an explosion of the cargo sufficiently severe to spread the material over large areas; and, therefore, the limited possibility of a large scale contamination spread to the immediate environment renders its potential for public health hazard small indeed. To repeat: while the nuisance value of a spill would be great, the public health hazard would be small.

The problem of transport of high level radioactive waste is an important one that must continue to receive attention and study. Preferably, fuel disposal areas should be in close juxtaposition to reprocessing plants, to minimize the extent of transportation of radioactive materials required. On careful consideration of the facts involved, most people would probably prefer trains or trucks carrying high-level radioactive waste to pass in the vicinity of their house, rather than trucks or railroad tank cars carrying gasoline, napthalene, other known flammables, or chlorine gas. Transportation of these materials do frequently give rise to severe fires that cause real and not hypothetical deaths of people in the vicinity.

§5.20 Fuel Reprocessing Plants. Our experience with possible hazards from reprocessing plants for waste from Light Water Reactors (LWR's) in the United States indicates that this is very small.[46,47] Essentially, our only source of data is from the West Valley reprocessing plant in northern New York State. Here the dose from the emission of gaseous radionuclides is small. There is always the potential for accident, with release of radioactivity. However, there is no possibility of anything like the "melt-down" that has been publicized as being possible in a reactor. Thus, any accident would be on a relatively small scale, with the consequences probably confined to the building in which it occurred or to the overall site. It is difficult to conceive of serious off-site consequences.

§5.21 A Comparison of Reactor Types. The foregoing discussions have dealt essentially with light water reactors (LWR) of the types known as boiling water reactors (BWR) and pressurized water reactors (PWR). Routine emissions of radioactivity from both plants are extremely small. The airborne emission rate from PWR's seems to be smaller than that from BWR's, but their releases of tritium in liquid effluents are greater. The potential for a severe accident is extremely small in each.

Although gas-cooled reactors date back to the earliest days (the Oak Ridge X-10 Reactor, the Brookhaven Graphite Research Reactor, the Windscale Reactor in England) these were not power plants. They were air-cooled and used metallic uranium fuel. The operating temperatures were low. Considerable experience has been gained, principally in England, with intermediate temperature, closed circuit, gas-cooled reactors.

Commercial high temperature gas-cooled reactors (HTGR or HTR) have been developed by private industry with government help, in the United States and in other countries. The first unit, Peach Bottom, was put into operation in Philadelphia in 1966 and has operated well. This was a small reactor rated at 40 MWe.

The first large HTGR in the United States was constructed at Fort St. Vrain in Colorado, and it went into operation in early 1973. The fuel is UC_2 (92% U-235 + Th-232 in the form of 100 to 300 micron pellets coated with carbon and SiC embedded in graphite). The moderator is graphite, and the coolant is high pressure helium gas. The reactor is rated at 330 MWe.

The advantages of HTGR's are the following: their thermal efficiency is higher than that of LWR's. It is in the order of 39.4%, which is quite comparable to that of the best fossil fuel plants. Routine emissions of radioactivity should be very low, although accurate data for large plants are not available. Extrapolating from the extremely low rates observed at the Peach Bottom plant, it is expected that the emissions from the Fort St. Vrain reactor will be much less than one percent of that from LWR's of comparable power rating.

The high thermal efficiency of this plant has obvious advantages. Not only does it reduce the problem of waste heat and its disposal, it conserves water as well. Thus the plant is suitable for use in areas in which the available water is inadequate for cooling of less efficient types of energy generators.

The accident potential for the HTGR appears to be less than the already extremely small accident potential associated with light water reactors. Several factors contribute to this conclusion:

(1) The possibility of uncontrolled power excursions is smaller because of the large negative temperature coefficient of the graphite moderator (the system responds very sluggishly to changes in reactivity).

(2) The graphite has a large heat capacity and can withstand extremely high temperatures. Thus a loss of coolant accident should not produce violent temperature transients followed by the necessity of rapid insertion of emergency cooling to avoid core melt-down. In HTGR, fission heat is absorbed by the enormous mass of graphite which rises in temperature very slowly. The design-base accident has been identified as "permanent loss of forced circulation" of primary coolant. This is an unlikely event which would require simultaneous failure of four independent redundant circulation systems. In the maximum credible accident, the calculated dose to any off-site resident is less than 1 rem.

Thus, although there has been less experience in the United States with HTGR's than in other countries, this type of reactor does seem to have definite advantages. Recently, several orders have been placed for larger units, and it is expected that the role of the HTGR in the overall energy production picture in this country will increase substantially over the next decade.

Breeder reactors are less developed in the United States than are those mentioned above. The obvious advantage of the breeder reactor is that it actually creates more fissionable fuel in the course of operation than it burns (conversion ratio greater than one), and thus the fuel supply becomes virtually inexhaustible. Most U.S. development effort has been expended on the liquid metal fast breeder reactor (LMFBR). The development program for the gas-cooled fast breeder reactor (GFBR) has received little federal support, but Gulf General Atomic has conducted extensive design studies.

Considerable experience has been accumulated with fast breeder reactors and liquid sodium cooling. Although the chain reaction with the fast breeder reactor has a smaller margin for control than do thermal reactors, the sophistication and reliability of modern control systems should adequately make up for this difference. The accident potential of the fast breeder reactor is much more difficult to assess than is that for LWR's or HTGR's. A number of factors have to be examined in more detail than they have been, and definitive statements on the subject are not possible at this time.

Small test breeder reactors on a federal location in Idaho have been in operation for years, with no apparent significant difficulties. Although the Fermi reactor near Detroit was not a fast breeder, it had breeder characteristics. Developed jointly by the Government and private industry, the Fermi reactor early experienced serious difficulties due to a blocked coolant channel which took years to repair. It has recently run into severe financial problems, and is no longer in operation. President Nixon requested continued attention to be given to the development of the LMFBR, and a sizable demonstration plant will be expected over the next several years.

The routine emissions from breeder reactors should be extremely low, perhaps on the order of those from the HTGR. Although the breeder will, of course, produce plutonium, none will be released into the environment in the course of normal operations.

§ 5.22 Plutonium toxicity.   The subject of Plutonium toxicity merits comment. Plutonium has been widely alluded to by reactor critics and even some Congressmen as "the most toxic substance known to man." Any statement as extreme as this must be examined in some detail. There is no doubt that plutonium is capable of producing cancer, as are other radioactive materials, and the amount required to do this is small compared to most other isotopes. However, the plutonium must first be inhaled or ingested by an individual in sufficient quantities before any such effects are possible; but barring the improbable accident, the plutonium in the reactor fuel is completely contained. One could equally well make such extreme statements with respect to the toxicity of a number of substances, e.g., botulinis toxin, some snake venoms, sulphuric acid, and the like and therefore such statements are absolutely meaningless unless the route by which they are expected to gain access to the human body is spelled out.

During the forties, 27 men at Los Alamos were exposed to plutonium via the inhalation route in the course of their work. All received appreciable amounts of plutonium, and of the 27, 20 received amounts equal to or greater than the present $0.04/\mu C$ body burden standard for plutonium. Several received two to ten times this amount. It has been possible to follow 23 out of the 25 closely and they have been examined quite recently.[48] None of these individuals has developed cancer or other clinical symptoms attributable to the plutonium that they inhaled. The elapsed time is 27 years. The numbers are small and the statistics are correspondingly poor. Nonetheless, the data are at least sufficient to show that if plutonium is "the most toxic substance known to man," it takes several decades for this toxicity to become manifest, even at relatively high levels commensurate with the body burden for radiation workers.

This is not to underplay the possible hazards of Plutonium exposure. It is a dangerous material that must not be released into the environment.

§5.23 Radiation Exposure of the Public. Sources of exposure to the public, and estimated dose rates per year, are given in Table 5.2. Also given are the guidelines for exposure of the public, 0.17 rem per year.

Note that by far the largest exposure is from diagnostic X-rays, which is roughly comparable to the dose rates received from natural radiation. Natural radiation varies from location to location and ranges in the United States from somewhat below 100 millirem per year in some areas to 300 or more millirem per year in other areas.[20] Of the approximately 130 millirem per year average, 45 millirem is derived from cosmic radiation, 25 millirem from naturally-radioactive materials in the body, and 60 millirem from naturally-radioactive materials in the soil, rocks, and masonry structures. It is perhaps of interest that the 25 mrem from naturally-occurring nuclides in the body derive mainly from potassium-40, and that some 330,300 disintegrations of this nuclide occur each minute in the average individual to yield this annual dose in mrem.

The recommendation of 0.17 millirem per year as a guideline was derived from recommendations made in 1956 by a National Academy of Sciences committee that reviewed the genetic effects of radiation.[7] This group stated that they thought that the development of peacetime uses of atomic energy could proceed at a satisfactory rate without exceeding these standards.

§5.24 Radiation Protection Guides. It is common practice with toxic substances to establish limits for exposure of the public which represent small fractions of those established for occupational workers. These limits are known as "standards" or "protection guides." A radiation dose of 0.5 rem per year (one-tenth of that which applies to radiation workers) was adopted for individuals; one-third of this value (0.17 rem per year) applies to groups. The 0.17 rem per year for the general public is approximately one-thirtieth of that set for occupational radiation exposure. Actually, however, the 0.17 rem per year has yet another important basis, which was put forth in the recommendations of a committee of the National Academy of Sciences.

In the mid-fifties the National Academy of Sciences established a series of committees, the Biological Effects of Atomic Radiation (BEAR) committees, to investigate the biological effects of atomic radiations, and in 1956 the committee on genetic effects issued its report.[7] This committee recommended that, on the basis of potential genetic effects, the total population should receive no more than 10 rem over a 30-year period, which was taken as the mean reproductive age of the human being. The 10 rem was intended to apply to exposure from all man-made sources including radiations used in medicine. Half of this value, or 5 rem over 30 years, was later allocated to all sources other than medical. This leads to 5 rem in 30 years, or 0.17 rem (average dose) per year.

The guideline of 0.17 rem obviously represented a value judgment. However, this value is equal approximately to the amount of natural background radiation that human beings receive, a fact which played a significant role in the derivation of this figure. Background radiation is discussed in the BEAR committee reports and in essentially all basic documents dealing with radiation protection.

Why does background radiation figure heavily in this judgment? The reason is that background radiation represents an exposure of human beings which has been experienced over eons. Living things evolved from the most primitive stages while being exposed continuously to background radiation levels that were probably higher than what we experience at the present time. We have evolved from Neanderthal man in the presence of this radiation and in the process have developed serious overpopulation problems. Further, the amount of background radiation varies considerably over the face of the earth. In large areas of France, the background radiation averages approximately twice what it is here in the United States. In some parts of India, very large populations of human beings have existed from the earliest known times in the presence of background radiation 10, 20, or more times that which is experienced in the United States with no noticeable detrimental effects. Thus, standards-setting groups feel confident about radiation protection guide numbers that are of the order of background radiation, and they feel less secure as exposure exceeds these levels.

The committee therefore set the number of 0.17 rem per year as essentially a bench mark, or an upper limit of exposure of the general population. In doing so, the committee made it clear that they were not necessarily saying that there would be no harm to the population at those dose levels or that such dose levels are "safe." They did say, however, that they felt confident that at levels near background exposure, the effect on the population, if any, would be quite small and that certainly the human species would not go "down hill" or disintegrate. Thus, it is quite clear that the 0.17-rem-per-year average dose to the total population does not represent a threat to the continued existence and propagation of human populations.

Although the BEAR committee provided a basis of 0.17 rem per year as an upper-limit bench mark, this was not their most important recommendation. Their most important recommendations for radiation protection guides and the reasons for it are as follows: Although they recognized that there may well be no harmful effects at low doses, they accepted the thesis that any amount of radiation exposure may carry some probability of harm to the population no matter how small that probability may be. Thus, the recommendation they would have liked to make is zero exposure. However, they realized that zero exposure is not only impossible but also impractical. Exposure from natural sources is inevitable, and some additional exposure is unavoidable if man is to realize the enormous benefits derived from uses of radiations and radioactive materials. They therefore made it quite clear that the population guides were provided with the idea of "stay just as far under that figure as you can." This idea is stated in many ways and frequently, not only by the BEAR committee but by standards-setting groups as well. The real recommendation is "keep it as low as practicable," and "it should most emphatically not be assumed that any exposure less than this figure (0.17 rem per year) is so to speak, all right."

Why is the real radiation protection guide "low as practicable" and not a fixed upper limit bench mark such as the 0.17 rem per year? It is realized that when one must assume some degree of effect in a large population even at low doses—when one cannot say with certainty that there will be zero effect in the population—then any number other than zero equates to some presumed degree of injury in man. To avoid this trap of saying, albeit indirectly, that some degree of injury to human beings is acceptable, the standards-setting groups introduced the "lowest practicable" approach as the real protection guide, and the numerical figures of 0.17 rem per year was introduced as an upper-limit guide.

Now let us examine what the standards-setting groups did with the recommendations of both the BEAR committee and the British study. The standards-setting groups are the International Commission on Radiological Protection (ICRP). The National Council of Radiation Protection and Measurements (NCRP), and the Federal Radiation Council (recently made a part of the Environmental Protection Agency). The recommendations of all these groups are essentially identical. Let us deal with the recommendations of the FRC, since this group has a more official status in the United States than do the others.

The FRC, in its first report,[28] said in essence, that the 0.17 rem per year recommended by the BEAR committee represents an acceptable bench mark, understood to be a barrier that is not to be approached or exceeded. However, it erected a much more restrictive limit or barrier in the form of a dose limit for the individual, i.e., 0.5 rem per year to the individual (still "not allowed" and "low as practicable" applies).

Why is the 0.5 rem to the individual more restrictive than the 0.17 average dose to the population? We can see this most easily by considering radiations from power reactors.

A principal AEC guide for radiation exposure from power reactors is identical to that of the FRC, or 0.5 rem per year to the individual at the site boundary. It is not the 0.17-rem-per year average dose. That the 0.5 rem is much more restrictive than the 0.17 average follows because of the rapid dispersion of material coming from the reactor stack and because of the rapid decay of radioactive elements. Thus, even if the dose falls off very rapidly with distance from the reactor site. Thus, even if the dose at the site boundary were 0.5 rem per year, the dose to most individuals and the average dose would be very much below this value. An individual 50 miles from the plant would then receive one one-thousandth of this amount, or 0.0005 rem per year. The average dose to the entire U.S. population is far, far below this figure.

Thus one could easily see that, if the 0.5-rem-per-year guide for the individual is not approached or exceeded, the average dose to the population will remain far below 0.17 rem per year. This principle holds not only for reactor radiation but for most other sources as well. For exposure from color TV, jet travel, luminous watch dials, etc., the 0.5 rem per year is an enormously more restrictive standard than is the average dose of 0.17 per year.

In the foregoing discussions the emphasis has been on providing information on radiation effects, on indicating areas in which there are bases for differences in interpretation, and on giving the substance of differing views. My own evaluation on several principal points that are subject to different interpretations are as follows:

With respect to dose-effect curves for neoplasia induction from exposure of man to low-LET radiations at high dose rates, there most likely are different shaped curves for different neoplasms. From my overall knowledge of neoplasia induction and of radiobiology, I would be very surprised if the most common dose-effect curve for low-LET radiation is not curvilinear with increasing slope, in the dose range of zero to perhaps 200 rads. Further it seems probable that for some types of tumors, e.g., skin and some bone tumors in man produced by radium, there may well be a threshold, or perhaps an "effective" threshold.

The above views are strengthened when dose-rate effects are considered. Such an effect with low-LET radiations is almost ubiquitous in radiobiology—when it has been looked for it has been found to exist to some degree. A marked dose-rate effect has been found for genetic effects in the mouse; perhaps a threshold dose rate effect in the female mouse. A marked dose rate effect has been demonstrated for radiation-induced myelocytic leukemia in mice. Thus, it is difficult to believe that low-LET radiations delivered at the very low dose rate commensurate with the radiation guidelines (0.17 rem/year) are not in general less effective per rad in terms of neoplasia induction than would be the same radiation delivered at high doses (tens to hundreds of rads) and high dose rates (tens of rads per minute or greater).

Although I feel strongly that the linear non-threshold hypothesis is conservative, probably to a considerable degree, I am not prepared to say by how much. The principal reason is that, at this time, we do not have enough data on radiation effects at low doses and dose rates to introduce a factor that could be satisfactorily, let alone rigorously defended. Thus for the present I am inclined to retain the linear model as a working hypothesis, and hope that through increasing educational efforts and by stressing the caveats that must be understood in using the hypothesis, the point can be made clear that estimates of risk at low dose and dose rates derived from the hypothesis are upper limit values that may exceed the actual risk by a considerable margin.

My principal overall conclusion is that extrapolation from low-LET radiation data obtained at high doses and dose rates most probably overestimate, perhaps by a considerable margin, the risk at low doses and dose rates. Thus the values for risk at low doses and dose rates, for low-LET radiation given in this paper most likely are indeed upper-limit values, and the true risk may well be appreciably smaller.

Despite the above-indicated views, I believe that radiation is potentially harmful and that exposure to it should continue to be carefully monitored and controlled. No unnecessary exposure should be allowed; the best radiation protection guide is, "as low as practicable." No exposure should be permitted without considering the benefits that may be derived from that exposure, and without considering the relative risk of alternative approaches.

§ 5.25 Internal Emitters.   Little space will be devoted to this subject here, not because of its relative importance, but because of its size and complexity, and time limitations. A wealth of experimental data are available.[3-6] As opposed to external radiation, internal radiation is delivered as a result of radioactive substances actually gaining access to the body and thus irradiating tissues during and after localization. Well known is the radium dial painter story. Young girls ingested radioactive material in the course of "pointing" small brushes with their lips in order to apply radium paint to watch dials. They later came down with bone tumors ascribed to the radium. This situation served to focus attention on potential industrial hazards associated with radiation exposure. Another somewhat similar situation is represented by the uranium (and other) miners, who in other countries as well as in the United States, inhaled radon in the course of mining operations. Here lung cancer, associated with radiation exposure, has been reported.

With internal emitters early clinical effects are rarely seen. This is because it is difficult for enough radioactive material to gain access to the body at one time to provide the high doses and dose rates necessary to produce early clinical signs and symptoms. Thus usually only late effects, principally neoplastic induction are of greatest concern.

Internal emitters to which the population at large may be exposed are usually "beta" or "gamma" emitters, i.e., low-LET radiations. Occupational exposure may involve alpha emitters, i.e., high-LET radiations (exposure of some families to isotopes from "mine tailings" in Colorado involved some high-LET radiations from radon). The eventual localization of isotopes in the body is highly dependent on route of administration (e.g., inhalation, ingestion), the chemical form of the isotope, the type of material with which the isotope is physically associated, and particle size. The spatial and temporal pattern of irradiation is highly dependent on the type and energy of radiation given off and the half-life of the isotope.

Thus, although there are no intrinsic differences between a given dose of radiation delivered internally or externally, the net effect of the above factors is that the spatial and temporal patterns of irradiation with internal emitters are usually highly complex and tend to be rather specific for a specific condition. Thus it is usually much more difficult to interpret "dose-effect" relationships with internal emitters in terms of possible mechanisms, or to generalize from specific situations. Because of the complexity, it is usually difficult to show that dose-effect curves obtained with internal emitters have any particular relevance to dose-effect curves for external radiations.

Although guidelines for the ingestion or inhalation of radioactive materials are oriented around radiation dose to tissues, they are usually given in terms of *Radioactivity Concentration Guides*, or RCG's, rather than in terms of radiation dose. This is because it is much easier to monitor amounts that might be inhaled or ingested, than it is to monitor the dose received. A great deal of attention is paid to nuclides in the environment.[4] It is well known that certain nuclides may be concentrated as they move up the food chain. This fact is taken into account in setting RCG's.

The Delaney amendment states that any food additive that is shown to be carcinogenic must not be allowed in food products for human consumption. The amendment says nothing with respect to route of administration, or dose levels at which a given material may have been found to be carcinogenic. Thus what is done routinely is to test possible carcinogenic agents at high doses, and assume that what is found there applies at low doses as well.

The Delaney amendment does provide a conservative approach to protecting the public; however, there are theoretical and practical problems concerned with its implementation.[25] These can be illustrated in light of experience with radiation, and many of the same problems encountered with radiation apply to hazardous chemicals as well. What is implied in the Delaney amendment is something akin to linear extrapolation from high doses to very low doses, which is usually considered to be a conservative approach. On the other hand, without the full dose effect curve, this approach could overestimate the hazard at low doses. It is quite possible that a number of such agents might show a threshold-type dose-effect curve, in which case the strict application of the Delaney amendment could unnecessarily prevent the use of some additives and agents that might be quite useful.

An additional difficulty comes in applying the amendment practically, and particularly with respect to radiation. This lies in the fact that zero is about as difficult to deal with in most instances, theoretically or practically, as is infinity. Often it comes down to the matter of how refined one's measurements are, and "zero" probably should be represented usually as "less than some number."

As noted above, human beings have been exposed for eons to background radiation, ranging from somewhat less than 100 millirem per year to several times this figure in parts of India and Brazil. There are a large number of naturally-radioactive materials, including potassium-40, which are a natural constituent in food. Also, fallout radiations are now ubiquitous, and can be found in very small quantities in many foods and some water supplies. Thus it is not within the realm of possibility to enforce the Delaney amendment strictly, with respect to radiation. All that could be done is to attempt to set guideline limits on exposure to any radioactive material or radiation exposure in excess of that which is inevitably encountered in the environment, in food or in water. This is precisely what the radiation guides, or the "radioactivity concentration guides" discussed above under Internal Emitters[5, 6] (see 5.25) are designed to do.

§ 5.26 **Risk to the Individual: Radiation and Other Hazards.** In the above discussions, only the possible effects of radiation on populations have been dealt with. From these estimates of risk to the population, one can develop the *probability* that an individual may develop a malignancy that he might not otherwise have had following exposure to radiation. As an example, consider leukemia. On the basis of extrapolation using the linear no-threshhold hypothesis, the excess incidence of leukemia in a population may be of the order of one to two cases per million people per year per rem, and one can use the average of 1.5 cases. Thus the chances of an individual developing leukemia following exposure to low-LET radiation, on this hypothesis, is 1.5 chances in a million per year per rem. The chances for any other dose can be obtained by simple multiplication, i.e., the chances, respectively, for two rem are 3, ten rem are 15, 1/3 rem are 0.5, etc.

The above risk figures are prospective, in the sense that they offer a basis for the estimate of an average upper limit probability of radiation-induced leukemia if one, in fact, becomes exposed, and before the disease develops, if it is going to develop at all. Different statistics can be presented for the probabilities that an individual who has actually developed a disease such as leukemia, with a documented history of radiation exposure, has radiation-induced as opposed to "spontaneous" leukemia.[21] This is best illustrated by an example. Consider that a man of a given age has developed myelocytic leukemia, and that he has a documented history of having been exposed to 10 rads of low-LET radiation several years earlier. Consider that the natural incidence of myelocytic leukemia for a man of his age is about 25 cases per million per year (1 chance in 40,000). On the linear hypothesis, and using the figures above, this 10R might result in a probability of radiation-induced leukemia of 10×1.5, or 15 chances per million per year (1 chance in 70,000). Thus the individual's total chances are 25 plus 15, or 40 (1 chance in 25,000). Of this total, the chances that the 10 rads might have been responsible for his leukemia, on the linear hypothesis, are approximately 15 out of 40. By analogous reasoning, corresponding probabilities can be worked out for leukemia at other ages, or for other types of tumors known to be associated with radiation exposure at high doses and dose rates.

An estimate of the relative hazard from different radiation sources to which the public is exposed can be obtained from the doses given in Table 5.2. If one adopts the linear no-threshold hypothesis, the risk is strictly proportional to dose; therefore the presumed hazard to individuals or to the public from exposure to different sources is strictly in relation to the average dose received.

Rather recently, in the United States and elsewhere, attempts have been made to evaluate the problems and risks associated with sources of power other than nuclear, in order to come up with a basis for realistic assessment of the role that each source should play under different conditions. These analyses are far from complete, and should be extended and refined over the next few years.

One fact that strikes one immediately is that the risks associated with nuclear power have been investigated in greater detail than have those associated with any other power source or environmental pollutant. With respect to fossil fuels and other types of energy sources, relatively little attention has been paid to the possible hazards, either from the investigative or analytical standpoint. Data on the sources of distribution and hazards of effluents from the use of fossil fuels are being investigated, however, and thus the assessments should improve with time.

Perhaps one of the earliest approaches is that of Stig Bergstrom of Sweden, who compared the overall impact of fossil fuel versus nuclear power plants in Sweden,[49] concluding that, "as to the choice between the oil and nuclear fuel alternatives the environmental impact from normal operations distinctly favors the nuclear plant." Similar studies have been carried out by Hill,[50] who came to the same conclusion, that nuclear plants produce far less air pollution than their fossil cousins, and that, "...the catastrophic potential of nuclear plants has been vastly over-exaggerated by their adversaries." The consequences of coal mining operations in the United States has been detailed in a recent mineral industry survey.[51] The annual toll from coal mining far exceeds that from any other kind of mining, and in the single year 1971, 342 fatal injuries and approximately 24,000 non-fatal injuries occurred. The cost of pneumonoconiosis (black lung) disease from a total of approximately 350,000 claims between 1969 and 1971 amounted to approximately $530,000,000 or approximately $340,000,000 per year, about $2,100 per case.

The health effects of electricity generation from coal and other nuclear fuels has been assessed by Lester Lave,[52] with the conclusion that "the analysis results in an unequivocal conclusion that nuclear reactors offer less of a health hazard than do coal burning generators." Mining coal has roughly 12 times the accident rate as mining and milling uranium, per megawatt hour of electricity. Chronic disability associated with coal mining is about 26 times as great as that of uranium mining, per megawatt hour of electricity generated. The routine effluents of coal and nuclear generators have been examined. Both release comparable amounts of gaseous radiation from routine operation. Thus radioactive emissions are not a reason to prefer coal generators. Finally, coal generators produce enormous amounts of air pollution which have been shown to be harmful to health.

Cohn and co-workers have studied the relationship between asthma and air pollution from a coal fueled power plant,[53] and have come to the conclusion that "significant correlations were found between reported attack rate and temperature, and between attack rate and pollution levels after the effects of temperature had been removed from the analyses. These…air pollution effects occurred at levels of pollution commonly found in large cities, and appeared greater at moderate than at low temperatures."

E. J. List looked carefully at energy requirements in California, and used as a "yardstick" EPA standards for air pollutants.[54] He came to the conclusion that "if we assume that all fuels (fossil) in these basins are burned at the minimum emission factors that appear technologically and economically feasible, the residual pollution is still such that the promulgated ambient air quality standards cannot be met. Thus, a simple policy of no-growth in these two basins still leaves the areas with significant air pollution. …continually lowering the emission factors will not attain clean air in the south coast and San Francisco area basins. …the problem is aggravated by the increasing consumption of fuel every year. The only other policies available for air pollution control are either the relocation of energy demands to those energy sources with zero emission factors, or the curtailment of the use of fossil fuels as an energy source. The only near zero emission energy source capable of accommodating the possible demand for energy at this time is nuclear generated electric power. Hence… the only way Los Angeles and San Francisco will attain ambient air quality satisfying promulgated standards is to replace fossil fuel consumption by nuclear power."

The results of such studies are shown in Table 5.3. Notice the enormous number of nuclear reactors that could be tolerated in the Los Angeles area, or the relatively few fossil fuel plants that could be permitted.

Table 5.3.   Tolerable Numbers of Power Plants as Implied by Curre nt Practices in Los Angeles County*

Plant Type	Critical Pollutant	Tolerable Number of 1000 Mwe Plants (exclusive of pollutants from other sources)
Oil	SO_2	10
Natural gas	NO_2	23
Nuclear reactor (LWR)	Radioactive gases	160,000

*Based on the following assumptions:

1. Unspecified mixture of radioactive isotopes released from nuclear plant. (most restrictive assumption based on 1 millirem).
2. Compliance with 0.5% by weight sulfur content for oil.
3. Air volume of Los Angeles County was assumed to be 3165 Km^3 which implies a mean inversion height of 300 M.
4. Ventilation of this volume requires 1 day.
5. Effluent volume rate for 1000 Mwe reactor is taken as 0.5×10^6 cfm which is an estimated upper limit.

Hamilton[55] has made comparative studies, and has stated in effect that, if one uses the same very conservative assumptions in assessing the hazards from toxic agents in air polluted by the burning of fossil fuels, as one does in assessing the hazards of radiation, the numbers of injured and dead will be so large as to make any possible hazard from nuclear power production pale by comparison.

C. Starr and M. A. Greenfield have made an extensive study on the public health risks of thermal power plants.[56] This detailed study consists of a summary report plus five appendices. Both the situations of normal operation and potential accident situations are compared in detail. They use an approach that has not been published in detail previously. In essence they state that, in assessing the hazards of any mode of energy production, one should use equally conservative assumptions throughout. Thus, if one is going to use extremely conservative assumptions in assessing the probability of the consequences of a severe accident in a nuclear power plant, then the same extreme assumptions must be used in assessing the probability and results of accidents associated with the use of fossil fuel for power.

They also make the interesting observation that the relationship between federal standards for different pollutants, natural background levels of that same pollutant,

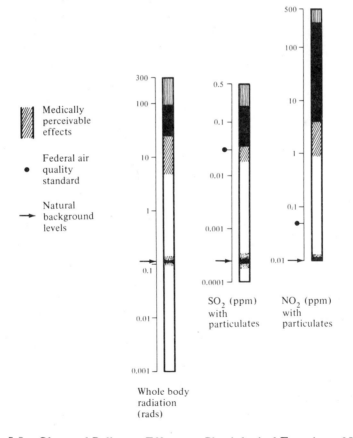

Fig 5.8 Observed Pollutant Effects on Physiological Function of Humans.

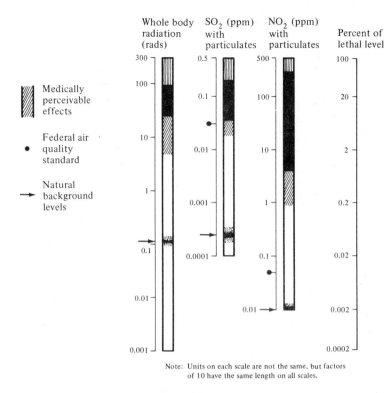

Note: Units on each scale are not the same, but factors
of 10 have the same length on all scales.

Fig 5.9 Observed Pollutant Effects on Physiological Function of Humans

and medically perceivable effects are vastly different for different pollutants (Figure 5.8 and 5.9). For instance, regulations or standards for radiation exposure of the public are at least two orders of magnitude lower than the levels of exposure at which one can begin to perceive medical effects. On the other hand, the federal air quality standard for SO_2 with particulates is set right at the level at which one can observe medical effects (Figure 5.10). It is ironic that even with a factor of 10^2 "cushion" in the radiation standards, there is still a great deal more concern about radiation exposure at levels even far below these already extremely conservative standards, than there is about the effects of SO_2 at or above the level of the standard set at the level where medical effects are readily perceived.

They have also pointed out that the severity of accidents in general varies inversely with their probability of occurrence. They have collected statistics on oil fires, and have compared these with the probability of accidental releases of radioactivity from PWR's. The results are shown in Figures 5.11 and 5.12. Note that the frequency of oil fires, real events, is much, much greater than the postulated frequency of the release of significant quantities of radioactivity in a reactor accident. In Figure 5.13 they have translated these data into cumulative accident mortality from oil-fired versus nuclear plants, as a function of distance from the installation. At all distances, the impact of nuclear plants in terms of death is considerably lower than that from oil-fired plants.

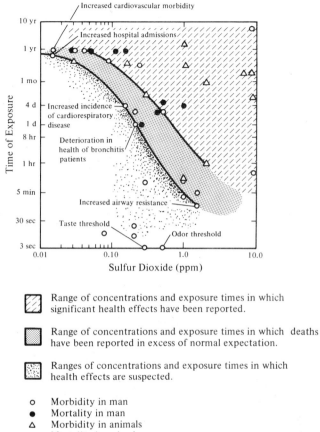

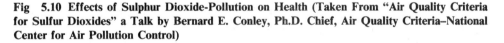

Range of concentrations and exposure times in which
significant health effects have been reported.

Range of concentrations and exposure times in which deaths
have been reported in excess of normal expectation.

Ranges of concentrations and exposure times in which
health effects are suspected.

o   Morbidity in man
●   Mortality in man
△   Morbidity in animals
▲   Mortality in animals

Fig 5.10 Effects of Sulphur Dioxide-Pollution on Health (Taken From "Air Quality Criteria for Sulfur Dioxides" a Talk by Bernard E. Conley, Ph.D. Chief, Air Quality Criteria–National Center for Air Pollution Control)

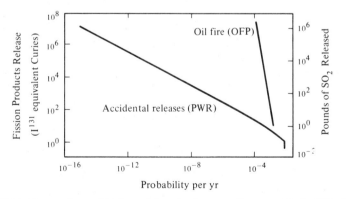

Fig 5.11 Comparison of Release Magnitudes on a Common Probability Scale

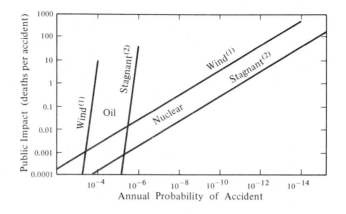

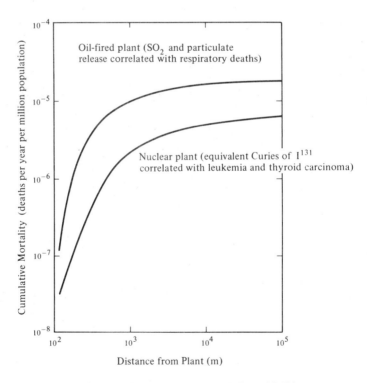

(1) The prevailing wind blows into a 30° sector at 2 m/sec under stable meteorological conditions (Pasquill F condition).

(2) The stagnant meteorological condition was assumed to prevail for 4 days with an inversion height of 300 m.

Fig 5.12. Comparison of Public Risk from Individual Accidents

Fig 5.13. Cumulative Accident Mortality with Distance

Table 5.4. Public Risk Comparison

		Expected Annual Averages (Deaths per 10 million population per 1000 Mwe plant per year)
Nuclear Reactor (Cancer Deaths)	1	Negligible (0.00006)
Oil Fired Plant (Respiratory Deaths)	60	Negligible (0.0002)

Their overall conclusions are given in Figure 5.11 and in Table 5.4.

No one has done truly exhaustive studies on the possible catastrophic results of a large oil fire occurring in an urban community (such as in New York where one large oil fire occurred very recently in the nearby community of Bayonne, New Jersey), making the extremely unrealistic and pessimistic assumptions used in a WASH-740 type of analysis. Under such circumstances one would postulate an extremely large fire, with the most adverse weather conditions such that the smoke spread out over a large urban area and was not dispersed by winds, and at a time when there was an extensive influenza epidemic. I am sure that if such an analysis were carried out in detail, the results would be horrendous to a degree comparable to those developed in the WASH-740 report.

§5.27 References and Notes

1. *Basic Radiation Criteria*, National Council on Radiation Protection and Measurements (NCRP) Report No. 39, January 1971, NCRP publications, P. O. Box 4867, Washington, D. C.
2. Victor P. Bond, Theodor M. Fliedner, and John O. Archambeau, *Mammalian Radiation Lethality, A Disturbance in Cellular Kinetics,* Academic Press, New York and London, 1965.
3. *Ionizing Radiation: Levels and Effects*, A report of the U.N. Scientific Committee on the Effects of Atomic Radiation to the General Assembly, No. E.72.IX.17, 1972.
4. *The Effects of Populations of Exposure to Low Levels of Ionizing Radiation*, Report of the Advisory Committee on the Biological Effects of Ionizing Radiations, (BEIR report), Division of Medical Sciences, National Academy of Sciences, National Research Council, Washington, D.C., 1972.
5. *Evaluation of Radiation Doses to Body Tissues, from Internal Contamination due to Occupational Exposure.* Report of ICRP Committee IV, ICRP Publication 10, Pergamon Press, 1968.
6. *The Assessment of Internal Contamination Resulting from Recurrent or Prolonged Uptakes*, Report of ICRP Committee IV, ICRP Publication 10a, 1969.
7. Summary Reports, *Biological Effects of Atomic Radiation* (BEAR) Committee Report, National Academy of Sciences, National Research Council, 1956 (and detailed genetics report).

8. W. L. Russell, *The Effect of Radiation Dose Rate and Fractionation on Mutation in Mice*, in *Repair from Genetic Radiation Damage*, ed. F. Sobels, Pergamon Press (Oxford), 1963, pp. 205–217; 231–235.

9. W. L. Russell, *Studies in Mammalian Radiation Genetics*, Nucleonics **23**, No. 1 (1965).

10. A. M. Weinberg, *Social Institutions and Nuclear Energy*, Science **177**, 27–34 (1972).

11. C. Edington and M. Kastenbaum (personal communication).

12. Alice Stewart and G. W. Kneale, *Radiation Dose Effects in Relation to Obstetric X-rays and Childhood Cancer*, Lancet **1**, 1185–1188 (1970).

13. H. Kato, *Mortality in Children Exposed to the A-Bombs while in utero, 1945–1969*, Am. J. Epidemiology **93**, 435–442 (1971).

14. A. Hull and F. J. Shore, *Sternglass: A Case History* (BNL # 16613). Presented at the Atomic Industrial Forum Information 3 meeting, Los Angeles, California, March 8, 1972.

15. A. C. Upton, et al., *Late effects of fast neutrons and gamma rays in mice as influenced by the dose rate of irradiation; induction of neoplasia*, Rad. Res. **41**, 467–491 (1970).

16. John W. Baum, *Mutation Theory of Carcinogenesis and Radiation Protection Standards*, presented at the Sixth Annual Health Physics Society Topical Symposium, Rivershore Motor Inn, Richland, Washington, on November 2–5, 1971.

17. R. D. Evans, Statement in *Radiation Exposure of Uranium Miners*, Hearings before the Joint Committee on Atomic Energy, Ninetieth Congress, First Session, Part 1, May, June, and July 1967, pp. 258–318.

18. A. M. Kellerer, and H. R. Rossi, *RBE and the Primary Mechanisms of Radiation Action*, Rad. Res. **47**, 15–34 (1971).

19. A. P. Hull, *Reactor effluents: As low as practicable or as low as reasonable?*, Nuclear News, Vol. 15, No. 11., November (1972), pp. 53–59.

20. *Estimates of Ionizing Radiation Doses in the United States 1960–2000*, U.S. Environmental Protection Agency document ORP/CSD 72-1, Rockville, Maryland 20852.

21. V. P. Bond, *The Medical Effects of Radiation*, NACCA Thirteenth Annual Convention, Miami Beach, Florida, 1959, pp. 117–129.

22. *Accident Facts*, National Safety Council, Chicago, Illinois, 1971 Edition.

23. W. H. W. Inman, and M. P. Vessey, *Investigation of deaths from pulmonary, coronary and cerebral thrombosis and embolism in women of child-bearing age*, British Med. J. **2**, 193–199 (1968).

24. F. P. Cowan, A. V. Kuehner, and L. F. Phillips, Final Report of an Interagency Agreement between U.S. Atomic Energy Commission and the Environmental Protection Agency, Brookhaven National Laboratory Report No. BNL 17291, October 2, 1972.

25. B. J. Culliton, *Delaney Clause: Defence against an uncertain threat of Change*, Science **179**, 666–668 (1972).

26. New York Times, January 8, 1973.

27. *Der Kolner Dom zerfallt zu Gips*, X Magazin, January 1973.

28. *Air Pollution across National Boundaries*. The impact on the environment of sulfur in air and precipitation. Sweden's case study for the U.N. Stockholm, Sweden, 1971.

29. I. Forbes, D. Ford, W. Kendell, and J. McKenzie, *Nuclear Reactor Safety, an Evaluation of New Evidence*, Nuclear News, September 1971.

30. V. Sailor, *Global Energy Needs*, International Convocation, Manhattan College, New York, October 14, 1972.

31. V. Sailor, *Costs and Benefits of Nuclear Power*. New Horizons in Physics Lectures, State University College, New Paltz, New York. October 17, 1972.

32. Newsweek, January 22, 1973.

33. Duane Chapman, Timothy Tyrrell and Timothy Mount, *Electricity Demand Growth and the Energy Crisis*, Science **178**: 703–708, 1972.

34. *Radiosensitivity and spatial distribution of dose.* ICRP Publication 14, Pergamon Press, 1969.

35. L. Rogers and C. Gamertsfelder, "*USA regulations for the control of releases of radioactivity into the environment in effluents from nuclear facilities.*" Environmental Aspects of Nuclear Power Stations. IAEA publication, 1971.

36. V. P. Bond, *Radiation Standards, Particularly as Related to Nuclear Power Plants*, Health Physics Society Newsletter, 18 February 1971.

37. N. Frijeno, "Cancer Epidemiology and the Radiation Background." Argonne National Laboratory, Argonne, Illinois. In Press.

38. Bernard Kahn, et al. "Radiological surveillance studies at a Boiling Water Nuclear Power Reactor." United States Public Health Service Report BRH/DER 70-1 March 1970.

39. "Theoretical possibilities and consequences of major accidents in large nuclear power plants." Report WASH-740, United States Atomic Energy Commission, Division of Technical Information, March 1957.

40. J.B. Horner Kuper, Testimony at the Shoreham Nuclear Power Station Hearings, Docket 50-322, May 25, 1970.

41. D.F. Ford, H. W. Kendall, and J.J. MacKenzie, A critique of the new AEC design criteria for reactor safety systems. Union of Concerned Scientists, Cambridge, Mass. October 1971.

42. *Letter to the Editor*, Nuclear News, November 1971, p 25.

43. Norman C. Rasmussen, *Nuclear Reactor Safety—An Opinion.* Nuclear News, January 1972, p 35.

44. H.J. Otway and C. Erdmann, *Reactor Siting and Design from a Risk Viewpoint.* Nuclear Engineering and Design **13**: 365–376, 1970.

45. F. Pittman, Management of commercial high-level waste. Speech, American Nuclear Society Meeting, Washington, D.C. 16 November 1972.

46. J.B. Knox, *Airborne Radiation from the Nuclear Power Industry.* Nuclear News, February 1971.

47. J.A. Martin, Calculations of environmental radiation exposures and population doses due to effluents from a nuclear fuel reprocessing plant. Environmental Protection Agency, Radiation Data and Reports, February 1973 Paper 59–75.

48. L. Hempelmann et al. "A twenty-seven year study of selected Los Alamos plutonium workers." In press.

49. S. Bergstrom, "Environmental consequences from the normal operation of an urban nuclear power plant." Presented at the Health Physics Society Mid-year Topical Symposium, Ramada Inn, Idaho Falls, Idaho, November 3–6, 1970.

50. A. Hull, *Some Comparisons of the Environmental Risks from Nuclear and Fossil Fuel Power Plants.* Nuclear Safety Vol. **12**, No. 3, May-June 1971.

51. "Injury experience and work time in the solid mineral mining industry, 1970–71." Mineral Industry Surveys, U.S. Department of the Interior, Bureau of Mines, Washington, D.C. 20240. F.T. Moyer, author, 1972 (August).

52. L. Lave, "The health effects of electricity generation from coil, oil and nuclear fuel." Presented at the Sierra Club Conference on Environmental Effects on Electricity Generation, Johnson, Vermont, January 1972. Graduate School of Industrial Administration, Carnegie-Melon University Report WP-93-71-2, April 1972.

53. A.A. Cohn, S. Bromberg, R.W. Buchley, C.T. Heiderscheit, and C.M. Shy, "Asthma and air pollution from a coal-fueled power plant," AJPH 62, 1181–1188, 1972.

54. E.J. List, "Energy use in California: implications for the environment." Environmental Quality Laboratory, California Institute of Technology, EQL Report, 3 December 1971.

55. L.J. Hamilton, personal communication.

56. C. Starr, and M.A. Greenfield, "Public health risks of thermal power plants." UCLA-Eng-7242, May 1972, University of California, Los Angeles.

Part IV

**ENERGY
AND
SOCIETY**

6

FABLE FOR ANOTHER TIME

Angelo J. Cerchione

To the amazement of their rude companion, three great magicians in turn caused a pile of bones to become a skeleton, then fleshed out the bony structure, and finally prepared to endow it with new life. Shaken, the simpleton cautioned, "Don't you realize that these are the remains of a tiger?!"

Caught up in the frenzy of artful invention, the magicians ignored the warning. From his tree, the simpleton watched the splendid moment of rebirth and—ultimately—the consumption of the three wise men.

Sanskrit *Panchatantra, fable.*

Prudence, mate! It is our only alternative when one does not understand the environmental effects of some invention.

Ian McHarg, environmental planner and teacher.

The new growth of leaves was still young enough to act as a soft green filter to the Spring sunlight and the park was suffused with a pale verdant glow. Beneath the trees, a spry, bearded figure made his way toward a group of young men and women who occupied the hillside just above the Schuylkill River, its banks specked with anglers. Soft spoken now in his later years, inevitably dressed in tweeds, he had long ago earned the respect and admiration of his young audience. Privately, they agreed that "he knows where it's at."

Settled on a favored bench, greetings and small talk exhausted, he felt a story coming on. The eyes twinkled, the adrenalin was up, and the storyteller magically shook off the accumulated years.

"As often as we have met here, none of you have ever mentioned or asked about this park. Did you know that this had once been an expressway which followed the Schuylkill through Philadelphia? It might still be an expressway but for the fact that many years ago our city fathers (with some prompting from the younger citizens) made a very crucial decision. But rather than beginning in the middle, let me take you back in time a bit.

"In the beginning men were content to move about the land on foot, to follow the migrating herds, to eat whatever berries and vegetables the land provided. It was apparent very early, however, that our people possessed a special inventiveness. We saw it first in the art of the shaman, the arrow maker, and the weaver of cloth. It is the ability to know a fact and to add to it a second thing which may be said to have been spun out of the wanderings of the mind. It is similar to the miracle of the spider which not only produces a filament from its body and knows how to anchor it but is able to conjure up a design for its web.

"This gift of creativity was everywhere in evidence in the works of our men and women. As a society we paid special honors to those who brought us pleasure through the arts, salvation through force of arms or religion, plenty through cultivation of the land, or wealth through the establishment of industry or business.

"Unfortunately, the same mind that is capable of conceiving of these things is also capable of perverse fantasies. We found that it is possible to become fascinated with a simple invention, to invest it with *mana*-like properties, then develop and further refine it, eventually to be consumed by our creation.

"By way of example, I could tell you of the better known creation histories— how from the chipping of flint came the arrow and its progeny, of the early use of iron and bronze and the great industries that followed, or of the first man-made spark and the fire that nearly bankrupted our planet's energy resources. Each of these and the final benefits and misery they conferred is well-known.

"I have chosen another because it is a tale of some foolishness, of tragedy certainly, and of redemption ultimately. Also because the land we are now sitting on is our city's monument to the balanced life.

"It begins in the shops of handymen and bicycle makers. Men trying to bring quixotic and pragmatic ideas into being. Busy men dreaming special personal dreams. At last from these shops the first motor cars appeared. We laughed, of course, to see the ungainly beasts shaking with self-importance and moving off at a rapid five to ten miles an hour to their own cacophonous salute. But *they did move and they were novel!* Eventually, there were dozens of small companies engaged in the production of a limited number of these primitive motor cars. Undaunted, these men stuck with their novelty, massaged its crudities, added improvements, and sought to convince us of the utility of their mechanical horse.

"At the turn of the century, there was an inventor and capitalist who developed a method of mass-producing his model. Once this production breakthrough had been achieved, other large companies got their version of the automobile onto production lines and each year more and more rubber-wrapped wheels hit the pavement. Why during the ten years between 1959 and 1968, Americans spent 700 billion dollars on cars, gasoline and accessories. In that same period of time, we used 670 billion gallons of fuel. What's more, in the Sixties, 14 million people were employed in industries directly concerned with motor vehicles.

"As often happens the owners of these new devices began to wish for a new way of organizing their lives and their communities—one more in keeping with their new sense of mobility. These wishes were not linked to what men really need to flourish but in a sense they seemed more the wishes of the automobile itself. Like a chimera, the auto was making its needs known through the minds and actions of the men whose soul it had come to possess.

"The auto cannot travel well without paved roads so these were built. The city restricted its movements and men soon bought homes in the suburbs or began to rebuild their cities. In some communities over 50% of the land was given over to the car. In many cases a weak individual was matched with a powerful vehicle and there were accidents: two million people a year were injured and another 55,000 were killed. Money given to the government in the form of taxes was set aside as a high-way trust fund and the presence of this extraordinary amount of money set in motion a frenzy of highway building in the 1950's and 1960's unparalleled anywhere in the world. Truly it could be said, the automobile was seeking a fit environment, was indeed making an adaptation, and its creative fitting with the land began to render the land fit only for the automobile. Other forms of transportation began to wither away and soon 89% of all intercity trips were accomplished by motor vehicles. Men seemingly forgot that there were alternative means to achieve mobility. Moreover, lovemaking, movie watching, banking, eating, home selection, and the collection and disbursement of tax monies—all of these activites were somehow becoming inextricably bound up with *the family car.*

"If the enthralled, busy knitters of this new cloth were working overtime and covering the land with their fabric, by the late 1960's, signs were appearing throughout the world that an unraveling had begun. At this critical point in time half of the 200 million cars loose in the world were being cared for in the United States (if in our country there was a car for every two people, it was encouraging to know that in Nepal the ratio was a single car for 2,500 people). Licensed, insured, groomed and repaired, this pampered pet of the American people consumed 88 billion gallons of gasoline a year—a great deal of which it wasted through inefficient combustion.

"The man called Ford brought the automobile to maturity as a transportation mode, but it was places such as the Meuse Valley, Donors, London, New York, and Los Angeles that challenged its preeminence. The thousands of people who died in these places drew the public's attention to the phenomenon which had snuffed out their lives: air pollution. While a variety of activities shared the responsibility for the corruption of our atmosphere, the automobile was discovered to be producing 60% of the nation's air pollution. In 1953, Dr. A. J. Haagen-Smit, discovered that the automobile was responsible for photochemical smog when oxides of nitrogen and hydrocarbons emitted by its engine underwent a chemical reaction in the presence of sunlight.

"It was this same smog that in 1949 made its presence known by harming plant life in the Los Angeles basin. In 1961, this smog damage accounted for eight million dollars in crop damage; and by 1969, many of these crops could not be grown in the Los Angeles basin at all. It was during this same decade of painful discovery that Americans came to learn that crops worth 500 million dollars were lost each year to air pollution of all kinds in the United States.

This in a world already hungry.

"In a land where lung cancer had been virtually unknown, some 4,000 deaths were reported in 1935; and by 1966, that number had risen to 43,000 a year. Emphysema became the country's fastest growing cause of death and had for handmaidens asthma and bronchitis. Cities such as Phoenix, Arizona, where, in years earlier, those suffering from respiratory problems were usually sent to recuperate, were being added to the list of the most polluted cities in the United States.

"American's love affair with the auto had not been disturbed by the car's influence on the shape of its cities, the alteration of personal habits or the high accident rate. However, the insidious pollution attack which was operating against the nation's health and the plant life which nourishes all living things was a physiological and psychological threat that no prudent man could ignore.

"As our ancestors learned more about the major causes of air pollution, their resistance to the car grew. Oxides of nitrogen, they were told, in sufficient quantities would cause pulmonary edema and the victim could drown in his own fluids. Smog, a product of hydrocarbons, oxides of nitrogen, and sunlight, was found to be linked with the destruction of the body's enzymes. Carbon monoxide was known to have an affinity for hemoglobin and would combine more readily with it than oxygen could, a process that can cause asphyxiation and death of the body's cells, especially the nerve cells. In low concentrations it led to nausea, headaches, mental confusion, vomiting, and unconsciousness. In many cases these same symptoms were contributing factors in auto accidents. Particulates, asbestos, dust, lead, sulfur dioxide, ozone—singularly and in combination—almost brought the nation to its knees.

"The impulse to improve the internal combustion engine with a host of anti-pollution devices was still strong, however. Valuable time and money were consumed trying to make the automobile a fit companion in a crowded world. Nationally, three billion dollars a year were committed to the air pollution fight by the Federal Government in the 1970's. Private estimates put the cost of controlling the emissions of the internal combustion engine at 100 billion dollars for the final quarter of the century. Even the president of the Society of Automotive Engineers in 1970 wrote that despite the cumulative improvements in the control of this pollution source, by the year 1985 another upturn in the overall amount of pollution was to be expected. It would have been caused by the continuing growth of the car population between 1970 and 1990.

"Faced with certain evidence of death and disease and the even more dire predictions for the last 30 years of the Twentieth Century, our people purposefully altered their habits and the legislation that governed them. Cities excluded the auto from their centers; mass transit which could move more people less expensively took over more and more of the auto's role. By 1985, steam turbine, electric and various natural gas engines completely dominated the market that had once belonged exclusively to the internal combustion engine. Where once a 700 cubic foot automobile dominated our roads, a much smaller personal module was put into service for those people who could not make use of the mass transit systems. Older cities found that they had only to remove the asphalt or concrete of the no longer needed freeways to create linear parks and to restore the land to more productive uses. There were cities that made a fine boast that their new mass transit systems were produced entirely from recycled automobile hulks. In some ways, we had as many laughs with the new contraptions as we did with the first motor cars. But we had learned our lesson.

"You see, at some point creativity bankrupt becomes monomania. The spider's display of web building so dazzles us that we miss the basic lesson in natural economics that is also being demonstrated. Men, too, produce marvelous devices and inventions but their needs for these things are not necessarily rationed by a natural thirst or hunger. Unfortunately, we were made to believe for many years that we were creating our own world rather than being an inseparable part of this one. By halting our exorbitant expenditure of natural resources, we proved that we were capable of serving as stewards to this good green earth. In a sense we owe our survival to the automobile and its internal combustion engine. Had we not pursued our inventiveness to the edge of self-destruction and discovered the enormity of our error just in time, we might have ended our existence on that darkling plain of the 1970's. It was a very close call for all of us my friends.

"Yes, those were exciting times to live in—to be among the saints and watch the ranks of the converts grow. And to think that the movement had its beginnings in this city." He smiled, "It's memories like those that make old men want to dance, sing and chase women." He chatted a while; finally rose and made his way through the trees. A cool breeze riffled the leaves and the group slowly broke up. Below, an angler and his son landed a fish—the shad were running once more. But then that is another chapter in the history of invention and prudent men.

*

7

CONSERVING ENERGY: OPPORTUNITIES AND OBSTACLES

David B. Large

§7.1. Introduction. The notion that the economic and environmental costs of our accelerating energy consumption are beginning to strain our ability to pay is now commonplace. Concern about "the energy crisis," an amalgam of economic, political, social and environmental issues, has produced a profusion of congressional hearings, conferences, lengthy studies, and much heated debate. (One industrialist was quoted recently as saying that the solution to the energy crisis is simply to burn all the energy studies for fuel.) Predictions about how much petroleum, natural gas, coal and uranium, the principal energy fuels, remain to be recovered in the United States vary widely, depending upon one's assumptions. But there is a concensus that unless growth slows, we are approaching a point at which most of the economically recoverable reserves of natural gas and petroleum in the United States will be exhausted within a very few decades. Recently, our response to shortages has been to increase imports — we already import one quarter of our petroleum, and the U.S. Bureau of Mines is predicting that this will be up to one-half by the early 1980's. However, both national security and balance of payments considerations have government planners extremely jittery about that degree of reliance upon foreign sources.

Predictably, the energy industries have responded to the spectre of a supply-demand gap by pushing for increased exploration and production, appealing for less federal regulation, more tax incentives for exploration, and a relaxing of environmentally motivated controls. Meanwhile the damage to the physical environment which necessarily accompanies the production, distribution and consumption of energy continues to increase, and battle lines have already been drawn over such controversial projects as off-shore drilling, the trans-Alaska pipeline, strip mine regulation, and the proliferation of nuclear power plants, to name just a few. Resistance to sacrificing our newly found environmental "ethic" is strong—organized citizen groups are fighting to see that it not be sacrificed to what they consider to be a wasteful and gluttonous rate of energy consumption.

In many quarters, there is an abiding faith that radically new technology will come to the rescue. Solar energy and thermonuclear fusion are often cast as the heroes—both call up the siren song of virtually unlimited, pollution-free energy. Unfortunately, these technologies have largely been neglected in the rush to expand our nuclear power capacity and to develop the breeder reactor (a scheme which increases the efficiency of uranium use by a factor of over one hundred). However, nuclear fission is fraught with environmental and safety problems which we are only beginning to appreciate, and several highly qualified scientists have spoken out against the whole breeder concept.

There is a fourth approach to the total supply/demand/environment dilemma now being promoted by many environmental groups, and it appears to be gaining recognition in parts of the federal bureaucracy, and to a lesser extent, in the energy industry. In its simplest form it means buying time by reducing waste—an "energy conservation" philosophy. Reducing our consumption of energy could of course be accomplished by drastically reducing our material standard of living ("returning to caves and candles," as one industrialist puts it), but several recent engineering studies, plus the common sense observation that we live in a wasteful society, lead to the conclusion that *there exist a variety of means for significantly reducing our rate of energy consumption, without compromising our material standard of living nor requiring radical changes in lifestyles.* Barriers to the implementation of these measures are political, cultural, and economic, rather than technical. Not all of the proposals being put forth are feasible, or necessarily socially desirable, but many do appear to offer the possibility of reducing environmental impacts while at the same time conserving fuels and enhancing the quality of our lives, and without the economically regressive effects of many pollution control measures. The following sections outline some of the most significant areas where energy is being wasted.

§7.2 Transportation. The day-to-day fueling of motor vehicles, trains and airplanes consumes one-fourth of our energy budget, largely in the form of petroleum. If we include the manufacture and servicing of these carriers, the portion increases to between 30% and 40%. The manufacture and servicing of the automobile alone, plus the construction and maintenance of highways, also consume fully one-fourth of our total energy budget. Thirty-nine percent of all the petroleum products burned in this country are consumed as gasoline in automobiles, and yet:

   The internal-combustion automobile engine is so inefficient that three-fourths of the gasoline burned is wasted.
   The average miles per gallon delivered has *decreased* since World War II from 13.5 mpg to 12.2 mpg.
   Eighty-two percent of all commuters travel by automobile, with less than one-

third of them carrying more than one passenger. The average urban passenger load during peak traffic hours is now only 1.2 people per car.

In cities, bus travel is more than twice as efficient (in terms of average energy expended per passenger-mile) as automobile travel. Commuting into a city by train is two and one-half times more efficient yet the number of revenue passengers on mass transit *decreased* by 48% between 1940 and 1971.

Shipping freight between urban areas by rail is four times as efficient (in terms of energy expended per ton-mile) as by truck; yet the percent of total tonnage shipped by rail has *decreased* steadily since 1950, while that shipped by truck has steadily *increased*.

Auto traffic in central cities moves at an average of about 12 miles per hour, the same speed achieved by horse-drawn carriages 100 years ago.

§7.3 Building Construction. The residential and commercial sectors of the economy together consume about 35% of our total energy budget, about half of that going to space heating and cooling. In 1970, 22% of all electric energy used by industry was consumed by the building construction industry, either directly in construction or indirectly through production of construction materials. Much of this energy is wasted:

Construction standards for commercial and public buildings are often grossly excessive: some structural engineers have estimated that 50% of the material used in constructing large buildings could be safely eliminated if more man-hours could be put into framework.

Readily available design practices could reduce the amount of energy commonly used for space conditioning and lighting in most large guildings by close to 50%. (The New York World Trade Center has a peak electrical demand of 110 megawatts, more than that required by Schenectady, a city of 100,000 people.)

Twenty-four percent of all electrical energy goes for lighting, yet ordinary incandescent lamps convert only 5% of the electrical energy they consume into useful light; fluorescent lamps convert only 20%. Lighting intensity standards have more than tripled over the past two decades, and are no longer based upon either physiological nor psychological criteria.

Increased thermal insulation in homes and buildings could reduce energy consumption by as much as 40% and still save the owner money by reducing fuel and electricity consumption. Insulating beyond even the recently revised FHA insulation standards is still economical in most cases.

§7.4 Air Conditioning. In all-electric homes, air conditioning follows space and water heating as a major consumer of energy. Air conditioning is particularly significant as a power consumer because of its contribution to (or in some cases, its cause of) the seasonal peak power load that occurs in the summertime in many localities. Eighteen percent of the growth in residential electricity consumption between 1960 and 1970 was due to the growing popularity of air conditioning. Yet the efficiency of these systems, expecially room units, varies widely:

Efficiencies of room air conditioners vary from 4.7 to 12.2 Btu of cooling capacity per watt-hour of electrical consumption. This means that the least efficient model consumes 2.6 times as much electricity as the most efficient one while accomplishing the same amount of cooling.

In 1970, the average efficiency for all room units was about 6 Btu per watt-hour. If the average efficiency had been 10 Btu per watt-hour, electricity consumed for air conditioning for that year would have been reduced by 15.8 billion kilowatt-hours, or 40%. That savings is the energy equivalent of 7.6 million tons of coal or the yield of approximately 1500 acres of strip mining.

The small increased cost of the more efficient models is more than compensated for by savings in operating costs. Even assuming the consumer buys his device on credit and pays an effective 18% annual interest rate, he would still be economically justified in paying up to $79 more for a high-efficiency unit, assuming electric rates applicable to the Washington, D.C. area.

§7.5 What Must Be Done. These are the most obvious areas where significant energy savings can be effected. There are many more, including the use of rejected heat from electric power generation, better design of heavy appliances, increasing the efficiency of many industrial processes, and utilizing the heating and recycling potential of the mountains of solid waste we produce each year. The above list was chosen because the savings potentials in those areas are great, and because the means to achieve those savings are technically, if not politically, available now. In simple terms, here are some of the things that need to be done:

- Dramatically increase funding for public transportation systems in urban and suburban areas, possibly by diversion of funds from the Highway Trust Fund.
- Discourage single-passenger automobile commuters by such policies as discriminatory parking taxes and bridge tolls, and the reservation of express lanes for commuter buses.
- Place "environmental impact taxes" on the sale of high horsepower, low-efficiency automobiles.
- Improve intercity rail networks and freight-handling procedures in order to create economic pressures to shift more intercity freight from trucks to railroads.
- Revise federal airline regulatory policies which in effect subsidize short inter-city air flights in order to shift more short-haul traffic to railroads and buses.
- Revise building codes to eliminate excessive construction and lighting standards.
- Develop standards for peak power demand and annual energy consumption for commercial buildings, and impose an impact tax on all buildings exceeding those standards.
- Tighten FHA and GSA insulation standards, and extend them to apply to more categories of construction (e.g., mobile homes).
- Establish standards for minimum allowable air conditioner efficiencies, and require that all units be equipped with thermostats.
- Disallow the inclusion of promotional advertising as a legitimate operating expense for electric utilities, especially the advertising of electric resistive heating.
- Alter rate structures for electric power to place a premium on use during peak demand periods.

Successful implementation of all these policies might reduce total annual energy consumption by nearly one-third from that now being projected for the next decade. Resistance by special interest groups to many of these changes is of course extremely effective—try telling the highway lobby to allow the Highway Trust Fund to be used for non-auto purposes, or the electric utilities that they shouldn't attempt to expand capacity, or the commuter that he can't drive to work any more. However, as the public becomes increasingly battered by the social and environmental costs of our profligate energy consumption, the pressure for change increases. Cracks are already appearing in the old guard—the fact that bitter controversy over highways vs. public transportation killed passage of the 1972 Federal Aid Highway Act by this Congress is highly significant. Of course the battle to open the Fund to public transportation has not been completed, but, in an observer's words, "it is no longer a sacred cow . . . it will be broken." Economic pressures in other sectors are also beginning to work in favor of energy conservation. The current shortage of natural gas, for example, has prompted the Gas Technology Institute (the research arm of the gas industry) to study ways of increasing the efficiency of industrial processes which consume large quantities of their product. Savings achieved in that sector would yield more gas to sell to the residential market, where profits are higher.

Unfortunately, the regulatory agencies in the federal government seem reluctant to turn away from their traditional, promotional posture toward energy. Both the Department of the Interior and the Federal Power Commission remain firmly entrenched in the concept that continued growth in per capita consumption is the only responsible public policy. And that growth must be fueled by more energy of all forms, especially electricity. The idea of pursuing conservation policies like those outlined here with the same vigor with which we are now pursuing ways of increasing *supply* gets short shrift from Hollis Dole, Assistant Secretary of the Interior for Mineral Resources: "We shall have to conserve all the energy we can, but energy conservation can never, by the remotest connection, be considered as an alternative to any measure designed to increase energy supply." "Between now and the end of the century," says Dr. V. E. McKelvey, Director of the U.S. Geological Survey, "we will need to build a Second America in the sense that we have to duplicate the entire U.S. plant—factories, homes, highways, and hard goods."

It would be naive to propose that total energy consumption could or should be strictly frozen at the present level. Some increase is inevitable. But do we *have* to double our total consumption in the next 28 years? (This is presumably what McKelvey means by his Second America.) What the federal regulators apparently have yet to grasp is that projections of past trends do not define the future—hopefully, we have some control over our destiny. Energy consumption does not grow independent of public policy. The means being vigorously pursued by the Interior Department to avert threatened energy shortages include development of the breeder reactor, research into coal gasification, further development of offshore oil, the letting of leases for oil-shale development, and efforts to bring in oil from Alaska's North Slope. Little more than lip service is being paid to measures for demand reduction. In its recent proposed policy statement on *Conservation of Natural Resources,* the Federal Power Commission says, "Overall, the Commission's basic purpose is to identify and articulate principles of prudent conduct which may be generally accepted on a voluntary basis in the further development of the Nation's primary energy resources, the conversion of those resources into electric energy and public consumption thereof." As a result of this emphasis on *supply*, government projections of future demand have historically been excellent examples of self-fulfilling prophecies—but the environmental impact brought to mind by a "Second America" can't help but make one skeptical about that pattern continuing much longer. (There are currently over 3.7 million miles of highways in the U.S.—Do *you* want 7.4 million?)

§7.6 The New Sacred Cows. Arguments stressing conservation as opposed to promotion are often countered with meretricious arguments such as "Those who want to reduce growth want to keep that man-in-the-ghetto in the ghetto" and "Here are 297 new ideas for cleaning up the environment—all would require electricity to make them go." The electric power industry (the fastest growing major consumer of primary fuels) has been especially ambitious in exploiting new symbols of social consciousness. A promotional campaign being run by the Edison Electric Institute claims that large increases in generating capacity are needed to:
- implement recycling programs
- clean up pollution
- keep up with inevitable population growth
- increase the living standards of the poor

Officials at Interior and on the FPC, and even the Environmental Protection Agency, have responded as hoped to these statements—one finds them mentioned in one form or another in a variety of recent government speeches. I find it quite disturbing that these arguments have apparently been accepted, at least tacitly, with little or no critical review. A few relevant studies have been carried out by public-spirited engineers in universities and research laboratories. Here are some of their conclusions.
- For steel, aluminum, copper, and paper the energy consumed in production from recycled scrap is considerably less than the energy required for production from raw ores. If just one half of the U.S. production of paper, steel and aluminum (common components of municipal trash) had been produced from recycled scrap in 1970, the overall electric energy *savings* would have been about 42 billion kilowatt-hours, 3% of the total electricity consumed that year.
- Currently, about 31% of our municipal sewage receives little or no treatment. Yet providing advanced secondary treatment for *all domestic and industrial* wastewater currently being produced would increase electric power consumption by only 1.3%.
- Over the last two decades, population growth accounted for only 18% of the increase in annual U.S. electricity consumption.
- The electric energy required to provide 23 million poor people with the U.S. average annual residential electricity (2200 kilowatt hours per person) would be only 3.6% of the total 1970 electrical consumption.

Any calculations of this type are necessarily approximate, and we don't have enough experience with many types of environmental control systems, including recycling, to make more than very tentative estimates of their energy consumption. But, until the energy industries produce data to back up their claims, there is no justification for incorporating them into government policy.

§7.7 Wise Use of Energy. Even a cursory study of energy consumption in America today reveals a great lack of prudence in the way we use our resources. This is not at all surprising, since energy has historically been cheap and abundant, and therefore there has been little motivation to use it efficiently. S. David Freeman, observes: "The past few decades can fairly be called a promotional era in energy growth. A variety of government policies have supported the promotional practices of the industry to make abundant supplies of energy available to Americans at the lowest possible price ... [Yet now] the shortages of energy and abundance of pollution point up the need for conservation, but the market place is still responding to the policies of promotion. If the existing trend of accelerated growth in energy use is not turned off, then the existing projections may even understate the future rates of growth. But such a course of action is likely to place society on a collision course with itself."

It appears that not only the market place, but also the federal government, is still operating from a promotional philosophy. Yes, Mr. Dole, we do need further, orderly development of some new energy resources. But at the same time, we need to inject a strong dose of the old-fashioned conservation ethic — i.e., husbandry of our precious natural resources. And this does not require great personal sacrifice, as some would have us believe, but only an elimination of the hidden waste that permeates our systems for energy conversion and use. In the long run, a vigorous pursuit of the energy conservation ethic might just buy us enough time to develop such clean, unlimited energy sources as fusion and solar power, and even aid in our eventual transition to a less-polluted, less material, more human-oriented society.

§7.8. Energy Conservation Measures. Many conservation measures can be put into practice in a relatively short period of time to provide for more efficient use of energy and reduce energy demand. Such measures involve decisions to manage existing buildings and urban service systems, transportation, and industrial processes in a way that minimizes energy demand without adverse effects. Measures which alter the present incentive structure through taxes and price increases have also been discussed. Medium range and long range conservation measures combined with short range steps can have a substantial impact on energy demand and reduce the need for new energy supplies.

The Subcommittee on Energy of the Committee on Science and Astronautics of the United States House of Representatives in June, 1973 published a program of individual action for energy conservation, pointing out that the use of energy for transportation and in homes accounts for 44% of our total energy consumption. Thus, by reducing personal energy consumption in these two areas, individuals like yourself can make an important contribution toward reducing the total United States energy consumption.

§7.8.1. Transportation. As can be seen from data, 25 percent of the nation's energy is used for transportation. Since most of this is consumed in private automobiles in the form of gasoline, it is clear that reducing the use of gasoline will have a significant impact in easing fuel shortages. Here are some actions you, the individual, can take:

Walk, take public transportation, or ride a bike for short trips.

Learn the schedules and routes of public transportation. Use it whenever possible to get to work, school, or shopping. Reduce your dependence upon automobiles.

If public transportation does not meet your needs, encourage additional routing and scheduling.

Encourage development of bike trails in your community.

Consolidate small tasks requiring an automobile into one trip, and thus reduce your total automobile mileage.

Create and support car pools for transportation to work, school, or shopping.

Driving an Automobile

Keep your engine tuned at all times.

Keep your tires properly inflated; under-inflated tires decrease gas mileage.

After starting your engine, drive slowly for the first mile instead of warming the engine up while standing still.

Drive slower. Increasing the speed at which you drive greatly increases fuel consumption. Driving at 70 miles per hour will increase the amount of gasoline you use by 33% compared to driving at 50 miles per hour, and 12% compared to driving at 60 miles per hour.

Anticipate speed changes and, where possible, allow your car to slow down before applying the brakes. Excessive braking increases gas consumption.

Drive smoothly. Changes in speed wastes gasoline.

Do not race the engine. If the automobile idles poorly, it may indicate the need for a tune-up.

Do not idle your engine for over three minutes while waiting.

While driving at highway speeds, check to see if the air conditioner is necessary. If possible, drive with it off. (You will need your air conditioner less if you avoid driving during the hottest hours of the day.)

Set your air conditioner to the warmest level that is still comfortable.

Purchasing an Automobile

When considering the purchase of an automobile, make fuel economy as a major consideration.

Purchase a car no larger or more powerful than you need. Try to eliminate unnecessary optional electrical features.

Remember that larger cars with more powerful engines consume more fuel than smaller ones, in direct proportion to their weights. (An automobile weighing 5,000 pounds uses over twice as much fuel as one weighing 2,000 pounds.)

Air conditioning units and automatic transmissions increase fuel consumption.

§7.8.2. Residential cooling. Shade windows from direct sunlight. It is best to shade them from the outside with trees, shutters, awnings, or roof overhangs. Be sure that this exterior shading does not trap hot air. (Deciduous trees give shade in the summer, but when they lose their foilage in the winter, the provide direct sunlight to windows and aid in heating.)

If the windows cannot be protected from the outside from direct sunlight, use light colored opaque draperies inside. These should be kept closed when the window is exposed to direct sunlight. In this way, you will reduce solar heating of your house through windows by 50%.

If you cannot shade a window from the outside, and do not wish to cover it with draperies, consider installing heat absorbing or reflecting glass. (This can reduce heat entering a house through the windows by 70%.)

Leave storm windows on windows that are not going to be opened during the summer months. This will reduce the transfer of outdoor heat into the house. Even with storm windows in place, it is important to shade windows from direct sunlight.

Make sure that your house is properly sealed so that the amount of warm air that can enter the house from the outside will be minimized. This should be done by checking areas of the house that could be sources of leaks. Check seals on windows and doors. Check and seal cracks in roofs and floors. Seal all exterior cracks. Use weatherstripping.

Close the fireplace damper. (This is important. It keeps hot air out in the summer and cold air out in the winter.)

Increase the insulation between the house and the attic to six inches of insulating material.

Allow for the ventilation of air through the attic. Reduce heat buildup by opening vents or windows. Using a small fan to exhaust attic air is particularly helpful.

If you are going to repaint or reshingle the house, use a light color. (Dark surfaces can become as much as 60° warmer than the surrounding air. Under the same circumstances, a light surface would only become 20° warmer.)

During hot weather, try to reduce the use of electrical or gas appliances within the house. These appliances give off excess heat during operation, and add to the load of cooling the house.

If possible, construct exterior vents for major appliances such as stoves and clothes dryers. Any excess hot air that can be expelled into the outside will mean less that will have to be cooled inside.

Minimize the use of hot water in your home. Wash in warm or cool water. Do not waste hot water when showering or bathing.

Turn off the lights when not in use.

§7.8.3. Residential heating. Check your house for insulation. It has been estimated that 15 to 30% of the heat required to warm a house is lost due to poor insulation. Check the insulation of your house.

If your insulation does not measure up to these standards, your energy output for heating will be greater than is necessary.

Install storm doors and windows.

Check for leakage to the attic and outside.

If possible, replace large glass areas with insulating or double pane glass. Close draperies in the evening or during exceptionally cold periods to reduce the heat loss through the glass.

Close the damper of the fireplace when not in use. If the fireplace is no longer in operation, provide an airtight seal in the chimney.

Have the furnace checked once a year and change the filters frequently during use.

Lower the daytime setting of the thermostat. Lowering the thermostat setting by one degree results in a 3 to 4% drop in fuel consumption; by 5 degrees, 15 to 20%

less fuel.

Lower the thermostat at night.

Close or reduce ventilation to rooms that are not in use or are used for limited periods.

§7.8.4. Home Appliances. All appliances should be evaluated for their usefulness when compared to the energy they consume.

When purchasing new appliances, decide how much you will use optional extras — they require extra energy.

Air Conditioners

Check your requirements before you purchase. An air conditioner with too large or too small a capacity requires more energy and will often not work as well as an air conditioner matched to your specific needs.

Determine the efficiency of an air conditioner before purchase. To obtain the efficiency, divide the rating of the machine in Btu's per hour by the number of watts required to operate the machine. Models on sale today have efficiencies (so measured) ranging from 5 to 12. The higher the number, the more efficient the machine; the more efficient the machine, the less energy required for a given amount of cooling power.

Keep your air conditioning system clean and in good working order. Clean filters are required for the machine to work at its maximum efficiency. Check all filters every thirty days during use and if necessary replace them. Before use, check and lubricate the bearings as recommended in the manufacturer's manual. Check for wear and maintain proper tension on all pulley belts.

Inspect the ducts in a central air conditioning system for blockage or leakage. The ducts should have a minimum of 1½ inches of insulation.

Gas Appliances

Since approximately 10% of all natural gas used in homes is consumed by pilot lights, consider switch operated electric starters instead of continuous burning pilot lights when purchasing new eqipment.

Extinguish all pilot lights on appliances that will not be used for long periods of time. Make sure no gas is flowing.

Television Sets

Turn off television sets when not in use. (If yours has the "Instant-On" feature, unplug the set when not in use.)

Refrigerators and Freezers

Frost-free refrigerators require 50% more energy to operate than a standard model. Side-by-side refrigerator/freezer models use up to 45% more energy than conventional models.

Decide whether you need a full size freezer before purchase: it can add up to $4 per month to your fuel cost. If you do use a freezer, keep in mind that a full freezer is more efficient than an empty one.

Washers and Dryers

When possible, wash dishes and clothes in warm or cold water. The water heater accounts for 15% of a home's utility bill.

If weather conditions are suitable, use outside clotheslines for drying clothes.

Check for dripping hot water faucets and fix if necessary. Such a leak wastes money, water and energy.

Set water heater to a lower temperature.

§7.8.5. Vacations. In the months ahead, the energy crisis may reach many of us while we vacation. Even while on vacation, we should consider measures that conserve energy.

Before You Leave.

Make sure that all gas outlets in your home are closed before leaving. If there are any that you want working when you return, have a neighbor turn them on the day before you reach home.

Use a timer (or neighbor) to turn the lights on and off in the evening rather than leaving them on while you are gone.

If you vacation in the winter, set the thermostat at the lowest setting. (Turn your heating system off if there is no danger of freezing your water pipes.)

Turn off your water heater.

On The Road

While driving, remember to slow down to save gas.

Try to minimize the use of your automobile air conditioner. Make sure that it is cleaned and checked before leaving.

If You Have A Trailer or Camper

Slow down. The speed at which you travel affects gas mileage even more than if you were in a passenger car.

Check propane or butane lines for leaks. Turn off all outlets during travel.

In the summer, choose sites for the trailer or camper that have natural shade. Open windows at night.

*

8

CONSERVATION OF ENERGY IN BUILDING

James A. Lowden

§8.1. Introduction. In broad terms we must consider the use of energy in real estate development in two areas:
1. The off-site use of energy;
2. The on-site use of energy.

§8.2. Off-site use of energy. Off-site use relates to the production of materials and their transportation to the site, and to the provision of the infra-structure to service the site.

§8.2.1. Production and transportation. Production of materials and their transportation to the site involves a knowledge of energy used in the production of various alternative building materials and their transportation costs to the site. A few examples are the use of steel -vs- aluminum; fibre board -vs- plywood; poured cement -vs- concrete block; reinforced concrete -vs- steel frame. Decisions on specifications, such as these, are usually made by the architect or engineer on a cost-use basis without consideration of the energy factor. Over-design of our buildings falls into a similar category and engineers contend that energy savings can be made in the production of materials that are over used (concrete). The time will come soon when we will no longer be able to afford the luxury of over-design.

§8.2.2. Location. The location of the property may create a large use of energy to provide for streets, sewers, water, power, communication cables, and for the construction of schools, churches, shopping and recreational facilities (new town in Arctic Circle). None of these may bear directly on the development *per se*, but each can be a consequence of it. The transportation of the buildings' occupants to and from their residences can waste substantial amounts of energy. Last year, cars in the U.S. burnt 90 billion barrels of gasoline—much of which went to provide daily transportation across town. There is no question that substantial savings in energy could be made if we lived closer to our work or could use rapid transit to and from work instead of the automobile. This is one of the big arguments used by the *In-City Renewal Group*. They say, and with some justification, that we should renew the worn out areas of our cities where the infra-structure already exists to support and maintain large populations, rather than abandon it for the suburbs where a new infra-structure must be created.

The whole question of off-site energy use is just beginning to be studied and I think we are going to reassess a number of our procedures as we learn more about this matter.

§8.3. The on-site use of energy. Major users of energy on-site are the heating, cooling, lighting, and ventilating systems. They are all related in their various functions and to a degree inter-dependent.

All these systems are wasteful but there are two kinds of waste. One is made consciously, the result of a trade-off between the capital cost of the system -vs- the cost of energy to accomplish the task. We may or may not be aware of the other kind of waste, and a decision to waste or conserve energy is not consciously made as we tend to go along with some historic procedure. This unconscious kind of waste can deal with the shape of the building, the extent of insulation, amount and type of fenestration, and the design of structural and mechanical systems to a capacity that will not be used.

Deliberate waste will decline quickly as the cost of energy increases in all but the most speculative buildings where the developer is not interested in the operating costs he passes along to the new owner. Unfortunately, the unconscious waste will take longer to control.

We have known for years that the building with the lowest perimeter to floor area ratio is the most efficient to heat and cool, but this factor is often disregarded in design, as are the savings that can result from improved insulation. Developers still make decisions based on thermal coefficients that applied when fuel oil cost 4¢ and 5¢ a gallon. At 15¢ a gallon and going up fast, these coefficients don't make sense any more.

Window design that makes most buildings a glass box is usually accepted without question on the basis that this is what the public wants. We have the technology to heat and cool these kinds of buildings no matter where they are built, and few concern themselves with the amount of energy consumed to make them habitable. Sure we all like to look out of our offices, particularly if there is something other than another row of windows 20′ away, but how many offices do you see where the occupant sits with his back to the window where drapes are drawn or venetian blinds pulled shut? In the last 10 years I have occupied offices with glass areas on one wall ranging from 95% floor to ceiling, to 95% desk height to ceiling. My present office which is 35% glass floor to ceiling I find the most attractive of the three. I believe that if the extent of glass area were pointed out to the public when it has to pick up the tab for increased energy costs a 35% glass wall would look pretty good.

The necessity to reconsider our use of energy because of its present and projected cost has been emphasized, but another factor that can't be ignored is the possibility of a limited supply. We could easily face energy restrictions in the 1980's and it would not be illogical to assume that maximum energy loads might be assigned to our buildings. Then builders would be forced to design to these loads.

Demand for electricity is projected to double every 10 years for the next several decades. If the United States now used about 1.7 million million KW hours a year, in 1983 this will be 3.4 million million KW hours and 20 years hence 6.8 million million KW hours. This sounds great from the demand side, but where are we going to get the energy to meet the demand? Perhaps we are not going to have 3.4 million million KW hours available in 1983 and to meet critical needs we will have to reduce the demand somewhat by being less wasteful.

§8.4. Conservation of energy in new construction. One of the first steps to conserve the use of energy in a new building is to do an energy profile or flow chart. This chart would indicate the expected energy demand of all the building systems and would facilitate an integrated approach to building design, (heating, lighting, cooling and ventilation, and other systems) by the various professional disciplines concerned with planning. Such a profile would ensure that specific design decisions would consider energy conservation and would generally encourage a greater efficiency in total design.

A second step would be the complete thermal analysis of the building to discover areas where both energy consumption and total costs could be reduced. Calculations which take into consideration heat from lights and people—as well as the heat storage effects of building furnishings—should be used. Trizec did a shopping center a few years ago where the architect and mechanical engineer discovered that the heat generated by the lighting and other equipment equalled the building heating requirements in the winter. Going to an all-electric system using heat pumps, heated the center with energy that would normally have been wasted at a savings in capital cost of 10%.

Two additional studies should be made before design is finalized.

The first is life cycle costing of various systems. Here, all of the costs of ownership throughout the life of the product are considered. The objective of this technique is to ensure that the building will result in the lowest overall cost of ownership. A combination of the optimum capital cost and operating cost can often result in substantial savings in energy.

The second would be a cost/benefit analysis. This is a process which compares the quantity and the quality of alternative means to accomplish a specific objective. (Capital cost of insulation against savings in heat and cooling.) It ranks the various alternatives according to the degree of economic efficiency with which they achieve the specific goal. It also sets out significant qualitative considerations in the ranking. In many cases the benefits flow toward a reduction in the use of energy.

The use of the computer has permitted these analyses to be done in a depth hitherto impossible and makes them a very worthwhile exercise in the conservation of energy.

Some of the ways energy can be conserved in the design and operation of individual systems are discussed in the following sections.

§8.4.1. Heating. Our heating systems are generally inefficient. Boiler installations are over-designed for safety factors and because they lack flexibility they seldom operate to maximum capacity. Trizec did a building recently in Calgary using a modular system of boilers fired by natural gas. There were a number of units that could cut in or out as the temperature dictated. Initial capital cost was about 20% less because design could limit over-capacity, and the costs of fuel are about 40% less than similar buildings with traditional installations. The method does work!

The type of building skin is a most important design factor capable of affecting heat and cooling loss. Heat transmission (and keep in mind that's in either direction) can vary as much as 3 to 1 depending on the amount of insulation used in the exterior wall, the amount of fenestration employed and whether this is single glazing, double glazing, heat absorbing, or heat reflecting, and the amount of solar shading in form of louvres, screens, blinds, trees, or other buildings. Engineers advise that a much improved thermal skin can be designed to save substantial amounts of energy at little or no additional cost.

As a society we have demanded indoor temperatures that will permit us to be comfortable in our shirt sleeves, regardless of the outdoor temperature. When I was growing up 67°-69° degrees was the standard but today 72°-74° is demanded. Those extra 4°-5° degrees use up much energy. There is no doubt in my mind that we can work efficiently in buildings that are a degree or two cooler in the winter and warmer in the summer, particularly if proper humidity levels are maintained.

The type of fuel used to generate heat has a bearing on the efficiency at which it is produced. For example, electrical heat, except when produced by hydro power, is about 1/3 as efficient as heat produced by burning fossil fuels. Yet more and more of our homes and buildings are heated by electricity. "An all electric house consumes about 5 times as much energy as that heated by fossil fuel—28,000 KWH -vs-5,600 KWH per annum", R. Stein states in the *Architectural Forum*.

One of the principal reasons for the switch to electric heating is due to the pollution associated with the burning of coal, oil, and gas—but another important factor has been a sales program on the part of the producers of electricity to help them fill in the troughs in their demand -vs- capacity profiles. It used to be that in Montreal summer air conditioning was encouraged to use up extra capacity needed for winter uses. Now it's reversed; the air conditioning load is so great that there is a surplus and electric heating is encouraged. With the demand for electricity increasing at such an incredible rate and our inability to increase supply at anything like the same rate, prices will rise and it will be interesting to see whether costs will reverse the present trends toward electric heat.

§8.4.2. Cooling. Much of what has been said about heating applies to cooling. In today's buildings the designs are so interdependent that you can't consider one without the other. Like heating, air conditioning is very wasteful of energy and largely for the same reasons. Building and mechanical system design use up energy at a fantastic rate. Fenestration and lighting are the biggest villains. Reduction in glass area and the use of reflective and heat absorbing glass can save large quantities of air conditioning energy. Lighting will be considered in a later section.

Air conditioning systems vary greatly in their use of energy. The most common system still feeds in continuous cool air which is then heated to the desired temperature. Energy is used first to cool warm air, then to warm the air that has already been cooled. This is known as the "energy fight". The more sophisticated dual duct induction system avoids this conflict by bringing warm and cool air in separate ducts to the area where the user can then make the selection. This system costs more to install, but with rising energy costs, developers should look at it more favorably. The introduction of cool outside air into the system, when it is available, can reduce the running time of the refrigerating machines and thus save energy. The central system, while being more efficient in the production of cool air, does lack flexibility because of the under use of selective controls. Our buildings should be zoned to recognize specific use of space — exposure to sun and prevailing winds. This is a flexibility we often don't have. Where a specific use requires a high air conditioning load, such as a computer floor, energy can be saved by installing extra package units in that area rather than by oversizing the central system to take care of this particular requirement.

The use of automatic controls on both heating and cooling systems can reduce the running period of the equipment at night and over weekends when buildings are not fully occupied.

Depending on design, the installation of multiple central refrigeration machines can work the same as multiple heating units. At least two machines are installed to provide the total cooling requirements in a building. One machine can be shut down when only partial or reduced cooling capacity is required, as often occurs in spring, fall, cool summer days, or when the building is partially occupied nights or weekends. This arrangement uses less energy than would be required by a single large refrigeration machine. It seems that electric motors running at or below half load are very inefficient, using 10% to 20% more energy than a smaller motor fully loaded.

§8.4.3. Ventilation. Ventilation is the third member of the triumvirate. While we think we are wasteful here as well, we are not so sure. Modern sealed type buildings have about 6 air changes an hour with 1 or 2 of the 6 being fresh air. Engineers are questioning this practice as being excessive. (The quality of the fresh air evidently is not considered.)

Ventilation represents about 10% of a building's total energy consumption, so it is an important element, and yet we apparently don't have adequate data on its real physiological requirements.

§8.4.4. Lighting. If there is consensus on any facet of energy waste, it's lighting. Standards have increased so dramatically in the past ten years that many must feel they worked in the dark prior to the war — literally. The effect of the 100 foot candles being designed into our new buildings, across the board, is blinding and what it is doing to the air conditioning load is something else. I understand that 2 excess watts of lighting require 1 excess watt of cooling, and when you consider that 50-60 foot candles are considered adequate for most tasks, the combined wastage is considerable.

Some experts think that by:
1. Reducing the light level to 60 foot candles;
2. Relating the light to the task and the specific area;
3. Providing proper switching;
4. Recognizing the existence of external light;
a saving of 50% in the energy required for lighting can be effected in new buildings and about half of that in old buildings. As lighting represents about 50% of all electrical energy consumed in a building, excluding heat and air conditioning, you are talking of a 25% reduction in the power bill. According to Richard Stein, writing in the *Architectural Forum,* this would be the equivalent saving 3% of the nation's total energy consumption. This would represent the capacity of thirty 1,000 megawatt generating plants. Lighting engineers seem to agree that lighting design in our buildings must be reassessed with more emphasis placed on the quality of light and less on the quantity. This would provide a more satisfactory light environment at substantial savings.

Author's note: I have selected only the most obvious ways in which the use of energy may be conserved in the design and operations of our buildings. There are many more—some being worked on, such as air conditioning systems that store ice crystals to help overcome peak demand. Others, such as the use of medium voltage service and distribution system when applicable, can save energy which would ordinarily be lost with a lower voltage system. Like it or not, it seems evident that we will have a serious shortage of energy between 1980 and the year 2000. We should do everything possible through design and operating procedures now to conserve it and to ensure its most efficient use in our buildings.

§8.5. Modular Integrated Utility System (MIUS). Alan R. Siegel, Director, Division of Community Environment and Utilities Technology, Office of Policy Development and Research, U.S. Department of Housing and Urban Development, at hearings before several sub-committees of the Committee on Science and Astronautics of the United States House of Representatives highlighted the modular integrated utility system—MIUS—program in the Department of Housing and Urban Development. This program was developed in response to the combined pressures of population growth and the increased mobility of society which have created a need for large-scale community development, including new communities, additions to existing ones, and large-scale redevelopment of our present population centers, while still maintaining environmental quality.

To make such developments livable, certain utilities are required: electricity, heating, air-conditioning, potable water, and waste treatment and disposal. The distribution networks which make these services available must also be provided. Increasing urban population is pressing the limits of environmental capacity, and the limited availability of resources and increasing unit cost and demand suggest the need to increase our productivity in supplying utility services.

The energy picture is bleak, particularly when the supply problem is coupled with the necessity to restore and preserve the environment. The cost of solid waste management is increasing at a rate faster than the population growth, due to the lack of close-in landfill and the need to improve current disposal practices. Liquid waste treatment facilities have not kept pace with the requirements of increased load and higher degree of treatment demanded by society today.

New approaches are needed to supply these needs of society in a manner consistent with conservation of our limited resources, and the need to restore and protect the environement.

The integrated utility systems are combined processing plants which generate electricity and use residual and recycled energy for heating, air conditioning, heating and water, treating water, and processing solid and liquid wastes.

The Modular Integrated Utility System (MIUS) can be located near appropriate users to minimize utilities service distribution infrastructure costs, which are traded against economies of scale. In addition, the modular approach maximizes the potential of factory assembly, thereby reducing the time from planning to delivery of services.

A MIUS must meet one or more of the following objectives:
- Conserve natural resources.
- Reduce primary energy requirements to satisfy the demands of the community being serviced.
- Minimize the environmental impact of providing services.
- Provide utility services on a schedule consistent with community development.
- Make land available for development which cannot presently be developed due to a lack of utility services.
- Eliminate the impact of local restrictions on sewer connections which may prevent construction of new housing for low- and moderate-income families.
- Provide a transportable system for emergency operations.

In order to place MIUS in perspective, a brief discussion of the conventional approach of providing community utility services, as contrast to MIUS, is in order.

Conventional practice, in nearly all cases, provides separate and dispersed community services for water, electric power, heating and cooling, and solid and liquid waste management. This approach of separated and dispersed facilities results in the inability to recover the wasted heat from power generation systems, or thermal energy from combustible waste products. It also does not present the opportunity to easily control the pollution caused by the outflow of liquid waste.

The MIUS approach on the other hand provides a combined utilities package, smaller than municipal-sized plants, to serve developments of limited size or portions of larger developments. It would integrate electrical power, heating, cooling, potable water, and liquid and solid waste facilities. An indealized MIUS might generate the electrical power and use the waste heat for living space heating and cooling, processing water or drying sludge. The solid waste produced by the development could be used as a supplemental energy source.

In the life support and utility system design for large, long-duration, manned space flight programs such as a space station, a higher degree of energy conservation was envisioned, and basic materials were recycled. In such a system, practically all water is reused: and solid waste materials are processed to produce additional energy, as well as reduce the volume of such waste. Residual waste heat from all sources is inventoried, and whenever possible, is used for supporting other systems. Water is used for cooling only when the heat connot be used or when passive, radiant heat rejection techniques will not be sufficient.

In 1971, NASA conducted a special study applying these technical approaches to the utility system of a 500 dwelling unit garden apartment complex. These concepts concentrated upon systems integrating engineering, energy management, life support provisioning, thermal design, and expendable materials conservation. In short, the utility system of a large manned space flight vehicle was conceptually designed for installation in a garden apartment.

The basic building blocks of any MIUS various process subsystems: are the power generation and distribution equipment; heating, ventilating, and air-conditioning units; the water and liquid waste process subsystem; the solid waste handling, transport and disposal components; the measurement and process control instrumentation.

Site characterization consists of establishing the utility system requirements for typical community facilities such as garden and high rise apartments, schools, hospitals, shopping centers, and office buildings. Questions must be posed such as: What are the water, heating, cooling, and power requirements? What are the characteristics of the liquid and solid waste produced and how much of such wastes will be produced?

HUD design depicts a 648-dwelling unit garden apartment which has been analyzed in the Houston environment. It was designed with the MIUS concept furnishing utility services and then with conventional utility services. Both designs satisfied identical requirements and included reasonable and available water saving devices.

In the conventional utility system design considered by HUD the electrical power is generated at a remote site with the residual heat being wasted. The liquid waste is transported through a sewage system to a remote site processing plant. The solid waste is transported to another remote site where it is incinerated and the heat energy is not recovered. Inflow water is treated at another remotely located central plant. All characteristics of conventional suburban development throughout the country in 1973.

By comparison, in the MIUS design system, it is shown the electrical power is generated onsite by diesel engines. Summer and winter air-conditioning is supplied by a combination of heat driven absorption and electrically driven compression chillers. Space heating and domestic hot water is provided by recovered heat supplemented by a fuel-fired boiler. Waste water is treated on-site by a physical/chemical process, followed by separation of solids by centrifugation, filtration, and disinfection. A cooling tower is operated with the process waste water, and the dissolved solids from the cooling tower are controlled by electrodialysis filtration. Solid waste is incinerated on-site, the heat is recovered, and the residue is hauled to landfill.

Too often, the first costs of an on-site powerplant have been compared directly to the capital investment requirements for a central station generating plant. This comparison can only be made after a detailed examination is made of the elements of the comparison. A central station powerplant by itself is not a system. In order for the consumer to utilize the electricity generated, vast and extensive transmission and distribution systems must be constructed.

The Modular Integrated Unility System (MIUS) integrates all utility services (power generation, liquid and solid waste treatment, etc.) into a single system that conserves resources and enhances the environment. A realistic cost comparison with conventional practice is necessarily difficult. A realistic cost comparison of each separate conventional utility service must be established (both capital and annual operation and maintenance costs).

§8.6. Building Code Standards. In compliance with the President's first energy message in June 1971, the Department of Housing and Urban Development (HUD) revised its insulation standards to require that all federally insured single family housing be sufficiently insulated to reduce the maximum permissible heat loss by one-third of that previously allowed. As a result of the action, fuel savings for a period of one to three years, depending on the climatic location, in a typical 1200 Sq. Ft. home, should at least equal the additional cost to install the insulating material.

Following the President's second energy message in February 1972, similar energy conserving measures were taken with multi-family structures. In this case, revised standards were issued calling for a reduction in maximum permissible heat loss up to 40% (in the northern regions) above that required under HUD's old standards. Due to the involvement of both insulating materials and glazing for these structures, a 5-year period in fuel savings is required to offset additional construction costs.

Section 101(c) of the Housing Act of 1949, as amended, established the Workable Program for Community Improvement which requires localities seeking renewal assistance to adopt nationally recognized codes, or state or local codes with comparable standards, unless local conditions exist which would warrant deviating from nationally recognized code standards. Where such conditions exist, such as soil, climatic or geological conditions, the policy provices for the granting of waivers upon submission of acceptable engineering data justifying such deviations.

The HUD code policy serves to broaden the choice of safe, nationally recognized materials. It employs nationally recognized codes and standards as a basis for evaluation the adequacy of local regulatory authority as provided in the context of Section 101(c). While the code policy appears not serve the interest of any particular individual or group of manufacturers, material suppliers or construction trade, nevertheless, by requiring modernization of codes to permit the use of accepted new materials and methods, it supports the right to broadest individual choice in the selection of any material or method which conforms to accepted industry standards sufficient to assure public health and safety. Further, the policy allows communities choice in the selection of code standards. While there is only one nationally recognized electrical code, communities may base their standards on any of four model building codes, three model housing codes, and three model plumbing codes.

§8.6.1. Background of HUD Policy. The HUD code policy was initially mandated by the language of Section 101(a) of the Housing Act of 1949, as amended, wherein Congress required the Secretary to *consider* the extent to which localities applying for certain federal renewal assistance encourage "the use of appropriate new materials, techniques, and methods," and eliminate "restrictive practices which unnecessarily increase housing costs." Current HUD code policy is based on the addition of Section 101(c) in 1954, wherein Congress required the Secretary to *determine* whether localities applying for renewal assistance have a Workable Program "effectively dealing with the problems of urban slums and blight within the community."

Both the Senate and House Reports accompanying the Housing Act of 1954 stated the effects of the Workable Program amendment to be as follows:

The requirements of Title I with respect to local responsibility and local action would be strengthened and increased. Existing provisions require the Housing and Home Finance Administrator, in extending financial assistance under Title I, to give consideration to the extent to which local public bodies have undertaken positive programs to prevent slums and blighted areas through the adoption and imporvement of local codes and to encourage housing cost reductions through the use of new materials, etc. *This requirement, which is applicable even at the initial stage of financial assistance,* i.e., at the making of advances for surveys, plans and other preliminary work for projects, would be broadened to include a consideration of code enforcement.

The Congress in 1949 recognized the relation of restrictive codes to slums and blight. No longer satisfied in 1954 with a mere *consideration* of the extent to which public bodies were undertaking programs to encourage the use of new materials and eliminate restrictive practices in local codes, Congress strengthened the Housing Act by requiring a *determination* the local programs are effectively dealing with the problems of slums and blight.

Further, in accord with the National Housing Policy, Section 2, Housing Act of 1949 as amended, Congress requires the Department to exercise its powers to encourage and assist:

(1) the production of housing of sound standards of design, construction, livability, and size for adequate family life; (2) the reduction of the cost of housing without sacrifice of such sound standards; (3) the use of new designs, materials, techniques, and methods in residential construction, the use of standardized dimensions and methods of assembly of home-building materials and equipment, and the increase of efficiency in residential construction and maintenance; (4) the development of well-planned, integrated residential neighborhoods and the development and redevelopment of communities; and (5) the stabilization of the housing industry at a high annual volume of residential construction.

Support for these goals has come from the Douglas Commission on Urban Problems, the Advisory Commission on Intergovernmental Relations, the General Accounting Office Workable Program Report, the Barrett Subcommittee Hearings on HR 13337, 92nd Congress, the courts, and the public sector.

Of particular interest among these is the Advisory Commission's Report A-28 which concludes as Finding #2:

Too many building codes contain unnecessarily high standards, prevent the use of economical methods and materials in building, and include provisions extraneous to the basic purposes and objectives of building controls. Local governments in the exercise of their building regulatory powers often include provisions that go beyond establishment of minimum requirements for public health, safety, and welfare. The cost of adhering to excessive requirements bearing only super-ficial relation to health and safety, limits the economic range of housing that can be made available within a community.

Similarly, a Federal court in *Radiant Burners, Inc.,* v. *Peoples Gaslight and Coke Company,* 364 U.S. 334, held that a standard may not properly be used to exclude a serviceable, but nondeluxe, product from the market-place by requiring that all the manufacturers of a given product conform to needlessly high standards.

§8.7. Pyrolysis: Approach to Solid Waste Disposal and Recycling. Solid waste management engineers are experimenting with a thermal process that can change a washing machine or television set into a small solid residue or molten slag. The process may also be used to convert solid waste into fuel gas and oil, and provide heat to make steam. Although this process may be used first as a replacement for incineration, its most important long-range value may be its ability to recover energy from solid waste materials. What is new about this process is that it features a method of producing tremendous heat in the absence or near absence of oxygen. The method is called pyrolysis.

Many municipalities and large developments, both residential and commercial, reduce the volume of their solid waste by burning it in incinerators. Suitable acreage for disposing of the wastes by sanitary landfilling (compacting it and covering it with earth) frequently is not available within their geopolitical jurisdiction. Sites outside of their jurisdiction are difficult to acquire because of political and citizen resistance.

Incinerators are not a complete system for managing a city's solid waste because: (1) about 20 percent of the average city's wastes are hard to burn in conventional municipal incinerators (refrigerators, sofas, and large tree limbs, for example), and this waste has to be pretreated or disposed of in some other way; (2) incinerators still leave a residue of about 20 percent of the original waste. This residue consists of such noncombustible items as glass and metal cans which must be buried in a sanitary landfill; (3) many incinerators pollute the air.

Pyrolysis is the thermal decomposition of materials in the absence, or near absence of oxygen. The high temperatures and the "starved air" situation present in the pyrolysis of solid wastes results in a breakdown of the materials into three parts: (1) a gas consisting primarily of hydrogen, methane, and carbon monoxide; (2) a "tar" or "oil" that is liquid at room temperature and includes organic chemicals such as acetic acid, acetone, methanol; (3) a "char" consisting of almost pure carbon, plus any glass, metal, or rock that may have been processed.

All of the pyrolysis systems presently under development require an added outside heat source, using a conventional fuel and burning process. However, pyrolysis-produced fuels may be used for this heat source in the future.

The heat is applied to a vessel containing the solid wastes. The vessel is sealed to eliminate virtually all oxygen. Once it is hot enough, most of the wastes are vaporized into gases. The gases are removed from the vessel for recovery or use. The remaining solids drop to the bottom of the vessel as char, which can be removed for recovery or heated further to produce a molten slag and then removed for recovery.

§8.7.1. Incineration and Pyrolysis compared. Incinerating and pyrolysis are both thermal processes. Incineration uses combustion to oxidize the wastes.

Pyrolysis in its purest sense just breaks down the wastes and does not oxidize them, although in every "pyrolysis" system under development some oxidation does occur. Pyrolysis and incineration both have the disadvantage of leaving a considerable residue. This residue must be disposed of in a sanitary landfill. Both processes achieve about the same degree of volume reduction of solids.

An exception to this is that some pyrolysis systems which use a cupola-type furnace lend themselves to melting of the ash (rock or silica) and other noncombustible materials (glass or metals). This is called "slagging" of the residue. Using this technique, it is possible to achieve about 50 percent greater volume reduction than in conventional incinerators.

Air pollution controls are considered to be less costly in pyrolysis systems than those for incinerators, since the pyrolosis systems generate about half the level of particulate air pollutants (soot and other particles) of usual incineration methods.

Resource recovery is possible with both incineration and pyrolysis. The heat from incinerators can be, and sometimes is, used to produce steam for heating, air conditioning, or electric power generation. Metals in incinerator residue can be retrieved and recycled, although this process is only practiced at a few facilities. Recovery of glass from incinerator residues is still in the experimental stage. In the pyrolysis process, these same recycling techniques can also be carried out. However, in addition, pyrolysis allows the option of producing either synthetic oil or to produce gas.

§8.7.2. Pyrolosis in municipal practice.

In Baltimore, Md., a full-scale intallation of *Landgard* developed by Monsanto Enviro-Chem Systems, Inc. will process 1,000 tons of municipal waste each day and simultaneously produce salable steam and other useful by-products.

The *Landgard* solid waste disposal process is a total and complete disposal system including receiving and handling solid wastes, preparation by shredding, pyrolyzing, residue processing, and gas purification. Minimum economical plant size is 250 tons per day with approximately 18 months as the estimated construction time.

In a typical installation, packer trucks discharge their loads in a receiving area from which it is fed to the shredders. The receiving and shredding area is housed in an attractive, ventilated building for weather protection and to control odors, noise, and blowing papers. Shredding reduces waste volume and produces a uniform particle size that is easily fed to the pyrolyzer.

The shredded waste is then conveyed to a completely enclosed storage hopper from which material is mechanically removed at a controlled rate. This hopper is sized to provide excess storage when trucks are not delivering.

Pyrolysis of shredded waste occurs in a refractory lined rotary kiln. Shredded waste feed and direct-fire fuel (oil) enter opposite ends of the kiln. Countercurrent flow of gases and solids exposes the feed to progressively higher temperatures as it passes through the kiln so that first drying and then pyrolysis occurs. The finished residue is exposed to the highest temperature, 1800°F, just before it is discharged from the kiln. The kiln is specially designed to uniformly expose solid particles to high temperatures. This maximizes the pyrolysis reaction.

The hot residue is discharged from the kiln into a waterfilled quench tank where a conveyor elevates it into a flotation separator. Light material floats off as a carbon char slurry, is thickened and filtered to remove the water, and conveyed to a storage pile prior to truck transport from the site. Heavy material is conveyed from the bottom of the flotation separator to a magnetic separator for removal of iron. Iron is deposited in a storage area or directly into a railcar or truck. The balance of the heavy material, now called glassy aggregate, passes through screening equipment and then is stored on-site.

Pyrolysis gases are drawn from the kiln into a refractory lined gas purifier where they are mixed with air and burned. The gas purifier prevents discharge of combustible gases to the atmosphere and subjects the gases to temperatures high enough for destruction of odors.

Hot combustion gases from the gas purifier pass through water tube boilers where heat is exchanged to produce steam. Exit gases from the boilers are further cooled and cleaned of particulate matter as they pass through a water spray scrubbing tower.

Scurbbed gases then enter an induced draft fan which provides the motive force for moving the gases through the entire system. Gases exiting the induced draft fan are saturated with water. To suppress formation of a steam plume, the gases are passed through a dehumidifier in which they are cooled (by ambient air) as part of the water is removed and recycled. Cooled process gases are then combined with heated ambient air just prior to discharge from the dehumidifier.

Solids are removed from the scrubber by diverting part of the recirculated water to a thickener. Underflow from the thickener is transferred to the quench tank, while the clarified overflow stream is recycled to the scrubber. Normally all the water leaving this system will be carried out with the residue or evaporated from the scrubber.

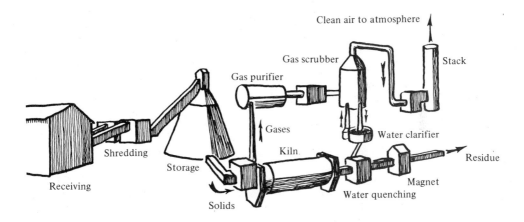

*

9

ENERGY CRISIS:
FACT OR FICTION?

Roland W. Comstock

§9.1. Introduction. Energy Crisis—Fact or Fiction? An energy crisis has already begun. It is real, not contrived (the FTC to the contrary notwithstanding). It's going to get worse before it gets better, and even if it does, it will not get better much before the end of this century. There probably are no technological miracles that will make the problem go away. There are only some hard choices—for you and for me as individuals, for this corporation [Northern States Power] and others , for government—and some of these choices must be made soon.

Although my comments are not an expression of either corporate or industry positions, I am going to share with you the work product of perhaps 25 or 30 people within Northern States Power (NSP) who have been studying the energy supply/demand problem for several months. My comments are based on the collective conclusions of this study group derived after examination of virtually all major studies and information sources. Incidentally, this is the first time [Midwest College Debaters Energy Symposium, Oct. 12, 1973] these conclusions have been publicly shared as the final report is still in draft form.

Finally, a word of caution. One of the concepts central to an understanding of the issue is the so-called "supply-demand gap." I will attempt to demonstrate that given a particular trend in supply capability and a particular trend in demand, there will be a gap *if we assume "business as usual."* Yet, it is apparent that in one sense there never can be a gap between energy supply and energy demand. Society can only use what is available and no more. If there is not enough energy to meet all of society's wants, then some wants will go unsatisfied. others will only be partially met, and some will continue as in the past. Demand, however compelling and strident, cannot receive that which simply is not available.

Why then discuss and illustrate supply-demand gaps? In fact, present trends cannot continue and it will no longer be "business as usual." The real question is what changes can we anticipate and what influence can we have over these changes. This paper suggests some of the possible futures.

In reality then, a prediction of energy deficits is a prediction of forces for change in consumption patterns and uses. Since energy is so basic, the "adjustments" to deficits will have physical, social, political, cultural and economic impacts. Severity of those complex and interrelated impacts is conditioned upon a variety of factors difficult to identify and impossible to quantify, each of which will vary over time. War for example, could be one "adjustment"—perhaps a fairly probable one—but one which can only be stated as a possibility.

Three important areas have been omitted from this paper for a variety of reasons. The Washington scene is already considered extensively in the media. The regional aspects of the energy problem are not considered because the United States is so large in area, with such widely varying geographic conditions, population concentrations, and economic activities, that the requirements for energy are by no means uniform throughout the nation. The nature of energy end-use, the sources of energy utilized, and per capita use all vary substantially from region to region. It is necessary therefore, to study regional needs on a detailed basis in the process of determining the future energy requirements for the nation as a whole. One of the few energy studies containing regional comparisons was done by the Chase Manhattan Bank. I have also omitted any discussion of what we should know that we don't. There are at least five major gaps in our knowledge about the national energy system that should be closed.

One prediction—most of you will feel discouraged when you finish reading this paper. There will be a strong need somehow to rebut the assertions—including the thought that well, anyone can look at only the dark side; but by golly, it's the American way to be optimistic. Be careful of such summary dismissal. The problem is too important for that. The basic report is not a doomsday, apocalyptic analysis. It's largely factual.

If you choose to interpret as pessimistic a statement of the energy crisis which sets out its real severity, that is both your choice and again, your problem. For my part, I don't think we can proceed toward solutions by pretending we only have a flat tire when in reality the engine block is cracked.

Finally, in the end it's probably not important whether any of us is optimistic or pessimistic. Far more important is what we *do* or don't do.

§9.2. Prologue—The riddle of the lilies. There are 2 obstacles to any presentation of the energy issue. The first concerns exponential or logarithmic growth, the second, fixed limits.

§9.2.1. Exponential or Logarithmic Growth. In June of this year, Dr. Jerome B. Wiesner, president of MIT, in testimony before congressional committees, warned:

We are victims of a rather deep human perceptual failing ... our inability to perceive intuitively the power of exponential growth. That is, the important idea that a phenomena that increases at a fixed rate will one day grow explosively.

There is a deceptively simple riddle which illustrates the apparent suddenness with which exponential growth approaches a fixed limit:

Suppose you own a pond on which a water lily is growing. The lily plant doubles in size each day. If the lily were allowed to grow unchecked, it would completely cover the pond in 30 days, choking off the other forms of life in the water. For a long time the lily plant seems small, and so you decide not to worry about cutting it back until it covers half the pond. On what day will that be?

On the twenty-ninth day, of course. You have one day to save your pond.

There is no single concept more central to an understanding of the energy issue than that of exponential growth. Full perception lies as much in the imaginative side of each person's nature as it does in the pragmatic, intellectual side. It must be deeply felt as well as understood.

Absent that there is the tendency to view papers such as this as vaguely interesting and academic recitals out of touch with the real world. Thus viewed, there are soon forgotten and seldom applied.

§9.2.2. Fixed Limits. The second fundamental obstacle to any serious study of the energy situation is to focus attention on an idea quite foreign to the western (and corporate) mind: the notion of finitude or fixed limits.

While the simple riddle of the lilies deals with the nature of exponential growth, it also deals with the notion of fixed limits. There is only *one* pond in the story.

More so than any nation in the world, the history of the United States has been characterized by endless frontiers. All of us (and all our institutions) have been nourished in a culture which largely accepted and encouraged unlimited expansion and growth. This is true whether thinking about land, natural resources, population, earnings per share, economics, technological achievements, outer space or ideas. We could always move on to other ponds not yet choked with lilies. We may now be running out of ponds.

Deeply embedded in all of us is the notion there are no limits to man's activities which somehow, someway cannot be overcome. Western cultures have particularly idolized the goddess of technology. Indeed, the great American dream and the good life were equated with limitless opportunity based on limitless growth. It is upon this often unspoken concept we have founded not only corporate and governmental policy but the very fabric of an economic system.

It is recognized that to ask any of us to consider the serious possibility of fixed limits is to ask for a shift in mind-set difficult, if not impossible, to accomplish. To make such a shift in values, even for the limited purpose of reading this paper, runs counter to our belief and experience. Yet such a request seems but the least of difficult tasks ahead. Eric Hoffer once said: "We can never be really prepared for that which is wholly new. We have to adjust ourselves, and every radical adjustment is a crisis in self-esteem. It needs inordinate self-confidence to face drastic change without inner trembling." Now we may have no alternative but to face the prospect of radical change. Our choice is whether it will be planned or accidental.

§9.3. The Global Scene. It appears highly probable an era of plentiful energy is nearing an end in our society ... globally and nationally. The future appears to be one of demand growing to exceed supply despite all reasonably probable technological developments and despite efforts by governments and industry both to increase supply and to limit demand i.e. to conserve. The question is not whether there will be a global energy shortage but rather how severe it will be.

The results of this depletion and shortage probably will be:

1. Extreme competition among nations as fuel supplies dwindle with consequent inter-nation tension and severe stresses upon all economic systems.
2. An increasing shift toward electric based economies among the affluent nations.
3. An accelerating gap between the have and have-not nations of the world.
4. The development of national energy policies based primarily on achieving maximum energy self-sufficiency.

§9.3.1. World energy uses. Overriding all other considerations seems to be the clear fact that world energy consumption rates are dramatically increasing with a resulting drain on finite fuel resources. Studies abound which document this simple fact over and over again. Figure 9.1 presents a typical summary of the past and projected global trends in world consumption of energy by fuel source. There are many

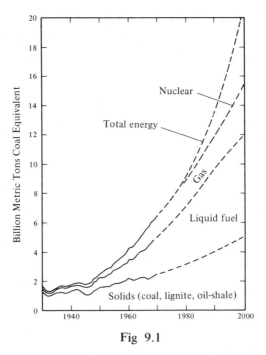

Fig 9.1

ways to scale this trend—per capita consumption, growth in gross national product, consumption in Btu's per year, etc. Regardless of the scale used and regardless whether one looks at energy consumption rates for the free world or by continents or for the United States, the same essential pattern emerges ... skyrocketing growth in demands placed on primary energy resources. While it is possible to haggle over precise numbers and estimates (and many do), the present trend seems clearly established beyond reasonable doubt ... worldwide growth in energy demands over the last 15 years of 5%+ per year coupled with every present indication that growth rate left uncontrolled will at least be sustained and probably increase.

§9.3.2. World energy resources. The question remains whether the world's supplies of energy resources are adequate to meet these rising demand curves. It is in the estimation of reserves that most divergence is encountered among various studies. However, even by doubling or tripling the most optimistic of reserve estimates and ignoring economic constraints, the same basic conclusion emerges: supplies of virtually every non-renewable natural resource—including energy resources—are inadequate to meet longterm demands if present patterns of consumption continue. This conclusion was voiced three years ago by the Council of Environmental Quality (CEQ) and is increasingly reinforced by subsequent studies.

These numbers raise some disconcerting problems. A projection to the year 2000 calls for a doubling of the coal supply in the next thirty years. While there are large amounts of coal reserves yet available, distribution and availability are another matter, quite apart from environmental considerations. The lion's share of coal is found in Russia, China and the United States. Getting coal out of the earth and transporting it in the required quantities pose tremendous logistic problems which will limit its utility in many areas of the world. However, as premium fuels, principally oil and gas, run in critically short supply starting in the mid 1980's, increasing effort will turn to converting coal to oil and gas. Obviously, it is vitally important that the conversion technology by commercially available by that time.

The global oil scene is not bright. Projecting the future oil supply is not just a matter of lining up points on a chart and then projecting infinitely into the future. That would result in the phenomenal figure of 100 billion barrels of oil per year required by 1993. A more likely estimate of what could be produced by that period is one-third that figure. No matter how optimistic one may be about petroleum reserves, it is evident that sometime in the 1990's, the world's oil production will taper off—probably rather sharply—and the gap in energy supply will have to be bridged by new sources of fuel.

Some scenarios seem obvious. The projection for the year 2000 cannot be fulfilled. Only if enough nuclear power is available to provide energy equivalent of about 6 billion tons of coal. Depletion of gas and petroleum reserves will inevitably shift the global energy scales to a coal/uranium balance. Given the problems of mining and transporting solid fuels, nuclear energy probably will dominate towards the close of this century. Whether from fission or fusion, the dominant energy form will be electric. The rich nations—because they have both the demand and the capital—will electrify their economies more extensively than underdeveloped countries.

This dichotomy between developed and undeveloped countries raises another scenario—a disconcerting one. The present energy disparity between the have and have-not nations needs little documentation. Figure 9.2 illustrates clearly the rela-

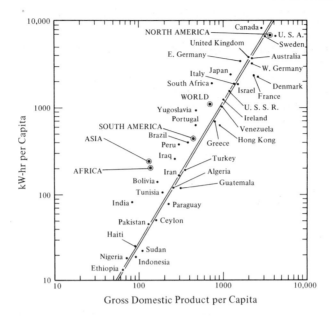

Fig 9.2

tionship of wealth to energy consumption. That the low-energy nations will be try-ing their best to shinny up the energy totem pole is apparent. This cluster of nations with hands grimly on the energy pole, is explosively ready to begin clawing their way upwards. The climb would be a slippery one! Even assuming an energy freeze of *no* increase in global per capita consumption of energy for the next few decades, even modest estimates of population growth   mean a total consumption by 1995 of energy equal to burning almost 1500 billion tons of coal or 6700 billion barrels of oil. If the poorer nations attempt to pattern their economies after the United States or Western Europe, they will be forced into the straitjacket of an electrified society for which resources, both fuel and financial, are quite inadequate. Long before other na-tions can approach, much less equal U.S. or European consumption figures in the energy sectors of their economy, planetary limits will prohibit such development.

There is then a high probability that the poorer nations of the world for the foreseeable future will be condemned to continuing poverty and the gap between the rich and poor countries will widen. Further, as the supplies of fuel begin to fall off beginning in the 1990's, the scramble among nations to secure their "share" will intensify. The combined European, Japanese, Soviet and U.S. energy consumption comprises about 70% of all fuels burned — leaving only 30% for the remaining three fourths of the world.

This scenario is hardly the blueprint for happiness on a planetary scale. The re-sentment over these islands of affluence cannot help but produce great global ten-sion. One can only speculate to what system breaks such tension may lead. Thomas Henry Huxley said in 1874:

> ... It is futile to expect a hungry and squalid population to be anything but violent and gross.

The root causes of this situation are not difficult to identify: more people using more and more. Added together, they involve our two friends introduced earlier: fixed limits and exponential growth. But why, one may ask, is it happening so fast? Answer: perhaps there is something inherent in human nature that makes the lilies seem innocuous until the 29th day.

A sense of the explosion of people upon the planet is conveyed in Figure 9.3. The explosion in consumption of these people is shown in Figure 9.4. Together they constitute a deadly combination.

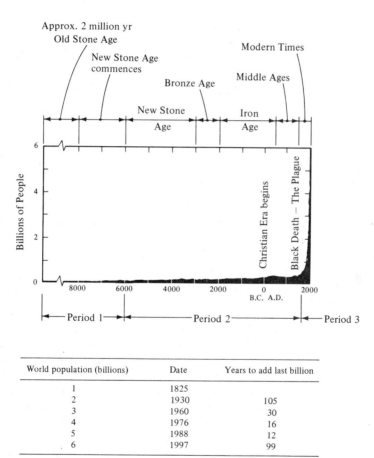

World population (billions)	Date	Years to add last billion
1	1825	
2	1930	105
3	1960	30
4	1976	16
5	1988	12
6	1997	99

Fig 9.3

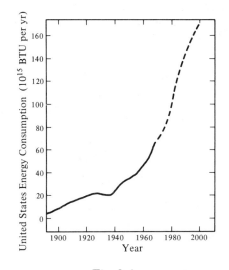

Fig 9.4

It is relatively easy to document the problem. To describe solutions — at least on a global basis — is far beyond the scope of this paper. However, for the United States, the path toward an appropriate energy policy seems clearly marked. While many nations have no other choice than a significant reliance on imported energy supplies, the United States has the potential to achieve virtual self-sufficiency through reliance on domestic resources provided there is significant reduction in growth rate.

As an aside, if I had to argue the negative of the proposition, "That the federal government should control the supply and utilization of energy resources in the United States, I would urge that no single nation — particularly the United States — should be allowed to manage unilaterly its energy supply and utilization patterns. It is a global problem requiring global management.

§9.4. The National Scene. A gap between energy supply and demand is now beginning and the gap will continue to grow at least until the middle 1980's. While the amount of the deficit may vary depending on various assumptions, the basic conclusion remains unchanged ... even though the most *optimistic* assumptions are made about supply and the most *conservative* assumptions are made about probable reductions in demand growth. Figure 9.5. Like the world scene, the only question for the near-term future in the United States seems to be how large will be the deficit, not whether there will be a deficit. The inevitable reliance upon fuel imports and the staggering impact upon balance of trade payments is well documented in the National Petroleum Council Study of 1972. Analysis of what stresses this may impose upon our economic ststem can only be raised and is better left to those more qualified. However, speculation can at least be made that it will be increasingly difficult to maintain an economic system based on narrow concepts of national sovereignty.

The demand/supply picture beyond 1985 is extremely difficult to predict ... primarily because of the uncertain impact over such extended periods of varying

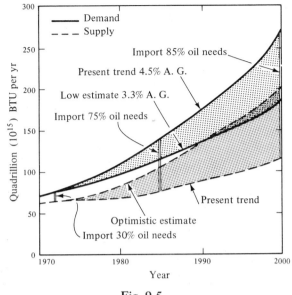

Fig 9.5

assumptions for which there is little or no historical experience or data. We are plowing new ground. However, some general conclusions can be tentatively offered.

Even under optimum conditions, the United States will continue to increase its dependence upon fuel imports if it is to avoid severe energy cutbacks. Further, it is unlikely that overall energy supply will meet demand until such time as fusion power arrives on the scene—many years from now. As will be discussed in §9.5 on Technology Assessment, even the most optimistic estimates for that development say not be until the late 1990's at the earliest. Meanwhile back at the ranch, there will be increasing growth in demand for fossil fuels by the non-energy sector: for petrochemicals, plastics, fertilizers, and the myriad other products which depend on fossil fuels for something other than heat. There are many (including some on the NSP Task Force) who feel part of our domestic hydrocarbon resources should be set aside *now* for these non-energy uses. In any event, it is probable there will be strong competition between the energy and non-energy sectors which probably will require congressional allocation legislation to resolve.

In the long term, there seems to be little alternative but significant changes in the national energy intensive life style.

In sum, the dimension of our future fuel availability problem over just this decade is as follows:

- Total energy consumption will increase about 50% between 1970 and 1980, but oil consumption in the same period will at least double.
- By 1980 the amounts of oil and natural gas imported will be triple the quantities imported in 1970.
- Probably half of our oil supply in 1980 will be imported.
- Probably 25% of our total energy consumed in 1980 will be imported.

The alarming statistics, however, are those of the last 2 years, the period 1970-1972. While total energy consumption grew by 6% and petroleum consumption grew by 11%, the growth in total petroleum imports was 38%. If the projected 1973 level of petroleum imports is realized (6,200 MB/D) the 1970-1973 growth in petroleum imports will be 80%. These figures indicate the NPC projections have been confirmed thus far, at least for the import projections.

§9.4.1. Discussion. The data graphically shown in the following figures indicate

- Continued high rates of growth will bring many aspects of the fuel supply problem to climax within the next decade.
- Moderate and intermediate reductions in growth rates may buy very little time — usually only 6 or 7 years delay in climax. Probably only drastic reduction (50-60%) in growth rates extends "intersect points" beyond the year 2000.
- Petroleum is most critical fuel resource with depletion of U.S. domestic supply possible under the worst case as early as the middle to late 1980's (Figures 9.6 and 9.7).
- Domestic natural gas supplies are in no significantly better position than oil. The middle to late 1980's will at least see significant reduction in domestic gas supply (Figures 9.8 and 9.9).
- Coal is a rich and abundant fuel resource. There is no alternative under any growth model but heavy reliance upon coal. It is critically important that coal gasification technology be sufficiently developed by the early 1980's. However coal alone cannot carry the whole energy load.
- Domestic supplies of uranium, although a function of price, are relatively plentiful (Figure 9.10). Here again, developing technology is important. Without the breeder reactor, uranium supplies may reach slightly beyond the year 2000. With

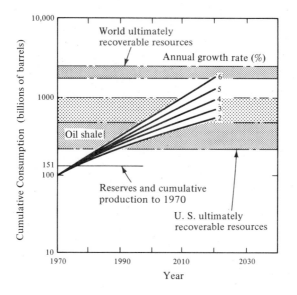

Fig 9.6

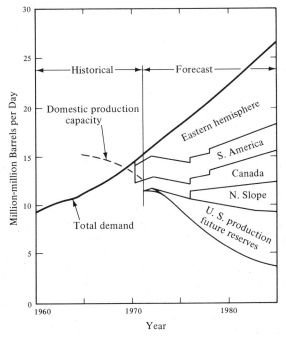

Fig 9.7

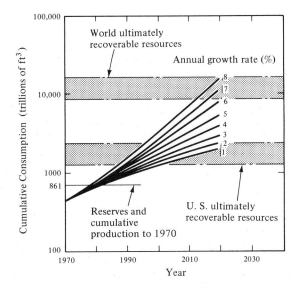

Fig 9.8

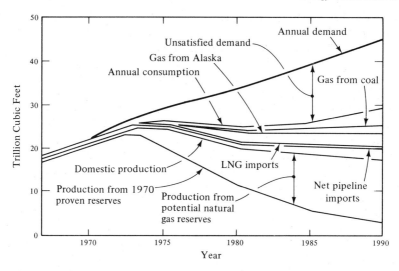

Fig 9.9

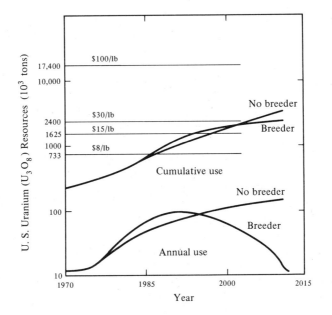

Fig 9.10

the breeder, the supply reaches farther into the future. While the administration is committed to early development of the breeder reactor, there is intense disagreement on all levels over the wisdom of this choice.

In summary then, the only manageable fuel future requires a reduction in the dominant dependence upon oil, sharply increased utilization of coal, rapid commercial development of coal conversion processes and some would argue for early development of the breeder reactor to carry us through the end of the century. In any event, coal versus nuclear fuel have been regarded in the past as competitive alternatives. That has changed, for we now must have both ... and even these must be accompanied by vigorous, effective energy conservation practices.

Essentially up to now the options of energy use and policy confronting Americans were *positive* options: choosing to substitute superior fuels for inferior fuels, because both were available; choosing to use energy more lavishly rather than to make our economic machine more efficient. They were not decisions forced by circumstances, but decisions prompted by opportunity.

Now, however, America faces new decisions. The options are more complex and may be *negative:* choosing to continue fueling important processes or to continue pollution control; or at least to accommodate these competing claims in the public interest; choosing to accept much higher cost for domestic fuels, or to allow our energy life blood to be increasingly pumped from overseas.

It is rare in any sphere of national activity that America has been forced to retreat. But it is clear that in the continued provision of fuels and energy America is encountering obstacles very difficult to negotiate. The policies selected to guide this nation and its institutions over or around these obstacles will be crucially important in determining the future development of our American economy and way of life.

§9.5. Assessment of Supply Technology. The cumulative contributions toward solution of the energy supply problem reasonably to be expected from new technologies not currently firmly established will be insignificant (less than 17%) during the remainder of the century. In other words, there probably will be no technological miracles and the nation (indeed the world) must rely largely on current technology and existing fuels for many years. This is not to say that research and development is not needed. Clearly it is. New technologies (e.g. nuclear fusion, solar power) *must* be developed now so as to make major contributions during the next century.

This conclusion rests on extensive and elaborate technology assessment studies. Because of time and space limitations, only the barest outlines can be offered in support of a conclusion which somewhat surprised the NSP Task Force and may well startle you. It is difficult to accept the idea that the next 25 years hold little promise of technological miracles in the field of energy supply. All of us have become incredibly inured to unending invention, titilated almost daily by reports of new "breakthroughs" in most every area of science. There probably will be few conclusions we might offer which will be more controversial.

There is a great danger in this faith (it is largely just that) western man has in the ability of science and technology. It might be called the "machine of the gods" syndrome. The ancient Greek playwrites often wrote plots so complicated, modern soap operas pale by comparison. The plays were long (5 to 6 hours). Sometimes, in order to extricate the characters, solve the problems and get the audience home, the playwright would use the *deus ex machina,* a machine of the gods, in the last scene of the last act. A box or chariot of some sort would descend from the heavens, a magical figure would appear and quickly put all things in order.

A blind, unreasoning faith in future, and therefore unknown, technology can be like that. We might wait until the last scene of the last act for a magical device. If it doesn't appear, we are worse off for the waiting. A far more prudent course is to act as wisely as we can on the basis of present knowledge and not expect miracles to come. If they do come, we are fortunate. If they don't come, we are prepared.

§9.6. A Technology Assessment Matrix.

To make this technology assessment, a simple matrix was used. Each technology was assessed against 7 constraints. It is analogous to a series of sieves vertically stacked, each a finer mesh.

Constraint of concept	1) Based on present information/knowledge, what is or seems to be workable? What survives because the basic idea is sound? Assume adequate research funding.
Constraint of time	2) Separate consideration of technologies that have little or no chance of development until after 2000 A.D. from those that could have near-term development possibilities.
Constraint of available natural resources	3. Of those which have near-term development possibilities, what are the relative conversion efficiencies in relation to fuel resource availability? Which ones do you reject because they are inappropriately wasteful?
Constraint of practicality	4. Of those which have near-term development possibilities *and* reasonable conversion efficiencies in relation to fuel resource availability, which ones pose unsurmountable operational problems?
Constraint of environmental impact	5. Of those which have near-term development possibilities *and* reasonable conversion efficiencies in relation to fuel resources availability *and* do not pose unsurmountable operational problems, which ones appear to pose unacceptable environmental impacts?
Constraint of contribution scale	6. Of those which have near-term development possibilities *and* reasonable conversion efficiencies in relation to fuel resource availability *and* do not pose unsurmountable operational problems *and* do not involve unacceptable environmental impacts, assuming maximum probable (or even maximum possible) development, which ones are excluded because their potential simply isn't large enough in relation to the size of the problem.

Constraint of money

To qualify, assume at least 1% of the nation's energy needs could be supplied by that concept.

7. Of those which have near-term development possibilities *and* reasonable conversion efficiencies in relation to fuel resource availability *and* do not pose unsurmountable operational problems *and* do not involve unacceptable environmental impacts *and* which can contribute significantly to satisfying the nation's energy needs, which ones—at this point in time—appear to be unacceptably expensive? (This should include consideration of cases where possible government subsidy results in the plant being economically feasible to private industry).

What's left is where solutions can be expected and where R&D priorities ought to focus.

Figure 9.11 summarizes the conclusions. Essentially the format of Figure 9.11 tests the principal technology concepts against the 7 constraints listed above. Since technology assessments become increasingly inaccurate the farther one looks into the future, the graph must be considered a conceptual analysis with the figures indicating order of magnitude, particularly those pertaining to the later years of the century and beyond.

We asked ourselves the question of whether these conclusions would still be true if there were a national energy research and development program on a scale comparable to the space program. This is difficult to answer as some research areas are sensitive to more money, more brain power, others are not. Further, analogy to

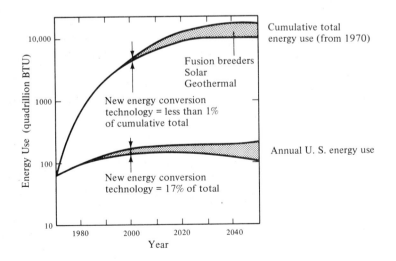

Fig   9.11

some international, some political, together with what appear to be some fixed limits i.e., there is only so much oil in the ground. Our response then would be that a sharply escalated R&D effort is warranted but don't bet all the blue chips on such an effort—maybe don't even bet too many—and in any event, hedge the bet.
the space program in some ways can be misleading. The space program limits were clearly technological. The energy problem involves other limits, some economic.

§9.7. Epilogue: Why is There An Energy Dilemma? The fundamental answer to the question of why America faces an imbalance in energy supply and demand is apparent.

Previously discussed are the Siamese twins of more people using more energy per capita. During the 20 years ending in 1966, United States population increased by 43% but energy consumption more than doubled. To the year 2000, United States population projections currently look roughly for a 30% increase. Yet projections of energy consumption four or five times 1970 levels are commonly found. The projected inability to supply energy demand at those levels already has been treated at length. (See Chapter 2.)

Why is our society so energy intensive? That it is energy intensive is beyond doubt. One evidence is that about half of everything transported in the United States today is fuel. It is a common statistic that the United States represents about 6% of the world's population, yet consumes about 30% of the energy extracted from the earth. Each American requires burning the equivalent of 24,000 pounds of coal every year. For comparison, each Swede 14,000 pounds, each Nigerian 71 pounds.

The phantom of our energy opera is affluence, wearing the cloak of increasing Gross National Product. But what does "affluence" really mean?

It means abundant food, nutured with high energy fertilizers and pesticides, cultivated, harvested and transported with energy intensive machines, processed extensively and convenience packaged (usually throw-aways), refrigerated and delivered to the shelves. All this despite the fact that per capita caloric consumption has not significantly increased over the last 50 years.

It means a wholesale shift from natural fiber textiles to energy dependent and energy intensive synthetic fibers (almost a 2000% increase between 1950 and 1968) manufactured from hydrocarbon fuel resources.

It means accelerated utilization of high energy plastics, aluminum and glass.

It means dedication to the automobile and private transportation rolling on synthetic rubber tires, to energy intensive truck transport rather than rail transport, to the single family detached residence, more and more of which are climate controlled and replete with energy intensive equipment.

It even means continued dedication to a cleaner environment through increasing use of energy intensive control equipment.

In sum, it is the way we live our lives that explains the energy dilemma. In sum, unless we choose to change the way we live our lives, the energy dilemma cannot be solved.

The reasons we choose to live an energy intensive life style are complex. Involved in part are basic values held by each person over which each of us presumably has some measure of choice and individual control. Involved in part are cultural values of which few are consciously aware, fewer still question and which are difficult, if not impossible to modify over the near term. Last, there are institutional policies (governmental, corporate, economic) which with a system of incentives and penalties establish a basic pattern of life style. No one is responsible, yet all are responsible.

That a new set of energy values is needed is clear. But some of these are for individuals to choose on a personal basis. Others are for individual institutions to choose on a quasi-public basis. Last, there are new energy values to be formed in the public arenas of social policy making. While the primary public arena is the Federal government, care must be taken not to neglect the "lesser" arenas in which energy values are also formed and choices made.

It may well be that the real solution (mitigation?) of the energy dilemma can only come as we gain new knowledge and new insights about planned modification of cultural values: the imposition of involuntary life styles. That being true, the path ahead may lie between equally hazardous choices: on one side, an impending energy crunch which threatens our national well being; on the other side, social engineering of new cultural values which could either threaten the democracy or which could open the frontiers of a new society with room for the first time for pluralistic values consciously chosen. To walk this razor's edge requires a delicate balance and the utmost in expanded social consiousness from individuals, from corporations, from government. In that setting, I can only wish you ... and me ... good luck. We shall need it.

*

10

ENERGY ECONOMICS: REAL AND UNREAL

Mike Morrison

§10.1. The Great Gasoline Give Away. For as many years as most of us have been driving, we Americans have pulled our automobiles into gasoline stations to receive a gift. Namely, the gasoline. And, for an equal number of years, almost none of us has been aware of this rather astounding largess. We remained unaware although its results stood unhappily all around us.

The great gasoline giveaway, now supposedly nearing its end, has been a primary constituent of a larger esoteric insanity that goes under the name of energy economics. These economics are the progenitor of an energy complex that is many times larger than other industrial components of national life. Energy is a colossus, and its effect on our lives is to scale.

As this giant ambles across the environment, it leaves gargantuan footprints stretching into the distance (like the endless march of high-voltage transmission lines). Oil spills, engine emissions, radioactive wastes, and lacerated coal regions are among the more familiar of these imprints.

Petroleum processing and marketing are sharply illustrative of how energy economics have worked in this country, especially since World War II. They have been linked for years to such urban problems as noise, air pollution, social decay, and physical deterioration—all the deformed offspring of the car culture and the freeway syndrome. Now they are reaching beyond the cities as motorized outdoorsmen surge into the wilderness aboard gasoline-powered recreational vehicles.

Paradoxically, many observers suspect that the "downstream economics" of oil marketing have not represented good business practice; that oil companies have been trapped by history and psychology on one hand and competitive and government actions on the other.

These observers, including senior economists, are capable of little more than suspicion. The best they can do is formulate informed theories and opinions from fragmentary evidence. The crucial sections of the jigsaw puzzle are still missing. How much, for example, does it cost to refine a gallon of gasoline? No one knows, despite efforts by legions of senators, congressmen, and committee counsel to find out.

Working in this twilight, oil industry observers—pro, neutral, and anti— generally have concluded that the companies give their gasoline away in the sense that they sell it at around cost. Given this broad conclusion, there remains a diversity of opinion about details, ranging from the surmise that the majority of large petroleum firms are actually willing to sell gasoline at a loss to the belief that a desire for some profit exists in gasoline sales, but is subordinated to the greater goal of getting rid of the stuff. To anyone who has ever run a business—even selling lemonade from a soapbox—this sounds absurd. It's true that higher volume with a lower price may equal a bigger overall profit. But how can you make money if making money isn't the primary aim?

The answer, which may seem even more absurd, is that the petroleum company earns its profit selling the original crude oil to itself. This Alice-in-Wonderland approach to commerce works for two reasons. First, the major oil companies are integrated. They produce, transport and market their product in a chain that stretches from a wellhead on a platform in the Gulf of Mexico to the pump at your local neighborhood service station. Second, the federal government has placed the tax advantages—on which so much annual passion is vented—at only one place in the chain: the wellhead.

It works like this. You drive into a service station and pay for the gasoline. Your payments should cover the costs of marketing, refining, transportation, and the price of the original crude oil. It may cover those items, barely, with a small profit. Or, in a competitive marketing area, your payment may not even meet costs, so the gasoline is sold to you at a loss.

The refining unit buys the crude oil from its parent company or from a sister producing subsidiary of the parent. As refinerymen are fond of pointing out, they must pay the "posted price" for the crude—the going rate which is reported in petroleum publications, just as stock-market prices are quoted in the daily newspaper. They must pay this price even though it is an intra-corporate transaction.

At this point, the oil company does indeed realize a profit, which is the difference between its costs of producing crude oil and the posted price for which it sells this oil to itself. Far from being fantastic or absurd, this intramural transaction culminates an ingenious process that moves the consumer's dollar upstream in such a way that profits are postponed until the point where they are no longer heavily taxed. This point is at the sale of crude oil, which is sheltered from the usual income taxes by the *depletion allowance* and *intangible drilling expense* provisions of the tax code. By foregoing profits until this point, the oil company retains much more of its profits than it could if they had been earned downstream at the filling station.

One question still remains: how do most service stations, which are franchised and presided over by nominally-independent, small businessmen, survive without a profit? The answer is that they are subsidized. The supplier grants a temporary dealer allowance that permits the dealer to sell gasoline at, or below, cost. Not only is the gasoline being marketed at giveaway prices, but the salesman may be subsidized to give it away. For many years, profits at the wellhead were so great that they supported in princely fashion virtually all oil company operations, including the gigantic downstream giveway.

The effect of this marketing system on American life should not be difficult to discern. Gasoline, which represents 50% of every barrel of oil produced, is not subject to the commercial checks and balances that set limits of possibility in other retail operations. Hence, oil companies have been able to go to impossible lengths to encourage Americans to burn prodigious quantities of gasoline (and, to a lesser degree, other petroleum products). Under existing law, they would have been remiss in their responsibilities to their shareholders had they done anything else. They can earn the most money and therefore are bound by their corporate charters to put their muscle behind whatever devices most quickly dispose of refined oil. So, they championed the most inefficient transportation mode known to man, fostering the level of combustion that has assisted so splendidly in achieving present levels of air pollution. They supported the severing of urban neighborhood cohesion by freeways and contributed mightily to blight by building millions of service stations which, under non-energy economics, would never have been dreamed of. Clearly, it was their duty as prudent corporate managers to do so.

This state of affairs may appear to be totally wanting in any base of reason. Not so. Energy economics may be perverse, but it is predictable, and like any other predictable phenomenon, it operates according to universal laws.

§10.2. The Second Law of Energy Economics.
The most inefficient purchase of fuels is the most efficient sale.
Translating the Orwellian into industry technese:
> Maintain the highest possible load factor by encouraging the highest possible level of consumption.
Reworked into common English:
> Move as much energy as possible along your system even if you must accept successively lower prices (including some that fall below cost) in order to get rid of successively larger amounts of energy. Since inefficient users will consume more energy, they are more efficient at helping to move the product along and should be encouraged.

While the oil chain is a stunning example of the second law in action, the law applies to other fuels and power sources, though often in rather different ways. Natural gas and electricity follow the rule with less Germanic obedience—differing from oil in that profits are pursued at all levels of sale—but here, too, power is often virtually given away in return for a high load factor. The consumer who uses the most natural gas and electricity pays the least for them. Critics claim that big users often buy natural gas and electricity below cost, with the utilities making up any loss by charging more to those who use the least.

Utilities are encouraged to subsidize big users by the bizarre system used to determine their allowable profits. Regulatory commissions allow utilities a level of profit computed as a percentage of their investment in buildings, operating facilities, equipment, and other costs of doing business. Taken together, these items constitute the utility's *rate base*. The utility cannot, with official sanction, earn more money by being more efficient. It *can* earn more money by expanding its rate base, so that its allowable profit becomes, say, 10% of $20 million instead of ten percent of $10 million. The utility thus is encouraged to get rid of its commodity as quickly as possible so that demand will exceed supply, thus requiring it to invest in new facilities and thereby realize larger profits. Critics of nuclear energy often cite this situation as a major stimulus of the electric industry's romance with the atom. Reactors cost more than steam boilers and therefore contribute more to the rate base.

By selling power to big users at preferential rates, the utilities encourage them to use much more than they would otherwise, thus necessitating an expansion of facilities with a corresponding increase in the rate base. Oil company charity at the service station is balanced by a large, lightly taxed return at the wellhead. But utility company charity to big power consumers is offset by the increased rates they can impose on the small consumer as a result of expanding their facilities and rate base.

§10.3. The First Law of Energy Economics. The bizarre nature of the second law of energy economics is exceeded only by that of the first law. Number one, which provides insight into how number two came into existence, specifies that:

Adherence to the principle of competitive free enterprise in energy requires rejection of the principle of competitive free enterprise.

In technese:

Energy must be supplied by private enterprises, but government must shield these companies from the constraints of private-enterprise economics. Government must be actively involved in energy to provide the "proper climate" that will permit private companies to fulfill whatever overdemand for underpriced energy exists at any given time.

In the vernacular:

Government must treat energy as a special case. It must go beyond the passive act of setting rules of the game through a code of prohibitions. Instead, it must actively engage itself in the energy business, granting special concessions (such as the depletion allowance) while imposing special burdens (federal control over natural gas, for one).

The pervasiveness of this first law is demonstrated by its hold on some critics of the energy industries as well as the industries themselves. Neither is willing to accept direction from Adam Smith's invisible hand. Neither believes energy industries would be guided to act in the public good within the usual business context by normal pressures of supply and demand on a finite resource.

The industry wants fuel imports blocked except where controlled by U.S. interests. It wants governmental subsidies and exemptions from such national social goals as diffusion of economic power and full public airings of public issues.

Reformers want more of the government regulation that has been such a failure, special taxes on windfall profits even if the profits result from socially desirable actions, and continuation of the most distorted (albeit popular) part of the mess: below-cost consumer prices.

The second law of energy economics sprang from the first and the first evolved out of history. It is one of history's great ironies, that energy has been deemed too valuable to be subject to the allocation constraints of the marketplace (the first law of energy economics). Without these constraints, energy was partially vested with the characteristics of a free economic good, and few among us give a damn about wasting a good that is free or nearly free.

The inevitable result was the second law of inefficient use to sustain efficient production; a rule whose enormous success and popularity reigned almost without challenge until just a few years ago. It has since come under intensifying assault from without, but it also is showing internal strains and these may ultimately be more important.

To function properly, the efficiency (second) law depends on energy companies, fossil fuel and electric, being capable of large production at low cost. Electric companies see the production half of their balance being crushed by the revolt against construction of new power plants. Failure to bring increased generating capacity on line, while demand continues to race ahead under the promotional spur, means shortages. Few people cared about inefficiency as long as everybody was getting all the power they desired. Gore the people's collective oxen with brownouts, shake in a few incendiary newspaper articles about waste and presto! utility executives who wish to keep their jobs become efficiency experts (save a watt, friends).

The fossil-fuel situation, especially in petroleum, is more complex, troublesome, and pregnant with implications for society as a whole. An oil company is concerned with capacity, but even more so with the cost-price spread at the wellhead. If profits are squeezed there, the firm is in trouble since it has that enormous downstream charity function to support.

The squeeze is precisely what has been underway for the last several years. Costs of finding and producing new crude have been rising faster than prices. The result: mergers to consolidate downstream operations and desperate attempts to abandon large-loss gasoline markets that sometimes spread across entire regions of the country. Most celebrated has been the dilemma of Gulf Oil, which recently announced a $250-million tax write-off and has been drenched in "raining red ink," according to an analysis in the financial journal, *Forbes*.

"Back in the sixties, Gulf used its gas pumps to extract profit from its ample crude," said *Forbes*. "Today, many of its gas stations make no profit at all."

If oil comapnies have not been compelled to operate under the rules that govern other business behavior, neither do they have access to remedies available to other businessmen in times of distress. The old castor oil of trade is to cut back unprofitable operations and raise prices. But retrenchment combined with price increases is doubly difficult to swallow under petroleum economics.

If an oil company tries to retrench to the part of its marketing system which is coincidently profitable, it first jeopardizes the large investment it has in marginal or unprofitable service stations and distribution apparatus. Secondly, it will lose volume, and crude sales will slow. The company has an even bigger investment in the crude producing and refining system and, if marketing operations are made efficient, the crude flow will drop further and further below capacity. This means the capital investment upstream becomes less and less rewarding, and the chain has become financially inefficient.

On the other hand, the oil company will have a difficult time raising prices because of the intense competition prevailing at the service-station level. It runs the risk of losing volume to other companies and the same denouement will proceed, starting at the refinery and moving back upstream, precisely as it would with retrenchment.

The interim answer, unsatisfactory to everyone, probably will be some price increase, some retrenchment, and a good deal of pain for little people involved in the business. A move already is afoot to try to force service station operators to carry more of the loss burden by cutting their allowances. (Money, it should be noted, can be made in gasoline marketing under normal economic rules. It has been done by a few companies of substantial size without access to large amounts of crude and by independent marketers. They merely approached their task with the intention of creating a profit-making market, not a spillway from the refinery.)

The anomalies of gasoline marketing, at least in their more inane forms, are well known to people with an interest in energy. For every hour petroleum industry critics spend fretting over the depletion allowance, oil company executives and trade journals spend five bemoaning the infernal problem of gasoline marketing.

One might wish that the two sides would get together, discover a way out of a situation which is becoming increasingly frustrating to them both, and arrive at a greater good. One might as well wish the Good Fairy would leave controlling interest in Standard Oil under his pillow, as the odds on both dreams are about the same. There is too much at stake: pride, power, money — the very fabric of American business life.

The energy industries and their critics have been enmeshed far too long in the commercial neurosis of contemporary energy economics to break free in one bound. The two great laws were fashioned in an evolutionary process. They now are showing the first signs of breaking down. But their disintegration and supersession are most likely to be evolutionary too. We may hope that the process will be quickened for, mean-while, a great deal of damage may be done.

11

NEPA, ENERGY,
AND THE ECONOMY

Irving Like

§11.1 Introduction. *The National Environmental Policy Act*[1] (NEPA) has provided a valuable tool for the assessment of all technologies which involve major federal actions significantly affecting the quality of the human environment. The Council on Environmental Quality (CEQ) has issued guidelines applicable to all federal agencies whose responsibilities subject them to the environmental impact disclosure requirements of the NEPA.[2] The federal activities which come under the scrutiny of the NEPA involve huge capital investments and depletion of natural resources and account for a significant portion of the Gross National Product (GNP). They represent a broad kaleidoscope of federally regulated or funded projects and programs dealing with highways, power plants, airports, dams, waterways, stream channelization, offshore oil drilling, pesticides, urban renewal, and national forest timber cutting.[3]

The potential interface of the NEPA with the American economy is as broad as the extent of federal intervention in the economic processes which govern the nation's output of goods and services and the magnitude of investment and consumption.

The cases in which the NEPA has been applied have usually involved the exercise of the federal power to regulate or spend money only after the point at which a particular project is perceived as necessary and beneficial and is defined in specific plans. For example, before a public utility submits an application to the appropriate federal agency for a permit to build a power plant, a corporate decision was reached that such a facility was needed and could be profitably undertaken. Similarly, before the Department of Transportation seeks federal funding to construct an interstate highway, it has decided that such project is justified by traffic needs and other economic considerations.

Other instances can be cited to document the point that before any project or program requiring federal authorization or appropriation is proposed, a preliminary determination has probably been made that it is necessary and in the public interest. It is also likely that substantial evidence can be marshaled to prove the benefits of the proposed federal action.

Thus, when the proposed federal action is placed upon the NEPA scale and subjected to the cost-benefit test, it is already weighted on the benefit side with the project's presumed need and advantages. If, for example, it is a nuclear power plant project, the Atomic Energy Commission (AEC) staff will already have concluded that the facility is needed to meet the power needs of the applicant utility's service area, and that the economics of building and operating a nuclear unit are more favorable than the fossil fuel alternatives.[4]

If it is a federally constructed or aided expressway project, the highway authority will already have decided that traffic counts and the volume of goods and passengers to be transported in autos and trucks in interstate commerce warrant the construction of the additional road network.

Whether it is a highway, an atomic power plant, or an Army engineers waterway that is subjected to the NEPA review process and its cost-benefit balancing requirement, it is likely that the proponents of the project will be able to demonstrate that it is needed and that it will serve progress and contribute to economic growth.

However, ruling that the NEPA is more than an environmental full disclosure law and that it creates substantive as well as procedural remedies, some courts have set aside agency action which arbitrarily and capriciously violates the substantive policies of the act, or which in striking a balance of needs, costs, and benefits, gives insufficient weight to environmental factors.[5]

It is fair to ask, however, whether the need and benefits (presumed or real) of a particular project have been generated by the prior exercise of federal economic power or policies which were themselves not subjected to the required NEPA evaluation.

§11.2. NEPA and Economic Powers. It is obvious that various federal agencies exercise economic powers which include credit, monetary, fiscal, and tax policies, and which qualify as major federal actions significantly affecting the quality of the human environment. As such they must comply with the NEPA's environmental impact statement requirements.[6]

The NEPA can and should be used to review the environmental impact of the federal exercise of credit, monetary, fiscal, and tax powers which significantly influence the size and composition of the GNP, and the quantity and quality of consumption and investment. It will be shown that economic policy-making by the federal government, although it avowedly serves various social objectives, fails to take environmental factors into account, as it is required to do by the NEPA.

The unforeseen result of this governmental omission may be the triggering of the chain of events which makes necessary the particular project, or supplies its justification. For example, if federal economic policies contribute to energy waste, or to unlimited growth of energy demand, they precipitate the claimed energy crisis, which in turn provides the cited justification or need for the particular power plant. If federal subsidy of critical stages of the nuclear fuel cycle gives the nuclear technology economic advantages over alternative energy technologies, then such prior federal action also becomes a causal agent which biases the cost-benefit balance in favor of nuclear power plants.

If federal economic policies operate so as to favor automobile production over mass transit, the concomitant increase in vehicular traffic becomes a factor which in part generates the need for new highways and may thereby bias the NEPA cost-benefit analysis of a proposed highway project.

It is essential then that the NEPA also be applied to the federal economic policies which precede and have a cause-effect relationship with projects further downstream in the planning or performance stages. If the NEPA cost-benefit analysis is applied at this earlier stage of the federal government's intervention in the economy, it will provide clearer insights on how to manage the problem of harmonizing growth and development with the protection of the environment. It will stimulate debate on how U.S. natural resources can best be allocated, what our national priorities should be, what the optimum size and composition of the GNP is, the broad alternatives to particular economic policies,[7] and what economic theories and models are best suited to managing the nation's productive machine to achieve the objectives of full employment, a high standard of living, and a sound economy, compatible with ecological constraints and resource limitations.[8]

§11.3. Federal Reserve Board. This article considers the Federal Reserve Board as a case in point to demonstrate how the NEPA can be applied to evaluate the exercise of federal economic power in credit and monetary matters. It is hoped that all federal agencies exercising economic power will be evaluated from the NEPA perspective.[9]

The Federal Reserve Board is an authority of the United States government[10] and as such is a federal agency subject to the NEPA and the Federal Administrative Procedure Act.[11] Its functions are subject to judicial review under the Administrative Procedure Act.[12]

The Board of Governors of the Federal Reserve System is charged with the responsibility of controlling the nation's money and credit through various powers including regulation of the banking system and the issuance of currency, Federal Open Market Committee policies, and determination of reserve requirements, discount rates, and interest rates. Each Federal Reserve Bank is an operating agency of the federal government and is authorized: (1) to make industrial and commercial loans; (2) to discount for, purchase, or lend money on industrial or commercial obligations; (3) to purchase and sell obligations of national, state, and municipal governments; (4) to purchase and sell acceptances of intermediate credit banks and agricultural credit corporations.[13]

The Federal Open Market Committee, which is composed of members of the Board of Governors and representatives of the Federal Reserve Board, carries out purchases and sales of securities eligible for open market operations to accommodate commerce and business with regard to their bearing upon the general credit situation of the country.

From its very inception, it was recognized that the original purposes of the Federal Reserve System—the provision of an elastic currency and facilities for discounting commercial paper, and the supervision of banking—were in fact parts of the broader objectives, namely to help counteract inflationary and deflationary movements and to share in creating conditions favorable to a sustained, high level of employment, a stable dollar, growth of the country, and a rising level of consumption. Yet the Federal Reserve System exercises this vast power without taking into consideration or disclosing the environmental impact of its economic policies.

Through its control of credit and money, the Federal Reserve helps to determine the level of production, employment, income, and consumption—in short, the level of economic activity and technology, and the size and composition of the GNP (the annual total output of goods and services). In this way, the actions of the Federal Reserve significantly affect the physical quality of the human environment.[14]

§11.4. Environmental Impact of GNP. In its third annual report (1972), the Council on Environmental Quality said:

> The amount of pollution will hinge on changes in total population, level of GNP, the stringency of abatement policies, the adoption of new industrial technologies, and other factors such as urbanization and hydrological cycles.... Because both the U.S. population and GNP will rise between 1970 and 2000, perhaps the key question in forecasting future pollution loads is the extent to which the pollution and GNP growth, will be counterbalanced by changes in industrial processes and by tighter pollution controls.

It is obvious that the GNP is the quantitative record of the aggregate of economic man's polluting and beneficent activities expressed in the flow of goods and services.

Recently, attempts have been made to document more precisely the relationships between changes in levels of specific pollutants and population size, environmental impact per unit of production, and the amounts of goods produced per capita.

> For it is modern technology which extends man's effects on the environment far beyond his biological requirements for air, food and water. It is technology which produces smog and smoke; synthetic pesticides, herbicides, detergents, and plastics; rising environmental concentrations of metals such as mercury and lead; radiation; heat; accumulating rubbish and junk.[15]

The relationship between the level of economic activity and pollution has also been described;

> Pollution can be said to be the result of multiplying three factors: population size, per capita consumption, and an environmental impact index that measures in part, how wisely we apply the technology that goes with consumption.[16]

and the impact of energy use upon the environment:

> The present flow of energy through U.S. society leaves waste rock and acid water at coal mines; spilled oil from offshore wells and tankers; waste gases and particles from power plants, furnaces and automobiles; radioactive wastes of various kinds from nuclear fuel processing plants and reactors. All along the line waste heat is developed particularly at the power plants.[17]

On comparing the changes in GNP and increases in environmental pollution since World War II, it appears that the predominant factor in our industrial society's increased environmental degradation is the increasing environmental impact per unit of production due to technological changes, particularly in the following general classes of production: synthetic organic chemicals and the products made from them, such as detergents, plastics, synthetic fibers, rubber, pesticides, and herbicides; wood pulp and paper products; total production of energy, especially electric power; total horsepower of prime movers, especially petroleum-driven vehicles; cement; aluminum; mercury used for chlorine production; petroleum and petroleum products.

Scientific and technical studies have been conducted which demonstrate in some detail how the size and composition of the GNP impose stresses on the environment through the specific impact of energy and materials consumption. For example:

> The effects of energy and materials consumption appear in the various forms of waste discharged to the environment: Air, thermal, and water pollution; radionuclides, noise, and solid waste.

§11.5. Energy Side Effects. The uses of energy fall within five major market categories, listed here according to their current size: (1) industrial, 32% of total; (2) electric utilities, 25% of total; (3) transportation, 24%; (4) residential, 14%; and (5) commercial, 5%.[19]

Diagrams of material flows in the economy have been published showing how fossil fuels (coal, natural gas, petroleum), nuclear fuel (uranium), agricultural products (food, wood, natural fibers), and minerals are transformed into various wastes through the processes of energy conversion and materials processing.

The use of fossil and nuclear fuels for thermal power, transportation, industrial, and commercial household purposes generates various particulates, gases, bottom ash, radioactive wastes, heat, and noise.

Materials processing of fossil fuels, agricultural products, inorganic chemicals and products, primary metals and products, and structural materials discharges slag, particulates, inorganic wastes, organic wastes, refuse, heat, and noise.[20]

§11.5.1. Air pollution. The pollutant by-products of combustion of fossil fuel in the form of particles and gases make up roughly 85% of the total tonnage burden of air pollutants in the United States, with the worst pollution resulting from the growing production of electric power. The particles include fly ash, soot, lead, other heavy metals, and hydrocarbon compounds. The gases include sulfur dioxide, carbon monoxide, and various oxides of nitrogen and hydrocarbon compounds.[21]

§11.5.2. Water pollution. Growth of population and the demand for water by cities, industries, and agriculture keep increasing the volume of pollutants while the water supply stays basically the same. Hundreds of different contaminants can be found in polluted streams, lakes, and bays: bacteria and viruses, pesticides and weed killers, phosphorus from fertilizers, detergents and municipal sewage, trace amounts of metals, acid from mine drainage, organic and inorganic chemicals, heat and radioactivity. All together, manufacturing activities, transportation, and agriculture probably account for about two-thirds of all water degradation.[22]

§11.5.3. Agriculture. The environmental side effects of the business of farming (an important component of the GNP) have also been investigated. Government farm price support payments and inducements for acreage limitations encourage further increase in per acre yields to obtain maximum benefit from land in cultivation. To capitalize on this situation, large farms use even more chemicals to boost yields from land allotted for cultivation.[23]

> It is ironic that the very success of programs to develop extremely productive crops to feed mankind's burgeoning numbers is now threatening the diversity in the gene pools of some of the most nutritionally important plants.[24]

§11.6. GNP and Energy Consumption. Statistical inquiries confirm the close correlation between the GNP and commercial energy consumption. The correlation is expressed in a nearly linear relation between the per capita consumption of energy for heat, light, and work, and the GNP.[25]

Since 1967, the ratio of total energy consumption to GNP has risen more steeply each year. In 1970 the United States consumed more energy for each dollar of goods and services than at any time since 1951.[26] United States energy consumption growth has outpaced the growth in population since 1900, except during the energy cutback of the depression years.[27]

Obviously, the economy is shaped by interrelations among resources, population, the efficiency of conversion processes, and the particular applications of power, and, in general, a high per capita energy consumption is a prerequisite for high output of goods and services.

Although there is a strong general correlation between per capita energy consumption and per capita output for a number of countries, it is not a one to one correlation. The relationship between energy consumption and output can differ, reflecting contrasting combinations of energy-intensive heavy industry and light consumer-oriented and service industries as well as differences in the efficiency of energy use.[28]

U.S. experience indicates that the ratio of energy consumption to GNP falls in response to increases in efficiency of energy conversion and utilization, the efficiency of electric power generation and transmission, and the increase in importance of the services sector of the economy in comparison to more energy intensive activities such as mining and manufacturing.[28]

If it is true that certain groups of production activities may be primarily responsible for the observed changes in pollution levels, it is important that a more detailed examination be made of the nature of the production activities that comprise the

GNP and of their specific relationship to environmental degradation. It is possible, through statistical investigation, to relate changes in different parts of the national economy to increased stresses on the environment.

Throughout the literature describing the operations of the Federal Reserve System, one encounters its stated policy of facilitating the expansion of money and credit necessary to continuing economic growth. This blind commitment to unlimited economic growth fails to consider the ecological consequences of such policy. It ignores the growing concern over what are the proper limits of growth in a country with finite resources.

The implications of unlimited growth have been graphically detailed and have sparked worldwide debate. Food and nonrenewable resources sustain the growth and maintenance of population and industry. Scientists have estimated the rate of growth in the demand for each of these factors and the possible upper limits to the supply. By extrapolating how much longer growth of each of these factors might continue at its present rate of increase, they have concluded that the short doubling times of many of man's activities, combined with the immense quantities being doubled, will bring us close to the limits to growth of those activities surprisingly soon.[29]

The heightening interest in the ecological imperatives restraining unlimited growth has stimulated the demand for comprehensive inquiry into the cost of unrestricted material growth and the consideration of alternatives to its continuation.[30] It has prompted a call for major restructuring of current economic theories, with reexamination of the concepts of "profit," and "economic growth" itself, and the introduction of "a system of energy accounting and simulation which could embrace description of how underlying energy-matter exchanges operate and how hidden energy subsidies or outflows obscure or prevent accurate accounting of the real costs, benefits and trade offs in human activities."[31]

Some environmentalists have focused on the quality rather than the quantity of growth. For example:[32]

> Environmental protection measures have an economic impact and raise specific issues that bear directly on international trade and investment, and that unless resolved satisfactorily, could damage international economic relations. The issues include how to prevent pollution controls and their costs from distorting international trade, what policy to adopt toward the movement of capital investment to pollution havens, how to reconcile real and imagined conflicts between environmental imperatives and economic development, and how to avoid damaging the export markets of less developed countries with environmental programs of developed countries.

> ...economic growth can be infinite, but it's the character of that growth that's creating the problem. We're just going to have to be producing a different mix of goods, and one of the goods is a clean environment.

The quest for a new economics which would assess the social cost of pollution has been taken up at the national and international level:[33]

Our price system fails to take into account the environmental damage that the polluter inflicts on others. Economists call these demands—which are very real—

"external social costs." They reflect the ability of one entity, e.g. a company to use water or air as a free resource for waste disposal, while others pay the cost in con- taminated air or water. If there were a way to make the price structure shoulder these external costs — taxing the firm for the amount and characteristics of the work discharged, for instance — then the price for the goods and services produced would reflect these costs. Failing this, goods whose production contributed to environmen- tal degradation are greatly underpriced because the purchaser does not pay for pollu- tion abatement that would prevent environmental damage. Not only does this failure encourage pollution but it warps the price structure. A price structure that took environmental damage into account would cause a shift in prices, hence a shift in consumer preferences and, to some extent, would discourage buying pollution- producing products.

In its third annual report of 1972, the CEQ discussed the international eco- nomic effects of environmental controls:[34]

The report noted that some industrialists worry that firms subject to strict en- vironmental standards will be put at competitive disadvantage with foreign competi- tors that are not. There is also concern that non-tariff barriers, such as frontier charges and export subsidies, may be established by nations with high environmen- tal standards to equalize environmental costs with trade competitors, and trigger a series of retaliatory trade actions.[35]

The confrontation of ecology and economics has led to various recommenda- tions designed to use the market system to penalize pollution and reward environ- mental protection activities by adopting the rationale that the costs of pollution must be internalized so as to be borne by the polluter.[36]

It has been suggested that through the use of monetary incentives and disin- centives it is possible to put a premium on durability and a penalty on disposability, thereby reducing the throughput of materials and energy so that resources are con- served and pollution reduced.   Various fiscal measures have also been proposed to accomplish the same goal — a raw materials tax penalizing resource-intensive indus- tries and short-lived products and favoring employment-intensive industries.

Also urged is an amortization tax proportionate to the estimated life of the pro- duct. For example, it would be 100 percent for products designed to last no more than 1 year, and would then be progressively reduced to zero percent for those designed to last 100 years. Obviously this would penalize short-lived products, especially disposable ones, thereby reducing resource utilization and pollution, par- ticularly the solid waste problem.[37]

It is also recommended that profound changes be made in the models by which economic performance is judged. Currently,

> ... the success of the economy is measured by the amount of throughput derived
> in part from reservoirs of raw materials processed by factors of production and passed
> on in part as output to the sink of pollution reservoirs. The Gross National Product
> (GNP) roughly measures this throughput. Yet both the reservoirs of raw materials and
> the reservoirs for pollution are limited and finite, so that ultimately the throughput
> from the one to the other must be detrimental to our well being, and must therefore not
> only be minimized but be regarded as a cost rather than a benefit.[38]

For these reasons, the economist Kenneth Boulding has suggested that the GNP be considered a measure of gross national cost and that we devote ourselves to its minimization, maximizing instead the quality rather than the quantity of our stock.   It has been noted that growth *per se* is not the environmental culprit.[39]

§11.7. Credit Regulation. Once applied to the actions of the Federal Reserve, the NEPA could require that the Federal Reserve authorities consider the adverse environmental impact of a policy of unlimited economic growth, and the alternatives, which might include, in addition to the suggestions already enumerated: (1) credit and monetary policies designed to stimulate the development of a "steady state" or recycling economy; (2) preferential treatment of industries which install pollution abatement equipment or develop technologies which are not as environmentally damaging; (3) higher interest rates on loans to businesses who use air and water as free goods and fail to assume responsibility for the external environmental cost of their operations;   (4) environmental protection standards and criteria governing all loans by financial institutions subject to the Federal Reserve's regulatory power.

The field of credit regulation provides good examples of how the Federal Reserve's regulatory power over the banking system can be used to implement the purposes of the NEPA. The Federal Reserve's general methods of regulating the flow of money and credit (discount operations, open market operations, and changes in reserve requirements) and its selective credit regulations can be employed with regard to the particular field of enterprise or economic activity in which the money or credit is used, and can be used to promote the purposes of the NEPA.

Examples of selective credit instruments are the Federal Reserve's special powers to regulate stock market credit, consumer credit, and real estate credit. In each of these instances, the Federal Reserve prescribes the terms on which lenders may extend such financing.[40]

The Federal Reserve authorities are enjoined by law to restrain the use of bank credit for speculation and to take restrictive action to prevent undue use of credit for the speculative carrying of or trading in security, real estate, or commodities.[41] Such regulatory power implements the legislative purpose of restraining undue speculation and controlling inflation. There is no statutory prohibition of the Federal Reserve fashioning selective credit regulations to carry out the public policy of protecting the environment. The NEPA authorizes the Federal Reserve to adopt regulations governing the use of bank credit to finance projects that significantly affect the physical environment, and the NEPA authorizes the Federal Reserve to prescribe the terms on which lenders may make such loans.

The history of regulation of consumer   and real estate credit by the Federal Reserve supplies analogues for new selective credit regulations designed to aid the objectives of the NEPA. In 1941, the Board of Governors issued Regulation W, prescribing terms upon which consumer credit might be granted. The objective was to curb the use of credit for the purchase of automobiles, certain appliances, and other consumer goods and services at a time when labor and materials were scarce and needed for the defense effort. Regulation W listed the articles subject to the restraints on credit and limited the amount of credit that might be granted and the time for repayment.[42]

In 1950, the Board of Governors issued Regulation X for real estate credit, prescribing the terms on which individual loans could be made, specifying the maximum amount that could be borrowed, the maximum length of time the loan could run, and the minimum periodic amounts that must be paid to amortize the principal amount of the loan.[43]

Under general NEPA authorization the Federal Reserve could devise credit regulations listing the types of projects deemed to have significant environmental impact and warranting particular restraints on credit. The Federal Reserve could then apply credit restraints or stimuli to discourage or favor economic activities or projects, depending on their environmental merit, by increasing or decreasing the amount of available financing and the terms and time of repayment. Credit bonuses could be granted for projects which incorporate the best available design and technology for abating or eliminating pollution. Since bank financing is the lifeblood of many, if not most major enterprises, the availability, or unavailability, of credit may often determine whether, when, where, and how a particular project will be built. Environmental criteria can be and have been incorporated into lending policy, in compliance with the NEPA.[44]

Federal Reserve policy requires that the benefits of selective credit regulation outweigh its burdens.[45] Selective credit regulations in which environmental protection benefits outweigh their cost burdens can be designed; they present no serious conceptual difficulty.

Regulations governing real estate or stock credit can take into account the external or environmental cost of the   activities dependent on such credit. They can be expanded to include environmental conditions. If consumer credit regulations can be used to curb the purchase of autos to further the defense effort, they can also be used to curb the purchase of automobiles which do not meet prescribed emission standards, or to favor the purchase of vehicles which incorporate the best available design and equipment for pollution control, or the Federal Reserve's power to "prescribe such lower margin requirements for the initial extension or maintenance of credit as it deems necessary or appropriate for the accommodation of commerce and industry, having due regard for the general credit of the country"[46] must now be interpreted, in the light of the NEPA, as permitting the Federal Reserve to exercise its power over stock credit to accommodate the objective of environmental protection.[64]

For example, the Federal Reserve could liberalize stock credit for the purchase of securities issued by corporations to raise capital to pay for new pollution abatement equipment. Conversely, it could reduce the amount of credit available for the purchase of securities of corporations issued to raise money for operations which utilize equipment not in compliance with effluent standards. Traditionally, lending criteria deal with the credit worthiness of the borrower and the appraised economic value of the loan's collateral. The environmental impact of the activity being financed can also be appraised. The air, water, and land affected by the project to which the credit is extended can be considered a species of environmental collateral, as important a part of the security for the loan as the physical property described in the security instruments. Since damage to this environmental collateral may also erode the economic value of the appraised security for the loan, the lender is justified in imposing loan conditions which insure that due precautions are taken to protect the loan environment.

The Federal Reserve Act provision regarding loans on forest tracts indicates some concern for environmental values:[47]

> Any national banking association may make real estate loans secured by first liens upon forest tracts which are properly managed in all respects. . . .

In short, it is reasonable for a lender to impose as a condition to making a loan, the borrower's obligation to build his project or conduct his activity in a manner least detrimental to the environment. Standards can be formulated by the Federal Reserve to implement the concept of a loan which is environmentally sound, or of credit which is used for purposes the environmental damage from which is minimized. A rule-making proceeding is a proper vehicle for shaping the environmental standards which govern loans or credit subject to the regulating power of the Federal Reserve.

Through their power over extensions of credit to member banks, the Federal Reserve authorities can encourage lending activity which promotes the goals of the NEPA and can restrain the undue use of bank credit for purposes inconsistent with those of the NEPA. Reserve banks are authorized to discount for their member banks, notes, drafts, bills of exchange, and bankers' acceptances of short maturities arising out of commercial, industrial, and agricultural transactions.[48] Obviously, the discount rate can be used selectively to favor or penalize transactions depending on their environmental impact.

The Federal Reserve may also exercise its power to change reserve requirements as a means of achieving environmental objectives. The Board of Governors, in order to prevent credit expansion or contraction injurious to the economy may, by regulation, change the requirements as to reserves to be maintained against demand or time deposits or both by member banks.[49]

The NEPA, together with the Federal Reserve Act, must now be deemed to authorize the Federal Reserve to regulate the uses of credit to prevent undue injury to the environment.

The Federal Reserve authorities have additional means of forcing member banks to consider environmental factors in their lending activity. The Federal Reserve Act limits the interest on loans, discounts, and purchases which any national banking association may receive to the rate allowed by the laws of the state or district where the bank is located, or to a rate of 1 percent in excess of the discount rate on 90-day commercial paper in effect at the Federal Reserve Bank in the federal reserve district where the bank is located — whichever is greater.[50]

By a selective application of its power to raise or lower the discount rate on commercial paper, the Federal Reserve can raise the interest rate chargeable for loans which finance activities injurious to the environment, and lower the interest rate for those borrowers whose projects provide special protection for environmental values.

The Federal Reserve's powers of inquiry can also be used to persuade member banks to incorporate environmental standards in their lending policies.[51] The Federal Reserve examines the affairs of member banks and may require reports as to the amount and nature of their loans and investments. Conceivably the Federal Reserve can identify any bank whose loan portfolio evidences little interest in protecting the environment. Or it can single out banks whose sensitivity to environmental concerns is deserving of commendation. The fear of bad publicity or the desire for good public relations, combined with emerging concepts of corporate social accountability and responsibility, may influence banks to adopt lending policies which take environmental factors into account.

It is not suggested here that the policies now followed by the Federal Reserve System are necessarily injurious to the environment. Nor can it be known at this time whether the Federal Reserve can develop worthwhile environmental standards by which to guide its operations. The NEPA, however, clearly obliges the Federal Reserve Board, as a federal agency whose actions significantly affect ecological systems on a macro and micro scale, to proceed to adopt regulations implementing the environmental impact provisions of the act.

The CEQ guidelines direct that "all federal agencies, to the fullest extent possible, direct their policies, plans and programs so as to meet national environmental goals."[52] The council has interpreted the phrase "to the fullest extent possible" to mean that each agency of the federal government shall comply with the requirement unless existing law applicable to the agency's operations expressly prohibits or makes compliance impossible.

The CEQ has identified a number of agencies which are to be consulted in connection with the preparation of an environmental statement.[53] These are agencies with jurisdiction by law or special expertise with respect to any environmental impact involved. CEQ states that Section 102(2)(c), which requires the filing of impact statements, applies to all agencies of the federal government with respect, among other things, to major federal actions significantly affecting the quality of the human environment.[54] It defines such actions as including federal programs, policy, and regulations.[54]

Congress has conducted hearings on the administration of the NEPA to determine how federal agencies have approached the task of assessing the environmental impact of the programs.[55] Among the agencies which responded was the Treasury Department, which reported on its procedures for preparing environmental impact statements.[56]

The Treasury included in the range of types of actions which may call for an environmental impact its administrative actions such as the issuance of regulations, procedures, rulings and decisions, and operations and continuing activities carried on directly by the department or supported in whole or in part by federal funds.[57] The department gave as examples of such operating decisions and actions, those relating to procurement, grants, subsidies, loans, or other funding assistance.[58] The Treasury's acknowledgment of its duty to comply with the NEPA supports the author's position that the Federal Reserve is also subject to the NEPA. As in the case of Treasury, the operations and continuing activities of the Federal Reserve, such as reserve, discount, open market, and credit operations are supported by federal funds, and involve policy, regulations, rulings, and decisions.

§11.8. Enforcement of Compliance. The NEPA, the Administrative Procedure Act (APA), and the *Calvert Cliffs* case supply the legal basis and remedial means of compelling the Federal Reserve to comply with the NEPA.

The NEPA duty as to disclosure and weighing of environmental impact is detailed in Section 102. Section 103 provides that all agencies must review "their present statutory authority, administrative regulations, and current policies and procedures for the purpose of determining whether there are any deficiencies or inconsistencies therein which prohibit full compliance" with the NEPA.

The APA provides:[59] "Agency action includes the whole or a part of an agency rule, order, license, sanction, relief, or the equivalent or denial thereof, or failure to act."

The APA says that in reviewing administrative action a court can "compel agency action unlawfully withheld or unreasonably denied."[60]

To discharge its duty under the NEPA, each agency must answer the following questions: (1) whether the agency action (in the case of the Federal Reserve, the exercise of credit and monetary power) significantly affects the physical quality of the human environment (NEPA, Section 102); (2) whether the present statutory, administrative, and regulatory authority contain deficiencies or inconsistencies which prohibit full compliance with the NEPA (NEPA, Section 103); and (3) whether the agency action can be lawfully withheld or delayed (APA, Section 706).

The agency's determination of each of the enumerated questions must be supported by specific findings of fact and is judicially reviewable.

The case of *Calvert Cliffs* v. *AEC*[61] makes it clear that a court may nullify an agency rule which fails to comply with the NEPA, and send the case back to the agency for further rule-making. It is reasonable to infer from *Calvert Cliffs* that an agency which fails or refuses to adopt a NEPA implementing rule may be directed to do so by the court. If a court can send back a case for further rule-making, it can in the first instance certainly direct the agency to initiate rule-making.

Where agency action is not deemed a final order subject to judicial review, an action under the APA or for declaratory judgment may be available. In *Natural Resources Defense Council, Inc.* v. *SEC (Securities and Exchange Commission)*[62] the court dismissed a petition for rule-making which raised a question whether the commission had delayed in carrying out its responsibility under the NEPA. The court found that the agency action was not a final order, and said:

> Whether the Commission has improperly delayed its action under NEPA or has improperly interpreted the act are issues which may be resolved in the United States District Court for the District of Columbia where petitioner may proceed by appropriate action under the Administrative Procedure Act or by way of an action for declaratory judgment.

Thus, a possible strategy for bringing the Federal Reserve into compliance with the NEPA, is the filing with the Federal Reserve of a petition for rule-making requesting the adoption of regulations implementing the NEPA. The petition could be accompanied by a proposed regulation and request its issuance or could simply ask the Federal Reserve to draft appropriate rules. This presentation has been an attempt to present legal and factual information of the sort that could be used in preparation of such a petition. If the Federal Reserve failed or refused to grant the petition or to adopt the NEPA implementing regulations, the remedy of judicial review would be available under the APA to resolve any Section 102 and 103 issues raised by the agency.

§11.9. Conclusion. Judge James L. Oakes, of the U.S. Court of Appeals, Second Circuit, recently observed:[63]

> Rather, the problem with NEPA that I perceive is the opposite side of the coin: it does not take into account that for the most part the agencies which must do the 'full good faith' balancing of economic and social costs against environmental costs are generally structured to be advocates of economic expansion. As long as agencies are left to do the balancing, and as long as they have a dual mandate of environmental protection and economic development in their particular field—for example, power growth for the FPC, nuclear development for the AEC, or flood containment for the Army Corps of Engineers— is not the environment bound to come out on the short end?

The *National Environmental Policy Act* applies also to those instruments of federal economic power—such as the Federal Reserve—which, as leading advocates of economic expansion, provide mission-oriented federal agencies with their pro-development and economic expansion bias.

The Federal Reserve System has shown itself to be a vital creative force adapting to the needs of the American economic system. It must lose no time in taking environmental factors into account in its policy-making.

The Federal Reserve Board of Governors should, as a first step, schedule rule-making hearings to assist it and its highly competent staff in drafting such regulations. The hearings should create a truly interdisciplinary forum with input from economists, environmental scientists, the various government agencies and special groups with relevant expertise, and the public.

It is hoped that this paper will stimulate research into the applicability of the NEPA to the operations of all federal agencies which exercise fiscal, monetary, spending, and taxation power contributing to a pro-development on economic expansion bias, and will thus lead to fuller disclosure of the rationale for, and environmental impact of, the economic growth premise of our political economy.

§11.10 Notes & References

1. 42 U.S.C., §4321, et seq.
2. Guidelines, *36 Fed. Reg.* 7724, 1971.
3. Examples of cases involving large-scale projects temporarily enjoined because of failure to file impact statements: *In re Cross-Florida Barge Canal,* 2 E.R.C. 1173, 2 E.R.C. 1796 (Cross-Florida Barge Canal across northern portion of Florida connecting Atlantic Ocean and Gulf of Mexico); *Wilderness Society v. Morton* 4 E.R.C. 1467 (trans Alaska Pipeline) 4 E.R.C. 1408 (waterway linking Tennessee and Tambigbee Rivers); *NRDC v. Morton,* 3 E.R.C. 1558 (offshore oil and gas leases).
4. Final environmental statement relating to operation of Shoreham Nuclear Power Station, Long Island Lighting Company, AEC, Docket No. 50322, Sept. 1972; Final environmental statement relating to operation of Midland Nuclear Power Station—Consumers Power Company, AEC, Docket No. 50-329, 50-330.

5. *EDF v. Corps of Engineers,* 4 E.R.C. 1721. *EDF v. Froehike,* 4 E.R.C. 1829, *Conservation Council v. Froehike,* 4 E.R.C. 2039. *Sierra Club v. Froehike,* 3 E.L.R. 20248.

6. The federal actions subjected to the impact statement requirement of the NEPA are usually agency-licensed projects with a direct physical effect on the environment. However, in a recent case, the NEPA was extended to an agency action stemming from the exercise of a rate-fixing power—an economic function with no immediate and direct physical effect on the environment, but likely to trigger other economic factors ultimately stressing the environment. (*SCRAP v. U.S.,* 4 E.R.C. 1312, preliminary injunction issued enjoining the Interstate Commerce Commission from imposing a freight surcharge on materials transported for recycling in the absence of an adequate environmental impact statement. It was shown by opponents of the surcharge that the railroad's rate structure discouraged the movement of recyclable goods in commerce, and that added disincentives to recycling would result in the increased degradation of the natural environment by discarded unrecyclable goods and in the increased exploitation of scarce natural resources. The U.S. Supreme Court refused to stay the injunction, 4 E.R.C. 1369.) See also *City of New York v. U.S.,* 3 E.R.C. 1570. A railroad abandonment proceeding, despite its potential for increase in the use of alternative modes of transportation with greater polluting effects, such as trucks, constitutes the type of federal action which requires a detailed environmental statement as prescribed in §102(2) (c) of the NEPA, 42 U.S.C. §4332.

7. The Courts will not permit agencies to shrink from their duty of examining the relative adverse environmental impact of alternatives which require a weighing of complex matters. See for example: *NRDC v Morton,* 3 E.R.C. 1558, 1562. The pertinent alternatives involved matters of economics, foreign relations, and national security, the court observing that when the proposed action is an integral part of a coordinated plan to deal with a broad problem, the range of alternatives that must be evaluated is broadened. The court ruled that the NEPA statement had to consider alternatives which raised a large number of key issues to be faced in the development of a national energy policy, including nuclear energy development, oil import quotas, natural gas pricing policies, offshore oil and gas lease production, and market demand prorationing.

8. Barry Commoner raises two fundamental questions which need the input of an NEPA assessment of federal economic planning and policies: (1) To what extent are the fundamental properties of the private enterprise system incompatible with the maintenance of ecological stability which is essential to the success of any productive system; and (2) To what extent is the private enterprise system, at least in its present form, inherently incapable of the massive undertakings required to "pay the debt to nature: already incurred by the environmental crisis—a debt which must soon be repaid if ecological collapse is to be avoided, in *The Closing Circle,* Bantam Books, 1971, p. 256. For a radical view that overproduction and undersirable growth, and their concommitant enviornmental disasters are endemic to corporate capitalism, see Willard M. Miller, "Radical Environmentalism," *Not Man Apart,* Nov. 1972, p. 14, and Harry Rothman, *Murderous Providence: A Study of Pollution In Industrial Societies,* Bobbs-Merrill, 1972. (The latter urges that environmental pollution must be understood by analyzing its social roots in the process of production and contends that it is the limitation placed by our society's social and economic structure on economic rationality that is the fundamental reason why our environment is polluted on such an enormous scale.)

9. The Council on Environmental Quality has not yet identified the numerous federal agencies which exercise various forms of economic power which proximately result in significant environmental consequences. It is under a duty to do so because the NEPA applies across the board to all federal agencies. (*City of N.Y. v. U.S.,* 337 F. Supp. 150, 164,

3 E.R.C. 1570—"NEPA is a new and unusual statute imposing substantive duties which overlie those imposed on an agency by the statute or statutes for which it has jurisdictional responsibility.") *Calvert Cliff v. AEC,* 2 E.R.C. 1779, 1782. ("... the §102 duties are not inherently flexible. They must be compiled with to the fullest extent, unless there is a conflict of statutory authority. Considerations of administrative difficulty, delay or economic cost will not suffice to strip the section of its fundamental importance.") *Kalur v. Resor,* 3 E.R.C. 1458, 1465, 1466. ("NEPA mandates a case by case balancing judgment on the part of federal agencies. In each individual case, the particular economic and technical benefits of planned action must be assessed and then weighed against the environmental costs; alternatives must be considered that would effect the balance of values.")

10. *Federal Reserve Bank of Minneapolis v. Register of Deeds for Delta County,* 284 N.W. 667, 228 Mich. 120, 12 U.S.C., S. 248J, 341.
11. The Administrative Procedure Act confers agency status on any administrative unit with substantial independent authority in the exercise of specific functions. *Soucie v. David,* 1 E.L.R. 10109 (holding the Office of Science & Technology to be an agency subject to the Administrative Act).
12. See for example *Gordon & Co. v. Board of Governors of Federal Reserve System,* 281 F. Supp 899.
13. 12 U.S.C. §352(a)(b), §355, §359.
14. The functions of the Federal Reserve Board clearly fit the definition of major federal actions subject to the NEPA. *NRDC v. Grant,* 2 E.L.R., 20185, 20186. ("A major federal action is federal action that requires substantial planning, time, resources or expenditure.")
15. Commoner, Barry, M. Corr, and P. J. Stamier, "The Causes of Pollution," *Environment,* 13(3):2, Apr. 1971. Commoner and his associates estimated the change in each of these factors for the period 1946-68 and submitted statistics relating changes in different parts of the national economy and particular groups of production activities to increased stresses on the environment.
16. Ibid., p. 3
17. Cook, Earl, "The Flow of Energy in an Industrial Society," *Scientific American,* Sept. 1971.
18. Makhijani, A.B., and A. J. Lichtenberg, "Energy and Well-Being," *Environment,* 14(5):10, June 1972. The authors made an analysis of the energy and materials consumption according to the goods consumed in the U.S. They calculated the amounts of energy needed to produce unit amounts of basic materials, for example, steel, aluminum, glass, cement, and so on (Table 1, p. 14) and estimated the energy content of various consumer goods and the total and per capita energy consumption represented by each finished product (Table II, p. 15).
19. "Outlook for Energy in the U.S. to 1985," The Chase Manhattan Bank, Energy Economics Division, June 1972, p. 8.
20. "The Economics of Environmental Quality," *Fortune,* Feb. 1970, p. 121.
21. "Some Burning Questions About Combustion," *Fortune,* Feb. 1970, p. 130.
22. "The Limited War on Water Pollution," *Fortune,* Feb. 1970, pp. 103-105.
23. "Plowing for Profit," *Environment,* 14(6):3, July/Aug. 1972.
24. Ehrich, Paul R., and John P. Holdren, "The Closing Circle," *Environment,* 14(3):24, 26, Apr. 1972.
25. Starr, Chauncey, "Energy and Power," *Scientific American,* 225(3):39, Sept. 1971. Makhijani and Lichtenberg, op. cit., p. 13.
26. Cook, Earl, "The Flow of Energy in An Industrial Society," *Scientific American,* Sept. 1971. p. 137.
27. Cook, ibid., p. 144.
28. Cook, ibid., p. 139.
29. Meadows, D. H., D. L. Meadows, J. Randers, and W. W. Behrens II, *The Limits to Growth,*

Signet, 1972, p. 97.

30. Ibid., p. 194.

31. Henderson, Hazel, "Economists vs. Ecologists," *N.Y. Times,* Oct. 24, 1971.

32. "Changing Times: Americans Engage in a Great Debate About Economy, Environment; Is Growth Good or Bad?" *Wall St. Journal,* Nov. 3, 1971.

33. *Environmental Quality,* First Annual Report of the Council on Environmental Quality, Aug. 1970, p. 12 See also the President's 1972 environmental message: "Our national income accounting does not explicitly recognize the cost of pollution damage to health, materials and aesthetics in the computation of our economic well-being. Many goods and services fall to bear the full costs of the damages they cause from pollution and hence are underpriced. The trade-off between economic progress and environmental quality is not encompassed within the standard indicators of economic development," Environmental Protection, House of Representatives Doc. 92-247, Feb. 8, 1972, p. 6.

34. *Environmental Quality,* op. cit., p. 92.

35. Ibid., pp. 92, 102-104.

36. "The Economic Impact of Pollution Control—A Summary of Recent Studies," prepared for the Council on Environmental Quality, Mar. 1972, p. 4.

37. "A Blueprint For Survival," *Ecologist,* 2(1):12, Jan. 1972.

38. Ibid., p. 10.

39. "Conservation of Energy, A National Fuels & Energy Policy Study," Committee Print, Committee on Interior and Insular Affairs, U.S. Senate, S. Res. 45, 92-18, 92nd Cong., 2nd Session, 1972, p. 5: "Much if not most of the environmental damage we associate with growth is a function of the way we grow—of the nature of our technology and the forms of production. By prohibiting ecologically deadly or dangerous activities and forcing producers to absorb the cost of using air, water, and land areas for waste disposal, we can redirect growth, technology, and production into environmentally more tolerable channels."

40. *The Federal Reserve System, Purposes & Functions,* Board of Governors of the Federal Reserve System, Washington, D.C., 1954, p. 56. (See Securities Exchange Act of 1934, Title I, §7 (a), which authorizes the Board of Governors of the Federal Reserve System to prescribe margin requirements for the purchase or carrying of securities.)

41. Ibid., p. 57.

42. Ibid., p. 61.

43. Ibid., p. 64.

44. See Agency for International Development Administrative Procedures, Manual Circulars 1221.2, 1214.1, and 1612.10.3 which set forth procedures for encironmental review of capital projects financed by AID. (NEPA, Appendix, pp. 1694-1763.) Environmental and health criteria to guide project developments have been prepared for the World Bank Group (the World Bank, its affiliate, the International Development Association, and the International Finance Corporation, which supports the private sector) whose basic function is to prepare and finance projects for economic development in the developing countries of the world. (See NEPA, Appendix, pp. 1765-1786.)

45. *Federal Reserve System,* op. cit. p. 57.

46. Securities Exchange Act of 1934 sec.(B) (1), *Federal Reserve System,* loc. cit. p. 57.

47. 12 U.S.C. §371.

48. *Federal Reserve System ... ,* op. cit., p. 32.

49. 12 U.S.C. §461.

50. 12 U.S.C. §85.

51. 12 U.S.C. §248(a).

52. "Guidelines for Federal Agencies under the National Environmental Policy Act. Issued by Council on Environmental Quality, Apr. 23, 1971," *Environment Reporter,* Federal

Laws 71.030.

53. Departments of Agriculture, Commerce, Defense, Health, Education and Welfare, Housing and Urban Development, Interior, State, Transportation, Atomic Energy Commission, Federal Power Commission, Environmental Protection Agency, Office of Economlc Opportunities. *Guidelines for Federal Agencies ...* , op. cit., 71.0302.

54. Ibid., 71.0301.

55. "Administration of the National Policy Act," 1972 Appendix to Hearing before the Subcommittee on Fisheries of the Committee on Merchant Marine & Fisheries. House of Representatives, 92nd Congress, 2nd Session, Feb. 17, 25, May 24, 1972, Serial No. 92-21 (NEPA, Appendix).

56. NEPA, Appendix, pp. 1345-1351.

57. Ibid., p. 1346.

58. Many federal subsidy programs involving, for example, agriculture, housing, natural resources, transportation, and economic development present significant environmental impacts and would require the agency involved to file NEPA impact statements. See "The Economics of Federal Subsidy Programs," a staff study, joint Economic Committee, Committee Print, 92nd Congress, 1st Aession, Jan. 11, 1972.

59. APA, 5 U.S.C. §551.

60. Ibid. §706.

61. *Calvert Cliffs v. AEC,* 2. E.R.C., 1779.

62. *Natural Resources Defense Council, Inc. v. SEC,* 3. E.L.R. 20154.

63. *Developments in Environmental Law,* 3 E.L.R. 50001, 50008.

64. Regulation A (12 CFR 201) as amended effective Apr. 16, 1970.

12

THE ORIGINS OF
A NATIONAL
ENVIRONMENTAL POLICY

Victor J. Yannacone, jr.

§12.1. Introduction.

If we can send men to the moon, we can clean our rivers and lakes, and if we can transmit television pictures from another planet, we can monitor and improve the quality of the air our children breath and the open spaces they play in.

The needs and the aspirations of the future generations make it our duty to build a sound and operable foundation of national objectives for the management of our resources for our children and their children. The future of succeeding generations in this country is in our hands. It will be shaped by the choices we make. We will not, and they cannot escape the consequences of our choices.[1]

This statement was made by Senator Jackson in support of the passage of the *National Environmental Policy Act of 1969*[2] (NEPA) which was signed by the President on January 1, 1970. Senator Jackson, the sponsor of the bill in the Senate, described the Act as "the most important and far-reaching environmental and conservation measure ever enacted by the Congress."[3]

The most important effect of the NEPA is that it provides all agencies and all Federal officials with a legislative mandate and a responsibility to consider the consequences of their actions on the environment.[4] Other provisions of less long range importance are the requirements that the President transmit an annual Environmental Quality Report to Congress,[5] and the establishment of a Council on Environmental Quality (CEQ) within the Executive Office of the President to advise the President on environmental questions.[6] The purposes of the Act are:

To declare a national policy which will encourage productive and enjoyable harmony between man and his environment; to promote efforts which will prevent or eliminate damage to the environment and biosphere and stimulate the health and welfare of man; to enrich the understanding of the ecological systems and natural resources important to the Nation; and to establish a Council on Environmental Quality.[7]

Congress enacted NEPA because of a belief that the policies and programs of the federal government, traditionally designed to enhance the production of goods and to increase the gross national product, were not designed to avoid environmental degradation and decay.[4]

§12.2. NEPA: Legislative History. *The National Environmental Policy Act of 1969* is a synthesis by a conference committee,[8] of bills introduced in the Senate by Senator Jackson,[9] and in the House by Congressman Dingell.[10]

Hearings on S. 1075 and two similar bills introduced by Senators Nelson[11] and McGovern,[12] were held before the Committee on Interior and Insular Affairs on April 16, 1969.[4]

S. 1075, after substantial amendment by the committee, was reported to the Senate, where the bill, as amended by the committee, was adopted on July 10, 1969.[13]

The Subcommittee on Fisheries and Wildlife Conservation of the House Committee on Merchant Marine and Fisheries held hearings on H.R. 6750, and similar bills introduced subsequently during May and June, 1969. At the conclusion of these hearings, the subcommittee unanimously reported a new bill, H.R. 12549, to the full committee.[14]

The new bill, which was in essence H.R. 6750, with amendments, was introduced into the House of Representatives by Congressman Dingell and coauthored by other members of the subcommittee. After extensive debate, the House of Representatives passed the bill, with amendments, on September 23, 1969. The House on the same day voted to amend S. 1075 by striking out all after the enacting clause and inserting in lieu thereof the provisions of H.R. 12549 as passed. The House requested a conference with the Senate insisting on its amendments. The Senate, after insistence by Senator Jackson, disagreed to the amendment by the House of S. 1075, and agreed to the request of the House for a conference. The compromise bill agreed to by the conference committee[8] was accepted by the Senate on December 20, 1969, and by the House on December 23, 1969. §§2, 101, 102, 103, 104, and 105,[15] are based on S. 1075 as passed, there being no similar provisions in H.R. 12549, as passed. The remaining sections of the *National Environmental Policy Act of 1969* are based primarily on H.R. 12549 as passed.[8]

§12.3. NEPA: A National Environmental Policy. §101 of the Act recognizes the substantial effects of man's activities on "the interrelations of all components of the natural environment . . . and . . . the critical importance of restoring and maintaining environmental quality to the overall welfare and development . . ."[16] To this end, the national policy of the federal government under the Act is, in cooperation with state and local governments and public and private organizations,

> to use all practicable means and measures, including financial and technical assistance, in a manner calculated to foster and promote the general welfare, to create and maintain conditions under which man and nature can exist in productive harmony, and fulfill the social, economic, and other requirements of present and future generations of Americans.[17]

§101 of the Act is a Congressional declaration of a national environmental policy predicated upon legislative recognition of the dependence of man on his natural environment for material goods and cultural enrichment. This national policy reflects the increasing pressures exerted upon the environment as a result of population growth, urbanization, industrial expansion, resource exploitation, and technological development.[4]

§12.4. NEPA: Federal Responsibility. In furtherance of this national environmental policy established by §101(a) of the Act, Congress declared in §101(b) that it is the continuing responsibility of the Federal Government to use all practicable means, consistent with other essential considerations of national policy, to improve and coordinate Federal plans, functions, programs and resources, to the end that the Nation may

> (1) fulfill the responsibilities of each generation as trustee of the environment for succeeding generations;
> (2) assure for all Americans safe, healthful, productive, and esthetically and culturally pleasing surroundings;
> (3) attain the widest range of beneficial uses of the environment without degradation, risk to health or safety, or other undesirable and unintended consequences;
> (4) preserve important historic, cultural, and natural aspects of our national heritage, and maintain, wherever possible, an environment which supports diversity and variety of individual choice;
> (5) achieve a balance between population and resource use which will permit high standards of living and a wide sharing of life amenities; and
> (6) enhance the quality of renewable resources and approach the maximum attainable recycling of depletable resources.[18]

§101(b)(1) is a recognition of the responsibility of each generation to enhance and maintain to the greatest extent possible the quality of the environment for the benefit of future generations. Congress has made it the duty of the federal government under §101(b)(2), through its planning and programs, to strive to protect the quality of the environment and to plan, design, and construct projects in order to protect and enhance every American's habitat. Each individual should be assured of safe, healthful and productive surroundings in which to live and work and should be afforded the maximum possible opportunity to derive physical, esthetic, and cultural satisfaction from his environs.[4]

Because this nation's natural resources must support larger populations in the future, the federal government must use these resources in the wisest and most efficient manner. In seeking intensified beneficial utilization of the earth's resources, the Federal Government must take care to avoid degradation and misuse of resources, risk to man's continued health and safety, and other undesirable and unintended consequences.[4] Though federal programs previously enacted are designed to preserve important aspects of our national heritage, many are single-purpose in nature and most are viewed as being within the province of a particular agency of Government.[4] §101(b)(3) was regarded by the Senate Committee on Interior and Insular Affairs as making clear that all agencies, in all of their activities, are to carry out their programs with a full appreciation of the importance of maintaining important aspects of our national heritage.[4]

The Senate Interior and Insular Affairs Committee regarded the maintenance of a natural environment providing the widest possible opportunities for diversity of experience and choice in cultural pursuits, in recreational endeavors, in esthetics and in living styles as an important aspect of national environmental policy.[4] §101(b)(5) is a recognition by Congress that uncontrolled magnitude and distribution of population underlies many of this nation's environmental and resource problems.

To insure that high standards of living are made available to all citizens and that Americans obtain esthetic enjoyment from a quality environment under the *National Environmental Policy Act of 1969* the Federal Government must strive to maintain magnitude and distribution of population which will not exceed the environment's capability to provide such benefits.[4] By §101(b)(5) Congress has thus implied that the rights of underprivileged American citizens to a high standard of living should not be sacrificed as a result of the national policy to protect the environment. This declaration of social policy may become the focus of extensive litigation and legislative activity as competition for fossil fuels as raw materials and energy as a means of social advancement increases between the haves and the have-nots. The intent of Congress under §101(b)(6) was to continue and intensify the efforts of the federal government to protect and improve renewable resources such as air and water. Through technology, and, if necessary, federal regulation, the recycling of depletable resources such as fiber, chemicals, and metallic minerals must be made more effective.[4]

§12.5. The Right to a Salubrious Environment. Congress has also recognized, under §101(c), that each person "should enjoy a healthful environment and that each person has a responsibility to contribute to the preservation and enhancement of the environment.[19] The Senate version of the Act originally provided: "The Congress recognizes that each person has a fundamental and inalienable right to a healthful environment and that each person has a responsibility to contribute to the preservation and enhancement of the environment."[14] The conference committee changed the language of §101(c) to its present form.

The managers of the bill in the House of Representatives stated that the language of the Senate bill was changed because of doubt on the part of the House conferees with respect to the legal scope of the original Senate provision."[8] The House conferees might well have thought that the original wording of the Senate provision might result in the issuance of injunctions against actions of the federal government that in any manner violated a citizen's inalienable right to a salubrious or healthful environment. In addition, the language of the original Senate provision might have been regarded as establishing an additional cause of action on behalf of every citizen against both governmental and private action that infringed their inalienable right to a healthful environment.[20] The change in wording by the conference committee was probably made, therefore, so that §101(c), *itself*, would not create a cause of action by which any citizen could restrain actions of the federal government that in any way adversely affected his natural surroundings.

§12.6. The Environmental Impact Statement. The most important provision of the *National Environmental Policy Act of 1969*, and the provision most debated by the members of Congress, is §102(1), which provides that "to the fullest extent possible: (1) the policies, regulations and public laws of the United States shall be interpreted and administered in accordance with the policies" of the Act. The meaning of this section is dependent on the meanings given to §§102(2), 103, 104, and 105.[21] §102 provides, in pertinent part, that all federal agencies shall, to the fullest extent possible:

(A) utilize a systematic, interdisciplinary approach which will insure the integrated use of the natural and social sciences and the environmental design arts in planning and in decision making which may have an impact on man's environment;

(B) identify and develop methods and procedures, in consultation with the Council on Environmental Quality ... which will insure that presently unquantified environmental amenities and values may be given appropriate consideration in decision making along with economic and technical considerations;

(C) include in every recommendation or report on proposals for legislation and other major Federal actions significantly affecting the quality of the human environment, a detailed statement by the responsible official on—

(i) the environmental impact of the proposed action,

(ii) any adverse environmental effects which cannot be avoided should the proposal be implemented,

(iii) alternatives to the proposed action,

(iv) the relationship between local short-term uses of man's environment and the maintenance and enhancement of long-term productivity, and

(v) any irreversible and irretrievable commitments of resources which would be involved in the proposed action should it be implemented.

Prior to making any detailed statement, the responsible Federal official shall consult with and obtain the comments of any Federal agency which has jurisdiction by law or special expertise with respect to any environmental impact involved. Copies of such statement and the comments and views of the appropriate Federal, State, and local agencies, which are authorized to develop and enforce environmental standards, shall be made available to the President, the Council on Environmental Quality and to the public as provided by Title 5 United States Code §552 and shall accompany the proposal through the existing agency review process;

(D) study, develop, and describe appropriate alternatives to recommended courses of action in any proposal which involves unresolved conflicts concerning alternative uses of available resources;

(E) recognize the worldwide and long-range character of environmental problems and, where consistent with the foreign policy of the United States, lend appropriate support to initiatives, resolutions, and programs designed to maximize international cooperation in anticipating and preventing a decline in the quality of mankind's world environment;

(F) make available to States, counties, municipalities, institutions, and individuals, advice and information useful in restoring, maintaining, and enhancing the quality of the environment;

(G) initiate and utilize ecological information in the planning and development of resource-oriented projects.[22]

§102 also provides that all federal agencies are required to assist the Council on Environmental Quality.[23]

§12.7. NEPA and Freedom of Information. The *National Environmental Policy Act* strengthened the right of public access to records and information of federal agencies under the *Freedom of Information Act.*[24] Senator Jackson, the author and sponsor of S. 1075, stated that §102(C) of the *National Environmental Policy Act* made the statements required with recommendations of proposals for major federal action that significantly affect the environment, available to the public under the provisions of the *Freedom of Information Act.* In addition, Congress has required all federal agencies to the fullest extent possible, to "make available to States, counties, municipalities, institutions and individuals, advice and information useful in restoring, maintaining and enhancing the quality of the environment***."[25]

These provisions, which should not be given a narrow construction[8] would enable individuals and private citizens, under the *Freedom of Information Act,* to obtain access to records of federal agencies that are relevant to the protection of the environment, in addition to the statements required to be filed by §102(C). Senator Jackson indicated that §102(F) required the federal agencies to give such advice and information "on environmental management as is available from their expertise and studies *** to other units of government."[26]

Members of the public may seek access to information relevant to the leasing of lands on the Outer Continental Shelf for off-shore oil drilling,[27] and to the leasing of lands under the Mineral Land and Mining laws[28] from the Department of Interior. Citizens can obtain information with respect to the hazards of thermal pollution and radiation from nuclear power reactors licensed by the Atomic Energy Commission, unless the information sought is clearly exempted from disclosure,[30] such as for reasons of national security, and they should be able to obtain equally detailed information on air pollution and water pollution from the Federal Power Commission for fossil-fueled electrical generating facilities. §103 requires all federal agencies to determine whether their present statutory authority, administrative regulations, policies and procedures prohibit full compliance with the purposes and provisions of the *National Environmental Policy Act.* If a federal agency finds that there are such deficiencies or inconsistencies, it is to propose to the President the measures that are necessary to bring their authority and policies into conformity with the Act.[31]

§12.8. NEPA: Agency Compliance. The specific statutory duties of a federal agency to comply with environmental quality criteria and standards, to coordinate or consult with other agencies of state or federal government, or to act or refrain from acting upon the recommendation or approval of other federal and state agencies, are not affected "in any way" by §§102 and 103 of the Act.[32] Finally, §105 provides that the policies and goals of the *National Environmental Policy Act of 1969* "are supplementary to those set forth in existing authorizations of federal agencies."[33] The effect of §§102, 103, 104 and 105 is to require all federal agencies and officials to consider environmental values in reaching decisions or in planning agency action, unless to do so would clearly conflict with their existing statutory authority and duties.

The Senate Committee on Interior and Insular Affairs indicated that §102 directs all federal agencies to follow the procedures and operating principles set forth in the section in carrying out their program activities.[4] The Committee also indicated that §101 would change the construction of authorizing legislation for some federal agencies which had been interpreted as prohibiting consideration of important environmental values.[4] NEPA provides such agencies with a mandate, a body of law and set of policies to guide such of their agency actions which have an impact on the environment. Though Congress has enacted numerous laws over the past decade that constitute congressional mandates on various aspects of environmental policy,[34] areas of federal policy and action have existed which have no environmental goals or policies or in which the conflicting operational policies of different agencies are frustrating and complicating the achievement of environmental quality objectives which are in the interest of all.[4]

Many older operating agencies of the federal government, for example, do not at present have a mandate within the body of their enabling laws to allow them to give adequate attention to environmental values. In other agencies, especially when the expenditure of funds is involved, an official's latitude to deviate from narrow policies or "the most economical alternative" to achieve an environmental goal may be strictly circumscribed by congressional authorizations which have overlooked existing or potential environmental problems or the limitations of agency procedures. There is also reason for serious concern over the activities of those agencies which do not feel they have sufficient authority to undertake needed research and action to enhance, preserve, and maintain the quality of the environment in connection with development activities.

S. 1075, as reported by the committee, would provide all agencies and all federal officials with a legislative mandate and a responsibility to consider the consequences of their actions on the environment. This would be true of the licensing functions of independent agencies as well as the ongoing activities of the regular federal agencies.[4]

§12.9. NEPA: Planning and Decision-Making. The *National Environmental Policy Act* is also designed to reemphasize the importance of existing statutory programs on the environment. Prior to the passage of NEPA existing legislation in some areas of federal activity did not provide clear authority for the consideration of environmental factors which conflict with other objectives. Many federal agencies have not given substantial and consistent consideration to environmental factors in decision making in certain areas of their responsibility. §102 is designed to remedy these shortcomings in the statutory foundation of existing agency programs by incorporating the policies and goals set forth in §101 into the actions and programs of federal agencies that carry out their responsibilities to the public. §102 establishes action-forcing procedures which will help to insure that the policies enunciated in §101 are implemented.

Prior to the passage of NEPA, planning and decision making that might effect the quality of the environment was too often the exclusive province of the engineer and cost analyst. The Act requires that federal agencies consider all relevant points of view and draw upon the broadest possible range of social and natural scientific knowledge and design arts in planning and decision making that may have an impact on the environment.

In the past, federal agencies have all too frequently ignored environmental factors in planning and decision making or omitted them from consideration in the early stages of planning "because of the difficulty in evaluating them in comparison with economic and technical factors. NEPA requires federal agencies and officials to strive to develop the methodology and techniques necessary for the determination of the total environmental impacts and full social costs of actions by agencies of the federal government. §102(C) requires that each federal agency which proposes major action, such as construction, new legislation, regulations, policy statements, expansion or revision of agency programs, determine the effect of that proposal in he environment. If the agency finds that the proposal will have a significant effect on the environment, the agency recommendation or report supporting the proposal must make findings with regard to the environmental factors set forth in §102(C)(i)-(v). These reports or recommendations must establish that the environmental impact of the proposed action has been studied and the results of the studies have been given consideration in the decisionmaking process which culminated in the proposed action. A finding must also be made that any adverse environmental effects which may occur cannot be avoided by reasonable alternatives and that these effects are justified by other considerations of national policy, which must be stated. The use of local, short-term resources must be consistent with the maintenance and enhancement of the long-term productivity of the environment. Proposals involving significant committments of resources that are irreversible and irretrievable under the conditions of known technology and reasonable economics must be environmentally responsible. It is in this area that conflicting claims to finite quantities of fossil fuels will be ultimately resolved. (see Chapter 3.)

The Senate Committee on Interior and Insular Affairs intended that §102(D) require federal agencies to develop information and provide descriptions of the alternatives in adequate detail for subsequent reviewers and decision makers, both within the executive branch and in the Congress, to consider the alternatives along with the principal recommendation. §102(E) directs all federal agencies which have international responsibilities to support international programs attempting to prevent a decline in the quality of the environment of the earth.

The statement of the managers on the part of the House at the conference state that under §102(C), the requirements for comment by other agencies should not unreasonably delay a federal agency from submitting proposals for major federal action.[8] The House managers indicated that the conference compromise bill anticipated that the President would prepare and establish by executive order a list of agencies that have "jurisdiction by law" or "special expertise" with respect to particular environmental matters within the meaning of §102(C).

The House managers felt that a federal agency need not seek the comments of those state and local agencies with only a remote interest and that are not primarily responsible for development and enforcement of environmental standards. In most cases, the requirement that state and local agencies comment on the proposals of federal agencies will be satisfied by published notice of the proposed action in the *Federal Register* and by supplying additional information on the proposal upon request of state and local agencies. To prevent unreasonable delay in the consideration of proposals by federal agencies, the House conferees recommended that the President establish a time limitation for the acceptance of federal, state and local agency comment.

The statement of the House managers supports the position of the Senate Committee on Interior and Insular Affairs that the *National Environmental Policy Act of 1969* requires all federal agencies to comply with the goals, policies and procedures of the Act unless the agency's statutory authority expressly prohibits such compliance or it would be impossible for the agency to fully comply.

"The conference substitute provides that the phrase 'to the fullest extent possible' [§102] applied with respect to those action which Congress authorized and directs to be done under both clauses (1) and (2) of §102 (in the Senate Bill, the phrase applied only to the directive in clause (1). In accepting this change to §102 (and also to the provisions of §103), the House conferees agreed to delete §9 of the House amendment from the conference substitute. §9 of the House amendment provided that 'nothing in this Act shall increase, decrease or change any responsibility or authority of any federal official or agency created by other provision of law.' In receding from this House provision in favor of the less restrictive provision 'to the fullest extent possible', the House conferees are of the view that the new language does not in any way limit the congressional authorization and directive to all agencies of the Federal Government set out in subparagraphs (A) through (H) of clause (2) of Section 102. The purpose of the new language is to make it clear that each agency of the Federal Government shall comply with the directives set out in such subparagraphs (A) through (H) unless the existing law applicable to such agency's operations expressly prohibits or makes full compliance with one of the directives impossible. If such is found to be the case, then compliance with the particular directive is not immediately required. However, as to other activities of that agency, compliance is required. Thus it is the intent of the conferees that the provision 'to the fullest extent possible' shall not be used by any federal agency as a means of avoiding the language in §102. Rather, the language in §102 is intended to assure that all agencies of the Federal Government shall comply with the directives set out in said section 'to the fullest extent possible' under their statutory authorizations and that no agency shall utilize an excessively narrow construction of its existing statutory authorizations to avoid compliance."[8]

The House managers also stated that §103 is to operate only where a federal agency finds a clear conflict between its existing statutory authority and the Act.

§12.10. NEPA: Supplement or Modification? The *National Environmental Policy Act* "supplements" rather that "modifies," existing statutory authority of federal agencies. §§103 and 104, and the use of the words "to the fullest extent possible" in §102, make clear that the Act is intended to supplement existing statutory authority of federal agencies to require them to follow the policies, goals, and procedures of the Act unless the existing law governing such agency's operations expressly prohibits full compliance with the Act's directives. If a federal agency finds that its statutory authority, regulations, policies or programs are inconsistent with the Act, the agency is required to propose to the President such measures as may be necessary to bring its statutory authority and practices into conformity with the Act. However, other actions of federal agencies must comply with the intent, purposes, and procedures of the Act.

The effect of the *National Environmental Policy Act of 1969* upon the existing statutory authority of federal agencies should be described as "supplemented" rather than "modified." The word "supplement" like the word "amend" is used where the purpose of a new statute is

> enforcement of the existing law, or to reach situations which were not covered by the original statute even though in its original wording the new statute does not purport to amend the language of the prior act. Whatever supplements existing legislation, in order to achieve more successfully the societal object sought to be obtained, *may be said to amend it.* (emphasis in original) Balian Ice Cream Co. v. Arden Farms Co., 94 FSupp 796, 799 (SD Cal 1950).[35]

Legislative history is an aid to the public and the courts in the interpretation of statutory law. The rules for the judicial construction of legislation allow the use of reports by committees of Congress in reporting a bill as an aid to determining the meaning of a statue.[36]

§12.11. NEPA and Administrative Law. The *National Environmentall Policy Act* is a "relevant statute" within the meaning of the *Administrative Procedure Act,*[37] thus giving standing to conservation organizations to obtain judicial review of actions and decisions of federal agencies that allegedly have not followed the policies, goals or procedures of the Act.[38]

In addition, private citizens, under NEPA, should have standing to challenge actions and decisions of federal agencies allegedly in violation of the NEPA where the action or decision threatens to have adverse effect on the Regional Ecological System in which they reside or upon which they may depend for recreation. §101(c) recognizes that each person should enjoy a healthful environment and that each person has a responsibility to contribute to the preservation and enhancement of the environment, and indicates that Congress intended to recognize the interests of individual citizens in the protection of their Regional Ecological Systems. Citizens who allege that decisions or actions of federal agencies are in violation of NEPA and have or will cause serious permanent and irreparable damage to the Regional Ecological System in which they reside or utilize for recreation, or any element of such system consequently have standing under the *Administrative Procedure Act* to seek judicial review of such action. Such citizens are within the class of persons whom the *National Environmental Policy Act of 1969* was designed to protect, and are thus persons adversely affected or aggrieved.[39] within the meaning of the *Administrative Procedure Act,* and entitled to judicial review of such agency action.[40] Indeed, NEPA, without the explicit statutory provision of the *Administrative Procedure Act* is sufficient to confer standing on such citizens if they are within the class of persons which the Act is designed to protect.

§12.12. NEPA: Goals and Policies. The national goals and policies for the protection, establishment, and enhancement of this nation's environment established by the *National Environmentall Policy Act of 1969* are

more than a statement of what we believe as a people and as a Nation. It establishes priorities and gives expression to our national goals and aspirations. It provides a statutory foundation to which administrators may refer for guidance in making decisions which find environmental values in conflict with other values.

What is involved is a congressional declaration that we do not intend, as a government, or as a people, to initiate actions which endanger the continued existence or the health of mankind: that we will not intentionally initiate actions which will do irreparable damage to the air, land, and water which support life on earth.

An environmental policy is a policy for people. Its primary concern is with man and his future. The basic principle of the policy is that we must strive in all that we do, to achieve a standard of excellence in man's relationship to his physical surroundings. If there are to be departures from this standard of excellence, they should be exceptions to the rule and the policy.[41]

§12.13. References and Notes.

1. 115 Cong Rec S 17452 (daily ed. Dec. 20, 1969).
2. 42 USC §§4321 *et seq.* (Supp. 1970).
3. 115 Cong Rec S 17451 (daily ed. Dec. 20, 1969).
4. S Rep No. 296, 91st Cong., 1st Sess. 14 (1969).
5. 42 USC §4341 (Supp. 1970).
6. 42 USC §§4332-4346 (Supp. 1970).
7. 42 USC §4321 (Supp. 1970).
8. Con Rep No. 765, 91st Cong., 1st Sess. (1969).
9. S 1075, 91st Cong., 1st Sess. (1969).
10. H R 6750, 91st Cong., 1st Sess. (1969).
11. S 1752, 91st Cong., 1st Sess. (1969).
12. S 237, 91st Cong., 1st Sess. (1969).
13. 115 Cong Rec S 7819 (daily ed. July 10, 1969).
14. H R Rep No. 378, 91st Cong., 1st Sess. 2 (1969).
15. 42 USC §§ 4321, 4331-4335 (Supp. 1970).
16. 42 USC §4331 (Supp. 1970).
17. 42 USC §4331(a) (Supp 1970).
18. 42 USC §4331(b) (Supp 1970).
19. 42 USC §4331(c) (Supp 1970).
20. Compare S 3575, 91st Cong 1st Sess (1969), discussed by its sponsor, Senator McGovern, at 116 Cong Rec S 3323-3325 (daily ed. March 10, 1970).
21. 42 USC §§4332(2), 4333, 4334, 4335 (Supp 1970).
22. 42 USC §4332 (Supp 1970).
23. 42 USC§4332(2)(H) (Supp 1970).
24. 5 USC §552 (1967).
25. 42 USC §4332(f) (Supp 1970).
26. 115 Cong Rec S 17455 (daily ed. Dec. 20, 1969).
27. 43 USC §1331 *et seq* (1964); *see* 30 C.F.R. Part 250 (1970).
28. Title 30, USC (1942).
29. 42 USC §§2011 *et seq* (1970).
30. 5 USC §552(b) (1967).
31. 42 USC §4333 (Supp 1970); see Chapter 12.
32. 42 USC §4334 (Supp 1970).

33. 42 USC §4335 (Supp 1970).
34. Some of the statutes that would be included in such a category would be the *Fish & Wildlife Coordination Act* 16 USCA §§661 *et seq* (1970); the *Multiple-Use Sustained-Yield Act,* 16 USC §§528 *et seq* (1970).
35. Another court has defined "supplemental" as "additional". *Texas & Pacific Motor Transport Co.* v. *United States,* 87 FSupp 107, 112 (ND Tex 1949). The word "modify", on the other hand, connotes "both by derivation and dictionary meaning, a limitation, not an extension of that which is modified." *Best Foods* v. *United States,* 158 FSupp 583, 589 (Customs Ct 1957). The word "modify" does not connote a major change, *State Airlines, Inc.* v *CAB,* 174 F2d 510, 514 (DC Cir. 1949), but alteration only of the thing modified." *Best Foods* v. *United States, supra,* at 589. To "modify" the change must "not be great and must not result in so transforming the original thing to be modified as to make of it something entirely new and different in substance. Volume VI of the *Oxford English Dictionary* (1933 edition) defines 'modification' as the action of limiting, qualifying or "toning down" (a statement etc.); a limitation, restriction, or qualification'." *State Airlines, Inc.* v. *CAB, supra* at 514-515.
36. *Wright* v. *Vinton Branch of Mountain Trust Bank,* 300 US 440, 464 (1937); *Duplex Printing Press Co.* v. *Deering,* 254 US 443, 474 (1921). Statements by the sponsor or by the committee members in charge of a bill, made during debates, are regarded as supplemental reports on a bill, and may similarly be resorted to for the meaning and purpose of a statute. *Wright* v. *Vinton, supra; Duplex Printing Press Co.* at 475; *National Woodwork Mfrs. Assn.* v. *NLRB,* 386 US 612, 640 (1967).
37. 16 USC §§1271-1287 (1970).
38. 5 USC §702 (1967).
39. 42 USC §4331(c) (Supp 1970).
40. 5 USC §702 (1967).
41. 115 Cong Rec S 17451 (daily ed. Dec. 20, 1969). (Senator Jackson).

*Portions of this chapter have been adapted from *Environmental Rights & Remedies,* by Victor John Yannacone, Jr., B. S. Cohen and S. G. Davison, published by Lauyeco Cooperative Publishing Co., Rochester, New York, 1972, in particular Chapter 5.

13

ENERGY AND CONSERVATION

Nicholas Muhlenberg

EDITOR'S INTRODUCTION

"To study Conservation, in its broadest sense, we must consider man in his total physical-cultural environement," said Nicholas Muhlenberg in 1952. His insight then, that "the total environment is more than the sum of its parts, because the parts as they exist in the whole are dependent and interrelated," led him to establish a system of interrelationships and to identify the limiting and controlling factors. Dr. Nicholas Muhlenberg in his Master's thesis chose energy as being the likely means whereby the sub-systems of the environment can be equated to the physical-cultural whole, and he threaded the total realm of conservation thought together with energy concepts. Noting that the trend of the non-living environment is toward increasing entropy and randomness, and that life systems reverse this trend, he demonstrates that cultural systems are motivated by energy availability, and that human, non-human, and inanimate energies are all significant in the evolution and sustenance of human cultures. In his own introduction to that paper, in 1952, Dr. Muhlenberg set forth his thesis:

> If energy provides a technique of interpretation enabling us better to understand the inorganic environemnt, the organic environment, and our cultural continuum, it seems to me that energy relationships can be used to interpret regional or national human communities in their environments. If our observations are well made, and the interrelationships of the human community well appreciated, we should be able to predict with warranted certainty what the effect of the in-

troduction of new techniques will be within any given human community. We should be able to find what the resource dependencies are and will be under given conditions. If human communities are successfully analyzed in these terms, we have a basis from which we can practice scientific conservation.

I have essayed to develop a total system of conservation thinking in energy terms and he did—twenty years ago.

§13.1 The Scope of the Problem. Both man and environment are complex and interrelated. Any number of approaches to the study of man and his total physical and cultural environment can be defined. Many aspects of these interrelationships have been investigated and considered in one or another context.

The consideration of saving or perpetuating parts of the human environment for future generations has become the *raison d'etre* of most conservation protagonists. Both saving and perpetuation imply time; and one of the chief values of the conservation approach has been the introduction of the public to an awareness of this factor.

We will, therefore, consider man and his total environment over time. This is, more exactly, a problem which involves spatial relationships (density, availability, etc.) over time. Thus, we must begin by first including the total system and then consider the parts, or sub-systems.

The first of the major sub-systems for our consideration is man. We are members of the articulate species *Homo sapiens*. Objectivity, therefore, is difficult, but it is absolutely essential for survival. While it is often postulated that the proper study of man is man himself, man lives within a physical-cultural environment, and individuals react to the environment as they are exposed to it; each within the context of his own total experience. The proper study of man is not just man, but man in his total environment. We must, therefore, attempt to find what the actions, reactions, and interactions are among man and the other systems of the environment.

Ecology is the study of actions, reactions, and interactions among plant and animal systems in order to interpret entire ecosystems. If man is included in the total sphere of thought we might, for convenience, call the study of Conservation applied human ecology. Human ecology, therefore, is the study of man in his total environment over time. The sub-system man is both a dependent and an independent variable in this context. The variation is due to environmental differences and man's dual role as initiator and responder, actor and reactor, to physical-cultural environmental stimuli.

The other major sub-system for our consideration is the environment. Human environment includes both physical and cultural components; however, the individual factors are so interdependent that it often becomes difficult to distinguish one from the other. For purposes of study a distinction must be made. The sub-system which constitutes the physical environment can be considered as being the available physical elements of the cosmos; our natural resources as defined by biological necessity; and our culturally conditioned wants.

Cultural considerations include media, technology, social situations, philosophical positions, and beliefs. These are functions of articulate speech, and when grouped can be called the cultural continuum. "Our present is the living sum total of the whole past." (Thomas Carlyle.)

Technology, a part of the cultural continuum, can modify man's physical environment; therefore, social systems are dependent upon technology, and a function of it.

> Social systems are secondary and subsidiary to technological systems. In fact, a social system may be defined realistically as the organized effort of human beings in the use of the instruments of subsistence, offense and defense, and protection. A social system is a function of a technological system.[20]

We have taken a brief look at the total realm of conservation or the system of human ecology and separated, for our convenience, several sub-systems. The objective of a more exhaustive analysis will be to find those controlling relationships among sub-systems which will enable us to interpret the whole system. Thus, we deal in a system of dependent variables over time. The vastness of our macrocosmos is so overwhelming at first glance, that we shall be obliged initially to approach the problem in terms of limiting factors and common denominators in order to construct a framework upon which we may have the answers to our initial questions.

§13.2. The Role of Energy.

"C'est dans l'energie que la realite s'encarne." Wilhelm Ostwald, 1910.

In searching for common denominators which are applicable to both the physical world and its life systems, including social organizations, there is little doubt that an understanding of the role of energy in this context renders intelligible, phenomena which are primary and fundamental to our study of human ecology or conservation.

In the non-living environment energy manifests itself in a variety of ways. The sciences of physics and chemistry have considered energy in its different forms and established the means whereby these different forms can be equated to each other. Once the fundamental unity of energy was recognized the laws of thermodynamics were formulated. In both physics and chemistry these laws have become recognized as being of fundamental importance in almost all considerations. Today, life-systems, ecological situations, and to a more limited degree, even social phenomena are being viewed in the light of these laws.

The First Law of Thermodynamics states that within any self-contained system the amount of energy present remains constant. "Energy can change its form but not its quantity."[4] For the systems with which we shall deal this holds true, but the law has been altered somewhat by Einstein's determination of the equimilane between mass and energy, ($E = mc^2$). The *Second Law of Thermodynamics* is more difficult to state in its broadest sense. Blum, however, summarizes the implications of the second law as follows:

1. All real processes are irreversible, and when all the changes are taken into account, they go with an increase in entropy, that is, toward greater randomness or less order.
2. The entropy of a circumscribed system into or out of which heat energy may flow, may either increase or decrease; but if the temperature of the system is

held constant the free energy will always decrease, unless energy other than heat is added from some outside source.

3. The Second Law of Thermodynamics points the direction of events in time, but does not tell when or how fast they will go.[2]

In brief, the above statements demonstrate that chemical reactions tend to run "down hill". The entropy or randomness of any system is, therefore, increased.

Plants are able to "trap" a small percentage of this available energy and the resultant synthesis by photochemical reactions decreases entropy and randomness on the planet Earth.[3] These photochemical reactions are the foundation upon which organic life, as we know it, rests. Life, therefore, is dependent upon energy, and life processes are influenced by the laws of thermodynamics.

These photochemical reactions, however, for practical purposes, are relatively inefficient.[6] The efficiency of these reactions imposes absolute limitations on the amount of life which the earth can sustain. Thus, we, *Homo sapiens,* are a function of the total life system on earth. As predators we are in a position to destroy great portions of it. Our objectives should be to attempt to sustain and improve these energy accumulating reactions, our allies against entropy and time's arrow.

We have seen that the constituents of both the non-living and living world are functions of energy and respond, to some degree at least, to the laws of thermodynamics. Human cultural systems appear to respond similarly.

Wilhelm Ostwald proposed that energy be used as a basis for interpreting social systems early in the 1900's, and the theory of social energetics has continued to evolve.

Leslie A. White uses energy in the interpretation of cultural systems and cultural evolution. He is primarily concerned with casual relationships, and in his essay, "Energy and the Evolution of Culture", he distinguishes

"...three factors in any cultural situation or system: (1) the amount of energy harnessed per capita per year; (2) the efficiency of the technological means with which energy is harnessed and put to work; and, (3) the magnitude of human need-serving goods and services produced. Assuming the factor of habitat to be a constant, the degree of cultural development, measured in terms of amount of human need-serving goods and services produced per capita, is determined by the amount of energy harnessed per capita and by the efficiency of the technological means with which it is put to work. We may express this concisely and succinctly with the following formula: $E \times T = C$, in which C represents the degree of cultural development, E the amount of energy harnessed per capita per year, and T, the quality of efficiency of the tools employed in the expenditure of the energy. We can now formulate the basic law of cultural evolution: other factors remaining constant, culture evolves as the amount of energy harnessed per capita per year is increased, or as the efficiency of the instrumental means of putting the energy to work is increased. Both factors may increase simultaneously of course." The energy spoken of here is the sum total of the animate (human and non-human) and inanimate energy which is harnessed by any cultural group.[20]

If one finds Professor White's premise acceptable, then the whole gamut has been run. Physical and chemical phenomena have been taken into consideration; life itself was equated to energy; and cultural systems are seen to be dependent upon animate and free physical energy for their growth and development.

Energy then is a common denominator and we might stop here, but then again we can go on to further considerations.

§13.3. Living Levels. Goods and services which serve the physical needs of humanity, and which are either already available in usable form directly from nature or can be made available by cultural means, are in effect those things which establish human living levels just as efficiency of utilization of food and energy in the plant and other animal communites establish trophic levels. The magnitude of these goods and services per capita varies directly with the amount of energy harnessed, and the efficiency with which it is put to use in a given situation; and with the relative accessibility of the physical components of the cosmos which are required to produce those products. The availability of these products varies inversely with the relative density of the population among whom the absolute final results of these efforts are to be distributed.

The things which are of importance in our consideration of living levels are food, the raw materials for clothing and shelter, and personal and community offense and defense. To these must be added the amenities considered essential by any particular cultural group. If the first four of these requirements can be obtained in adequate amounts man can live and reproduce. The amenities which very often are elaborations of these four essentials are however, extremely variable, and are present in myriad forms throughout the many cultures of mankind.

Those components of the cosmos which are essential for maintaining living levels are our natural resources, both inorganic and organic. Certain elements and compounds as they are found in the earth, on the earth, and in the atmosphere are basic non-renewable resources, and left alone represent frozen assets. The organic resources resulting from the photochemical reactions discussed earlier, are renewable, and with proper management a balance can be established between the production and use of these resources and the needs of human population.

Fossil fuels and water are examples of renewable and non-renewable resources. The fossil fuels are a fund of carbon compounds which have energy potential due to their organic nature, but are present on earth as non-renewable resources. Water is a renewable resource from the point of view of man. It is reconstituted in different forms throughout the hydrologic cycle, and its position and state determine its availability and usefulness to mankind.

We can consider much of the inorganic, or non-living world, plus the fossil fuels then, as being essentially non-renewable or fund resources. They are finite, and are wasting resources if they are irretrievably destroyed or dissipated with use. The objective of Conservationists should be to eliminate waste in the utilization of non-renewable resources, in order to extend their availability over the longest period of time possible.

The other organic resources and water are our renewable resources, and with proper management a balance can be established between their supply and the demands of human population. The establishment of this balance should be the other primary objective of conservationists. Man can control, to a degree, the species and varieties which grow in a given habitat, and can by cultural means increase somewhat the efficiency of productivity.

Making non-renewable resources available and overcoming some of the inherent environmental resistance to utilization of renewable resources requires expenditure of energy in one form or another. The successful management of these resources, then, is dependent upon energy. If the amount of energy expended and the efficiency with which it is utilized can be determined for any given region (assuming that the resistances remain constant) we have an index of potential productivity.

If there is an adequate index of potential productivity for any given region and the number and structure of the population is known, the average living level potential can be determined. If the productivity index is in units of energy, then that index divided by the number of people gives the potential living level index in energy units expended per capita per year. This has to be further qualified however, by the degree of resistance to resource utilization presented by the environment.

Basically, such resistances are of two general types: (1) the physical-chemical resistances of time-space and entropy which determine the accessibility and state of natural resources, and (2) the resistances imposed by the relative limitations on the productivity of plants and animals. If the amount of energy requried to overcome such resistance in order to yield a given end product is known, we can establish a qualified index of living level.

There is one more major consideration. As new sources of energy become available two things can happen: (1) the population can increase and continue to live at more or less the same living level which had existed formerly, or, (2) if the population grows less rapidly than does the potential productivity, the living level can increase. After new sources of energy are accommodated by populations, the energy production must be sustained, and energy sources replenished. If they are not, there will be either a regression of living level, and/or a population catastrophe. As new sources of energy are introduced into a region, therefore, careful consideration must be given to the trends which will probably occur in the population, and steps taken to minimize the rate of population increase beyond the rate of renewal or maintenance of energy production.

We can, therefore, analyze any given situation in these terms and obtain a result. If the approach is valid the result should be valid, too. Conservationsits must concern themselves with differences in human living levels, and attempt to establish and maintain adequate living conditions without destroying or unduly wasting the resource base.

§13.4. Standard of Living and Aesthetics. The phrase "standard of living" has been used to encompass a multitude of concepts. Everyone who uses it seems to have his own definition, and these definitions vary greatly.

A clear distinction must be made between living level and standard of living. Living levels are largely determined by materialistic and quantitative considerations of apparent natural processes. Standard of living is the product of broad aesthetic relationships as they are experienced by any individual or group of individuals who may share the same or different living levels. (Aesthetics is defined in this context as the capacity of any individual or group of individuals for bringing order to a given state of complexity—the capacity for equating one's self to one's environment and one's environment to one's self).

Experience is equated in individual contexts in the light of all other experience. The experiences themselves are defined by physical relationships which are a function of resources (including time), technology, and the social systems which integrate human beings within the context of their physical-cultural environments.

Once again resources and technology are primary considerations. As energy per capita has increased, so has the complexity of the physical environment due to the intricacies of technological manipulations, the resultant specialization of labor and function, and the increasing interdependence of area social group relationships to the widening of resource regions from which physical resources are drawn.

Church and state have attempted to bring order to the complexity of environmental processes—the state by controlling men's bodies and actions, welding them into reasonably effective mutual aid groups, and the church by ordering and subjugating their minds.

Individuals then are subjected to at least three main forces which are constantly influencing their state of mind. The primary one is the complexity of the physical-technical environment. This increases as energy available per capita increases. The increase in energy utilization per capita is responsible for increased specialization and interdependence. The second force is a socio-political-ecclesiastical one, which attempts to maintain order in any given primary or natural environment. Abstract thought represents third environment and is responsible for the philosophical, artistic, and scientific concepts which are instrumental in affecting the other environments. These abstract thoughts are a function of other environmental stimuli, however, and the three are therefore dependent variables. Abstract thought attempts to order environmental conditions but at a different level from the church-state.

Within Western civilization the amount of energy available per capita has increased rapidly since the advent of the steam engine. Roughly, the increase has resembled a geometric progression or an exponential curve. The incidence of technological invention during the same period has been staggering. The resultant industrialization has introduced social forces heretofore unknown in the world. All of these contribute to complexity. The physical resource complexity factor has been increasing in the same order of magnitude as has energy. The multiplication of interactions has been bewildering.

The church-state influence is more difficult to consider accurately. The church alone is at best holding its own. Much ecclesiastical dogma is being reconsidered and appears untenable and even repugnant in the light of modern science. It can be safely generalized, therefore, that as energy consumption per capita increases, the influence of the church on the general population tends to decrease. This loss of influence is replaced to some degree by the rise of the state and the elaboration of certain scientific concepts to the status of religious dogma.

The size of the effective state tends to increase with increases in energy availability due primarily to increased efficiency in communication and transportation. Thus the trend is toward a single world state achievable several ways. The increase in the control exercised by the state, either as single states or cooperatively as international unions, over larger effective areas and resource bases tends to increase order in the sense that larger social groups are at least externally attempting to achieve the same ends. International anarchy may thereby decrease, and the predictability of individuals and groups increase, thus increasing order.

Philosophy based on scientific premises has increased encouragingly and is becoming a means whereby individuals can organize their concepts into a private or group cosmology, essentially replacing ecclesiastical dogma, while still retaining certain tenets of the institutional church.

Alfred North Whitehead states in his essay "The Origin of Modern Science"

> The thesis which these lectures will illustrate is that this quiet growth of science has practically recoloured our mentality so that modes of thought which in former times were exceptional are now broadly spread thru the educated world. This new colouring of ways of thought had been proceeding slowly for many ages in the European peoples. At last it issued in the rapid development of science; and has thereby strengthened itself by its most obvious application. The new mentality is more important even than the new science and the new technology. It has altered the metaphysical presuppositions and the imaginative contents of our minds; so that now the old stimuli provoke a new response.[21]

At this point I would like to put forward a deliberate extension of a quantitative theory of aesthetics proposed by George D. Birkhoff in his book, *Aesthetic Measure*. He essayed, with a high degree of success, to obtain a direct quantitative means of measuring the fundamental aesthetic problem. His basic objective was the following: *within each class of aesthetic objects, to define the order,* O, *and the complexity,* C, so that their ratio $M = O/C$ yields the aesthetic measure of any object of the class.[2] M here is the aesthetic measure. Professor Birkhoff applied this theory to polygonal forms, ornaments and tilings, vases, diatonic chords, diatonic harmony, melody, and the musical quality of poetry. While these limited examples are far from being equatable to whole cultural systems, I choose to apply the same premise to cultural systems, for it withstands testing and justifies further consideration.

Professor Birkhoff qualifies his statement with the clause, "within each class of aesthetic objects", and modifies his original formula by making aesthetic measure equal to a *function of* order over complexity. Thus, for our purposes we can say that our class of aesthetic objects are civilizations themselves, and, using the elements of order and complexity outlined earlier, we can then empirically arrive at a graphic representation of our problem.

If we plot the elements of order and complexity over time and have the ordinate represent increasing complexity order and aesthetic measure we find that the elements of complexity (the amounts of energy available per capita and technology) are increasing as an exponential curve. The effective ordering achieved by the church-state is at best an arithmetic progression, cumulative to some degree over time. The progress of scientific philosophy is too evasive to plot, but any speculations concerning this or any other elements of order are left to you, patient reader.

Acknowledging the gross over-simplification of the following graphic representation, the following elements are plotted:

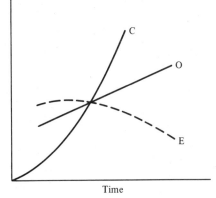

Time

AESTHETICS WITHIN THE CLASS WESTERN CIVILIZATION SINCE THE ADVENT OF THE INDUSTRIAL REVOLUTION

If the elements of order and complexity are equated according to the $M = O/C$ formula, the M curve becomes a mirror image of the C curve over the O line. This over-simplified graph is oppressive and arbitrary enough to be completely ignored if one chooses; however, it is intriguing and I have been unable to put it aside. Its implications have been corroborated by the writer's empirical observations and research. Alfred North Whitehead in his essay, *The Nineteenth Century,* states,

> What is peculiar and new to the century, differentiating it from all its predecessors, is its technology. It was not merely the introduction of some great isolated inventions. It is impossible not to feel that something more than that was involved ... It was a peculiar period of hope, in the sense in which, sixty or seventy years later, we can now detect a note of disillusionment or at least of anxiety.[21] I have assumed here that the complexity of the social force, industrialization, was the factor which disrupted the aesthetic status quo of the century, and caused the disillusionment and anxiety.

It appears that this same principle may be applied to socio-political situations, within limits.

We now have a system for analyzing level of living and standard of living. There is one element which is used in both considerations, the energy-technology-industrialization complexity factor.

§13.5. Another Definition of Conservation. The study of conservation should deal with human ecology. It is concerned with both the physical and the cultural environments of human populations and the ways in which these parts are articulated with the whole. The findings of researches into these relationships should ameliorate the struggle for survival and maintain living levels, utilizing human and natural resources in ways which will minimize the disruption of aesthetic relationship by establishing means whereby populations can accommodate change and articulate themselves to it. The nature of people and resources must be respected, establishing and maintaining a balance between the two which will minimize the degradation of either or both by one or the other.

§13.6. Some Statistical Corroborations. The world's population has been growing steadily and can be expected to continue to grow. We must assume that each individual person in the world is receiving at least the minimum amounts of food, clothing, shelter, and protection required to sustain life at the level that individual differences in the availability of the basic essentials of life in various parts of the globe are easily observable.

If the amount of physical energy available per capita is a limiting factor in these availability patterns, the statistics of energy consumption should be most revealing.

Based on the statistical data available in 1952, Dr. Muhlenberg proceeded to consider energy in the context of his thesis, satisfactorily demonstrating that the insight was correct and the technique sound.

We have taken a brief look at some statistics. Much more nearly complete analyses would be possible utilizing the existing statistics; correlations could be drawn using other statistics, and complete country or regional studies could be developed using this approach. ... Levels of living could be determined with a high degree of accuracy, and the relationships established within which aesthetics could be ascertained.

Once this was done, the basic differences which exist between the living conditions of the world's millions of people would be known with much more accuracy, and the socio-political problems which are contingent upon these level-of-living and standard-of-living relationships could be approached much more directly.

§13.7. Some Socio-Political Implications. Toynbee in his *Study of History* states,

> Whenever the existing institutional structure of a society is challenged by a new social force three alternative outcomes are possible: either a harmonious adjustment to force, or a revolution, which is delayed and discordant adjustment , or an enormity.[17]

If an enormity is considered as being complete social breakdown, which, of course, is to be avoided if possible, there remain but two possibilities for adjusting to the new social forces of our age of fuels and industrialization.

Only a small area of the world has undergone the changes wrought by the fuel revolution. Social upheaval has followed the introduction of industrialization. The history of the Western world during the 19th century was one of political adjustment to these new social forces, the main trend being toward the democratic republican form of government.

We in the West are currently proposing to aid in the development of the areas of the world which have been largely unaffected by the rise of industrialization and all that it implies. Our motives are manifold, and everyone concerned hopes to derive a just return from these efforts. I personally have a few reservations about the way in which certain proposals have been outlined and would like to speculate what the outcome of the "development" of "backward" areas might imply.

If we "Westernize" these more primitive areas by exporting our techniques, we shall introduce essentially the same social forces which were unleashed in Western Europe and North America in the 19th century.

One of the ostensible motives forwarded by proponents of the development idea is to alleviate the "misery" of the indigenous peoples in these more primitive areas. This would, of course, raise the buying power of the nations in question and by so doing would contribute to an expanding world economy.

The inherent dangers of population releases have to be considered here. If there are population explosions the beneficial effect of such attempts could never be realized.

By introducing the social force *industrialization* into these underdeveloped areas we must expect social and political unrest to follow. If this unrest manifests itself in "uncoiled" populations we will have created social chaos where we had planned to bring order.

If we equate this to Mr. Toynbee's three possibilities—harmonious adjustment, revolution, or an enormity—and assume that harmonious adjustment is the solution we would most like to see, we must be completely ready to do everything in our power to bring it about.

The USSR will be more than ready to aid and abet any revolutionary situations which present themselves, and it is possible that we shall find in these areas being developed the same situations now present in Korea, Indo-China, et. al., manifesting themselves all over the world.

I believe, however, that much can be done to aid these underdeveloped areas without totally disrupting the indigenous cultures. A development program must constantly take into consideration certain aesthetic relationships; must realize the dangers inherent in Westernization *per se;* and must do what can be done to minimize potential population releases.

If technical collaboration is carried out with discretion in an open-minded way, perhaps we can ameliorate levels of living and standards of living simultaneously, and at the same time make available some of the resources of these areas for the betterment of the well-being of the peoples of the world.

§13.8. References and Notes.

1. Ayres, Eugene. "Major Sources of Energy", *Proceedings American Petroleum Institute,* Vol. 28, 1948. pp. 109-142.
2. Birkhoff, George D. *Aesthetic Measure.* Cambridge, Masschusetts: Harvard University Press, 1933.
3. Blum, Harold F. "Life: a Photochemical Steady State", *Science,* Vol. 86, No. 2230, 1927.
4. Blum, Harold F. *Time's Arrow and Evolution.* Princeton, New Jersey: Princeton University Press, 1951.
5. Clark, Colin. *The Economics of 1960.* London: Macmillan Co., 1943.
6. Curtis, Otis F., and Daniel G. Clark. *An Introduction to Plant Physiology.* New York: McGraw-Hill Book Co., Inc., 1950.
7. Daubenmire, R. F. *Plants and Environment.* New York: John Wiley and Sons, 1947.
8. Deevey, Edward S. Jr. "Recent Textbooks of Human Ecology", *Ecology,* Vol. 32, No. 2, April, 1951.
9. Ebenstein, William, ed. *Man and the State.* New York: Rinehart and Company, Inc., 1948.
10. Guyol, Nathaniel B. *Energy Resources of the World.* U.S. Department of State. Washington, D.C.: U.S. Government Printing Office, 1949.
11. Hoffman, Paul G. *Peace Can Be Won.* Garden City, New York: Doubleday and Co., Ind., 1951.

12. Hubbert, M. King. "Energy from Fossil Fuels." *Science,* Vol. 109, pp. 103–09.
13. International Development Advisory Board. *Partners in Progress.* New York: Simon & Schuster, 1951.
14. Mumford, Lewis. *Technics and Civilization.* New York: Harcourt, Brace and Co., 1934.
15. Osborn, Henry Fairfield. *The Origin and Evolution of Live* New York: Charles Scribner's Sons, 1918.
16. Ostwald, Wilhelm. *L'Energie.* (French tr. by E. Philippi from the original German.) Paris: Felix Alvan, 1910.
17. Toynbee, Arnold J. *A Study of History,* abridged edition. New York: Oxford University Press, 1947.
18. United States Department of State. *Point Four.* Washington, D.C.: U.S. Government Printing Office, 1950. Publication 3719, Economic Cooperation Series 24.
19. Wells, H. G. *The Outline of History.* Garden City, New York: Garden City Publishing Co., Inc., 1949.
20. White, Leslie A. *A Science of Culture.* New York: Farrar, Straus and Co., 1949.
21. Whitehead, Alfred North. *Science and the Modern World.* New York: New American Library of World Literature, Inc. (Mentor Books), 1949.
23. Zimmermann, Erich H. *World Resources and Industries.* New York: Harper & Bros., 1951. Rev. ed.

EPILOGUE

CRISIS ...

> "We have met the enemy and he is us."
>
> Walt Kelly, 1969, *Pogo*

> "Energy Crisis! The bell rings and we stagger to the center of the ring, touch gloves with Pogo's "us" again and have at it.
>
> "Bite down on the rubber mouthpiece to ease the pain. Keep moving ... keep swinging ... what round is this, anyway? When did this fight start?
>
> "It seems that we first came out of our corner in the '50s against a savage club fighter called discrimination. But there's been so many rounds since then ... civil rights ... peace ... environment ... women ...
>
> "See the drawn faces on television saying the same things they said at the beginning of all those other rounds. Hear the tense voices of the commentators raising the same old questions. Hasn't this all happened before?
>
> *Energy Crisis,* Angelo J. Cerchione

The energy crisis is largely a rerun of any other historical crisis in human society. In the fifties it was America practicing the lonely isometrics of Cold War vigilance alternating between fits of neurasthenia and noisy machismo that prepared the People for the effort to reach the moon when President Kennedy, publically and dramatically, took exception to Sputnik.

Once the bell sounds and a crisis is recognized, the People spring into action. When the people have been stung into action, they always ask, "How could this have happened?" Investigations are launched, Congressional Committees and Subcommittees meet in emergency sessions. The media finds the right words, and eventually some kind of public perspective emerges, as many people make a sincere effort to tune in on discussions that were once the conversational white elephants of a few specialists.

Unfortunately, just when the public becomes most interested, the thinlipped spectre of secrecy begins to stalk the halls of government, and the blind, stubborn refusal to divulge necessary information makes the people edgy. The irony of being uninformed citizens in a republic where freedom depends on information encourages sensational speculation in the media, and bizzare nightmares among the people.

It was the public reaction to environmental degradation beginning with the attack of a New York woman on the indiscriminate use of DDT for mosquito control in Suffolk County during 1966 and culminating with Earth Day, 1970, that prodded first the Courts, then the legislatures into action. Unfortunately, since the great majority of the American People became aware of the environmental crisis on Earth Day, and the Energy Crisis as winter approached in 1973, there has been a concerted effort to move the planning for resolution of these crises back into official-bureaucratic-academic channels where the public is simply forgotten or deliberately ignored. Having been whipped into an educational and public-spirited frenzy, no-one is really ready to go back to sleep. It will not be possible to return to business-as-usual. Many of the current problems faced by our lawmakers and corporate executives are the result of the public desire to participate in the planning process, particularly in planning those activities which will affect the lives of all the people of this generation and those generations yet unborn.

It is the sudden need to know about energy and the energy crisis that has brought you to the closing pages of this book. Let us now look at the energy crisis secure in the knowledge that most of the crises which spawned the mass movements that have wracked America during the last two decades seem to follow the same evolutionary track: prelude, crisis, confrontation, public education, official recognition of the problem, legislative remedy, public recriminations.

For years, as Dr. Hubbert so precisely demonstrated in Chapter 2, men have known, or with the exercise of reasonable prudence should have known, that at some point in time, all our fossil fuels: coal, oil and natural gas would eventually be consumed. Nevertheless, during those same years, the public has been led to believe that when coal, oil and natural gas were no longer available as energy sources, other sources of cheap, convenient energy would be available. (Plucked from the ether, perhaps, by the nimble technological fingers of our scientists and engineers.).

Satisfied, mankind dozed—warmed and cozened by the petrochemical fire in the basement and illuminated by the electrical fire in the lamp—"fat headed in fossil fueldom."

This book is being written during the public education phase of the energy crisis. The industrialized nations of the world have embarked on a massive attempt to educate the people about energy and its wise use. Whole countries are reading the primer of conservation and the magnitude of public concern is creating a spiritual heat the likes of which has already burned holes in the history of mankind. The public, those undefined masses of the body politic, may even come looking for the Salem witches of the energy crisis and the public apparently still hopes that some technological prophet will come forth to lead them into a promised land full of cheap, clean, abundant energy.

"In 1933, F. Scott Fitzgerald described the heroine of *Tender Is The Night* as a lovely lady genie whose birth was signalled by the Industrial Revolution. To the applause of capitalism, the silent nod of obeisance from workers everywhere, and a Disney-like swirl of sparkling, clattering, bank specie, reaching its crescendo in a Ziegfeld-Hurok spectacular, Scott's Nicole was born.

"If F. Scott Fitzgerald were still with us today, his heroine might be a Tinker Bell grown to starlet size, whose existence is even more magical than Nicole's could ever be. The consort for this generation descends into our midst in a cloud of restless electrons, delights us with an overpowering attention to detail, and does many things for us with intoxicating speed. Yet our technological Tinker Bell is really a puppet and the strings that control her lead far from the scene of her efforts.

... To the resources stored within the earth

... To that constant source of life-giving energy, the sun

... To corporate boardrooms where economic policy is fashioned

... To the offices of government agencies with limited statutory missions to advance

... To that Janus-like figure of technology that promises to deliver so much comfort for the ransom it demands

... To the environmental nursery and graveyard of all those other aspiring civilizations of the past

... To the power plants which energize our Tinker Bell from a distance.

And now, just at the height of our fascination, strings snap threatening the handmaiden of our way of life."

Energy Crisis, Angelo J. Cerchione

DANGER ...

There is already a clear and present danger of government interference with the free enterprise system in the name of energy conservation and pollution control, yet there is still time for business and industry to respond to our most critical problem—survival. Not merely the survival of man as just another animal species, but rather the survival of those uniquely human characteristics which transcend the mere biological heritage of mankind.

Correlation between pollution of air and water and the incidence of poverty, social disease, chronic disease and social unrest has been established. The conclusion is inescapable.

The urban pressures resulting from dense packing of human beings act synergistically with any degradation of environmental quality to encourage social disorder, the manifest symptoms of which may be disease, social unrest or simply chronic, hopeless poverty. It has become a demonstrable fact of life that our core cities, if they are to survive, demand the most sophisticated environmental engineering of which man is capable.

If the many recent advances in basic science are to be made relevant to social problems and if the nation is to tap the enormous and tragically underutilized reserve of engineering, business and industrial talent in order to meet the very real crisis facing most of our metropolitan centers, at least one fundamental question must be answered:

What levels of Environmental Toxicants are tolerable?

There is a substantial probability that by the year 2000 we must establish an economy in which most of our production is recycled. The choice is whether this reordering of our national resource economy will be accomplished by massive government intervention or by the orderly evolution of Business and Industry within the free enterprise system.

The limited experience of scientists with the analysis of stability and productivity in natural ecological systems clearly demonstrates that there is a great deal to be learned before we can safely recommend balanced environmental control measures. However, unless we dramatically change the structure of our quest for cheap, abundant energy and environmental quality; and unless we improve the access of our legislators and the voting public to systems techniques and socially relevant environmental data, the free enterprise system will be seriously compromised.

Unfortunately, the leadership responsibilities for the restructuring of our national natural resource economy are now scattered among federal and state agencies, universities, business corporations and industrial associations. It is clear that this continued diffusion of leadership and dilution of initiative cannot meet the timetable imposed by the reality of our environmental and social crises.

OPPORTUNITY . . .

American Business and Industry can lead the way toward national mobilization for the war on energy waste and environmental degradation. Once before, Industry, Science and the American People joined together to meet a common enemy. During World War II, we put aside many individual professional differences to unite against a common threat. The technological revolution which followed this effort—radar, racons, shoran, loran, sonar, RDF, rangefinders, operations research, reconstructive plastic surgery, dried blood plasma, protein fractionation of whole blood, dynamic testing techniques, jet propulsion, rockets, insect repellents, magnetic airborne detection, aerial reconnaissance and remote sensing, atrabrine, chloroquine, and the other antimalarials, advances in psycho-acoustics, and psycho-physiology, rodenticides, anticoagulants, the sulfonamides, insecticides and many more—that technological revolution has yet to be duplicated.

The pressing need to mobilize the rich resources of scientific, technical and administrative skill traditionally separated from each other by the differing priorities of university, industry and government affords the opportunity to bring together the human and technological resources of these disparate communities and mount a concerted effort to restore and maintain our beneficent environment while assuring sufficient energy to maintain social order and civilization. The very act of bringing together talented, concerned individuals who would otherwise be working on energy and environmental problems in isolation or not at all would, of itself, represent an immediate improvement.

In spite of the frenzied flurry of environmental legislation at all levels of government, there is still no mobilization of those human and physical resources already available in Business and Industry, at the Universities and hidden away in the many pidgeon holes of government.

Consider the choices available to Business and Industry. They can lobby against the proliferation of environmental legislation and legislative authorizations for citizen originated environmental litigation. They can defend the many lawsuits already filed and under consideration throughout the country, winning some and losing others in the courts and before administrative agencies, but on the whole losing the respect so hard won from the American public.

Or, they can recognize the existence of a new challenge and a new opportunity — the opportunity to participate in the development of new business, new industry — environmental rehabilitation — the new business and new industry which can provide a solid economic base for the next generation.

Any such national mobilization should, of course, be influenced by

● Surveys of our remaining non-renewable resources, the extent of waste buildup, air and water quality degradation.

● Natural constraints on the timetable for the present decade during which the initial changeover to a recycling resource economy must be accomplished.

● Continued progress in the integrated study of complex ecological systems.

● The existing data, manpower and techniques available in both the public and private sector which could be combined in meaningful remedial action programs to rehabilitate the environment and restructure our natural resource economy.

Let us assume that we do mobilize to protect the environment and meet the energy needs of civilization. What are the short term, real time efforts that such a national mobilization should make?

● Analysis of regional processes such as urbanization, land utilization, land abandonment, atmospheric contamination, changes in water quality and population dynamics. Such studies must take advantage of the modern remote sensing techniques developed by the military, government and industry to furnish much of the data needed for meaningful simulation of regional ecological systems.

● Detailed description of the variable elements of existing social, biological and environmental systems and the processes by which their state values change. These studies would lead to descriptive conceptual models at first, and finally operational models permitting computer simulation of environmental processes for the remainder of the century.

● Development of strategies for monitoring the secondary effects of environmental toxicants.

● Determination of the energy needs of civilization and the limits of mankind's tolerance of environmental alterations.

● Examination of the psychological interaction of man with his environment and the adaptability of mankind as a species to technological change.

● Establishment of planning methods which take into account the ecological, social, economic and political factors involved in the development of real property — a quality of life approach to regional planning.

● Implementation of data collection, storage and information retrieval methods, systems modeling and optimization techniques already available in order to develop criteria for choice upon which elected officials and the voting public can act.

● Promotion of public educational efforts incorporating the results and experience of the national mobilization effort against energy waste and environmental degradation.

There is little need to review the environmental litigation docket, or the announced policies of many organizations to "go after business and industry." There is even less need to review the many bills pending before the Congress of the United States and the 50 state legislatures.

There is, of course, no need to remind anyone that *The Law* is the framework of civilization and the ordering program for society; that our adversary system of litigation is the civilized alternative to bloody revolution; and that so long as the door to the courthouse remains open, the door to the streets can remain closed.

There is, apparently, some need to point out that a very real opportunity still exists for American Business and Industry, individually and collectively, to take direct action to restore and maintain the quality of our environment on which the quality of our lives depends.

Many of the concepts expressed in this epilogue were first presented on February 19, 1970, at a Conference in Washington, D.C., sponsored by the Public Affairs Council and the U.S. National Committee for the International Biological Program, *ENVIRONMENT: The Quest for Quality; Mobilizing Science, Industry and Government.* in a paper and proposal entitled *Project Eagle and a National Trust for the Environment,* prepared by an *ad hoc* committee which included, Dr. Robert Cancro, research psychiatrist, The Menninger Foundation, then visiting Professor of Computer Science, University of Illinois at Urbana-Champaign; Dr. Orie Loucks, Professor of Botany, Coordinator, Lake Wingra Project, IBP, University of Wisconsin, Madison; Dr. Ian Marceau, Agricultural economist, leader, large-scale planning group, than associated with the ILLIAC IV Project, University of Illinois at Urbana-Champaign; Professor Ian McHarg, Chairman of the Department of Landscape Architecture and Regional Planning, University of Pennsylvania, Philadelphia; Dr. John Rankin, Professor and Chairman of Department of Environmental Medicine, University of Wisconsin School of Medicine, Madison; Dr. Lawrence Slobodkin, Director, Evolution and Ecology Program, State University of New York—Stony Brook; Dr. Daniel Slotnick, Professor of Computer Science and Director, ILLIAC IV Project, University of Illinois at Urbana-Champaign; and Victor John Yannacone, jr, attorney.

The fate of Project Eagle was accurately described in a newspaper feature a few weeks later.

Project Eagle, a crash program to save the environment, took wing at a national conference on environmental problems last month and immediately flew into heavy flak.

The Washington conference was a three-day talkathon at which some of the nation's biggest industrialists and most important environmentalists gathered to tell each other just how critical the problem of preserving the environment had become.

Project Eagle was the only concrete proposal put before the conference and it gave the assembled scientists, government officials and industry publicists something to chew on. And chew they did.

Some of the scientists wanted to know just who would administer the program and how the funds for research would be distributed, not to mention just how much money, finally, the project would require.

The attack of some of the industry public relations men was a bit more personal— the men behind Project Eagle were described as everything from 'henchmen' to 'Commies.'

"Several industry representatives, meanwhile, said that they were awaiting the creation of a National Trust [for the Environment] and a more specific proposal from the Project Eagle people. The result, at this point, seems to be a which comes first, the-eagle-or-the-nest-egg impasse that is keeping the Eagle grounded."

Pearson, Harry, *"Clipping the Wings of Project Eagle,"*
Newsday, March 5, 1970, Environment/II, 13A

*

INDEX